Diamond Deposits: Origin, Exploration & History of Discovery

Diamond Deposits: Origin, Exploration & History of Discovery

Editor

Minh Chi Ngo

scitus
academics

Diamond Deposits: Origin, Exploration & History of Discovery
Edited by **Minh Chi Ngo**

ISBN: 978-1-68117-573-7
Library of Congress Control Number: 2016943008

SCITUS Academics LLC,
www.scitusacademics.com
Box No. 4766, 616 Corporate Way,
Suite 2, Valley Cottage,
NY 10989

Printed in United States of America on Acid Free Paper

Preface

The diamond is the hardest natural substance known. It is found in a type of igneous rock known as kimberlite. The diamond itself is essentially a chain of carbon atoms that have crystallized. The stone's unique hardness is a result of the densely concentrated nature of the carbon chains. Like other igneous rocks, kimberlite was formed over the course of thousands of years by volcanic action that occurred during the formation of the earth's crust. It is thought that diamonds were first discovered in India about 6,000 years ago in the riverbeds of the region. India was the only known source of the rocks for over a thousand years, until they were unearthed in Borneo around A.D. 600. During the Middle Ages, the diamond was overshadowed by some of the more colorful gems like the ruby and emerald. In the 18th century, diamond deposits were discovered in Brazil in small quantities, and later in Australia, Russia, and the United States.Block caving is the most commonly used method in excavating diamonds from kimberlite deposits. This method offers the highest yield and thus is the most cost effective.In the crushing operation, which occurs in the below-ground mining facilities, large chunks of kimberlite are broken up into more easily transportable segments. Diamonds will continue to be used in industry and high-technology enterprises, but synthetically produced facsimiles—first manufactured in 1953—may accomplish some of the tasks originally the exclusive province of the real stone. These "manufactured" gems have the same properties of hardness and durability, and while they will never be as popular as the real diamond for adomment purposes, they are well suited for industrial applications.

This book provides comprehensiveoverview on origin, exploration, and history of diamond.

Table of Contents

Chapter 1

Diamond Formation Due to a Ph Drop During Fluid–Rock Interactions

Dimitri A. Sverjensky[1] **& Fang Huang**[1]

[1]*Department of Earth and Planetary Sciences, Johns Hopkins University, Charles and 34th Streets, Baltimore, Maryland 21218, USA*

ABSTRACT

Diamond formation has typically been attributed to redox reactions during precipitation from fluids or magmas. Either the oxidation of methane or the reduction of carbon dioxide has been suggested, based on simplistic models of deep fluids consisting of mixtures of dissolved neutral gas molecules without consideration of aqueous ions. The role of pH changes associated with water–silicate rock interactions during diamond formation is unknown. Here we show that diamonds could form due to a drop in pH during water–rock interactions. We use a recent theoretical model of deep fluids that includes ions, to show that fluid can react irreversibly with eclogite at 900 °C and 5.0 GPa, generating diamond and secondary minerals due to a decrease in pH at almost constant oxygen fugacity. Overall, our results constitute a new quantitative theory of diamond formation as a consequence of the reaction of deep fluids with the rock types that they encounter during migration. Diamond can form in the deep Earth during water–rock interactions without changes in oxidation state.

INTRODUCTION

Diamonds record fluid migration events over billions of years in the deep carbon cycle in the Earth's interior[1, 2, 3]. Major advances in our understanding of the origin of diamonds in the deep Earth have come from studies of these fluids[4, 5, 6, 7, 8, 9, 10, 11, 12]. Aqueous fluids, melts and/or supercritical mixtures of the two are suspected to have been involved in the origin of diamonds in both the cratonic lithospheric mantle[1, 9, 11, 12, 13, 14, 15] and the ultra-high pressure metamorphic environments[4, 5]. However, previous quantitative models of the fluids involved in diamond formation have been so simplistic as to severely restrict the scope of theories about the origin of diamonds.

Traditional models of the fluids involved in diamond formation have long treated the fluids as mixtures of neutral gas molecules such as CO_2, CH_4, H_2 and H_2O (that is, COH fluids) without consideration of aqueous ions or species derived from silicate rock components[4, 16]. Consequently, the only link in the models between the fluids and their silicate host environments is the fugacity of oxygen (f_{O_2}). For this reason, the cause of diamond precipitation has typically been attributed to redox changes[1, 2, 3, 4, 13, 17]. Addressing alternate possibilities, such as the possible role of evolving fluid chemistry during fluid–rock interaction as a cause of diamond precipitation, has not been possible. Modelling diamond formation during silicate alteration reactions involving aqueous ions and aqueous species derived from silicate rocks could not be attempted, because COH models at high pressures do not contain any ions or silicate rock components.

Recent advances in theoretical and experimental aqueous geochemistry resulting in the Deep Earth Water (DEW) model enable calculation of equilibrium constants involving minerals and aqueous ions, metal complexes and organics to 6.0 GPa and 1,200 °C (refs 18, 19, 20, 21). In the present study, we use these advances to address the potential role of evolving aqueous fluid chemistry in the origin of diamonds. We first calibrate our model at 5.0 GPa and 900 °C using experimental data (see Calculation Methods and Supplementary Table 1 and Fig. 1), then we model diamond formation during the irreversible chemical mass transfer associated with aqueous fluid–rock interactions[22, 23], to investigate the role of pH changes in driving diamond precipitation. The model uses a conceptual scenario in which fluids from the oceanic mantle lithosphere in a subducting plate react with eclogitic mineral assemblages at 900 °C and 5.0 GPa, as a generic scenario of subcratonic lithospheric diamond formation (Calculation Methods). The aim of the present study is not to model the formation of specific diamonds and their fluid and solid inclusions. Instead, we demonstrate how it is now possible to monitor evolving fluid chemistry during diamond formation and the potential importance of pH drop as a new mechanism for diamond formation.

RESULTS

Mineral products during fluid–rock interaction

An initial fluid chemistry (Supplementary Table 2) was set by reaction of water with a carbonated mafic rock[24] resulting in a fluid very close to equilibrium with a Mg-rich carbonate solid solution and diamond (Fig. 1a). The fluid was then permitted to react with a model metasedimentary eclogite[25,26] consisting of clinopyroxene (mole fractions of jadeite 0.7, diopside 0.2 and hedenbergite 0.1), garnet (mole fractions of pyrope 0.6, almandine 0.3 and grossular 0.1) and coesite. The final fluid chemistry is given in Table 1 (see also Supplementary Note 1 for the full reaction path). The reactions produced abundant diamond,

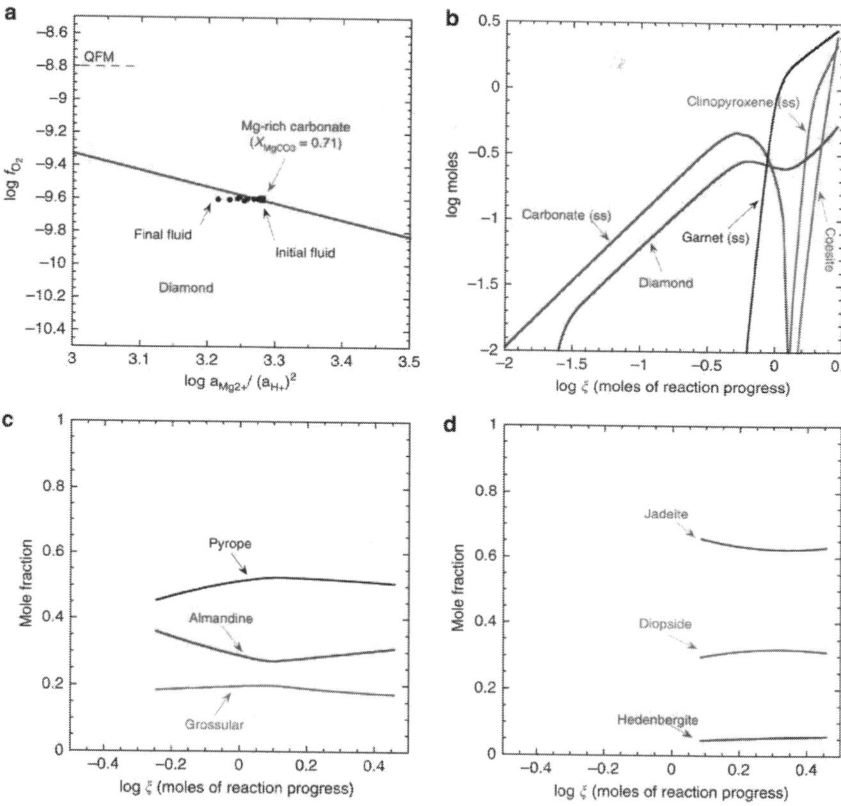

Figure 1: Predicted reaction path and mineral products of the fluid–rock interactions forming diamond.(**a**) Reaction path of the fluid (black dots) relative to the equilibrium boundary between a specific magnesite component composition in a carbonate solid solution, diamond, aqueous species and $O_{2,g}$. (**b**) Minerals produced during the reaction of the fluid with eclogite (Supplementary Table 2). (**c**) Composition of the garnet solid solution produced during the reaction of the fluid with eclogite (Supplementary Table 2). (**d**) Composition of the clinopyroxene solid solution produced during the reaction of the fluid with eclogite (Supplementary Table 2). All calculations refer to 900 °C and 5.0 GPa. In **b,c,d**, the x axis represents the logarithm of the reaction progress variable ξ, which is equal to the number of moles of each reactant mineral destroyed during the reaction progress.

while reactant silicate minerals were destroyed (Fig. 1b). As the initial fluid was in equilibrium with Mg-rich carbonate solid solution (magnesite 0.71, siderite 0.12 and calcite 0.17), secondary Mg-rich carbonate was predicted to precipitate and replace eclogite at small extents of reaction progress (logξ values of −2.0 to about −0.30; Fig. 1b). However, at large extents of reaction progress (logξ values greater than about −0.30, Fig. 1b), carbonate and eclogite are replaced by diamond, secondary garnet, clinopyroxene and coesite. The predicted secondary garnet and clinopyroxene compositions (Fig. 1c,d) could serve as models of the solid inclusions typically preserved in natural diamonds[13]. Future studies could include more detailed comparisons of predicted evolving solid solutions in garnet and pyroxene during diamond formation with specific inclusion compositions in natural diamonds.

Table 1: Comparison of the predicted final fluid composition after reaction with eclogite at 900 °C and 5.0 GPa with the measured composition of a highly siliceous fluid inclusion*.

Final model fluid	Parameter value†	Siliceous fluid inclusion*	Parameter value†
Na	0.94	Na	0.84
K	1.69	K	1.69
Mg	0.24	Mg	0.78
Ca	3.07	Ca	1.62
Fe	0.14	Fe	0.87
Al	0.68	Al	1.0
Si	5.18	Si	5.2
Cl	0.10	Cl	0.7
S	0.18		
C	33.7		
pH	4.74		
$\log f_{O_2}$	−9.60		
Mole ratio of $C/(C+H_2O)$	0.4	Mole ratio of $C/(C+H_2O)$	0.1‡
wt.% H_2O in fluid	32	wt.% H_2O in fluid	Up to 40‡
wt.% CO_2 in fluid	47	wt.% CO_2 in fluid	Up to 11§

*Ref. 55, inclusion number Sp.BR-1(inner) normalized to the K content of the model eclogitic fluid given above (1.69 m).
†Concentrations given in molality (m) unless otherwise noted.
‡Ref. 6.
§Calculated from the given mole ratio of $C/(C+H_2O)$ equal to 0.1 and the given upper limit of the wt.% H_2O in fluid equal to 40%.

It is interesting to note that the first part of the reaction path shown in Fig. 1b (logξ<−0.30), corresponding to the smallest extents of fluid–rock interaction, could correspond to the formation of diamond in a Mg-rich carbonate-metasomatized eclogite. It might be expected that the carbonate phase could disaggregate during transport in a kimberlite magma[1], which would result in xenocryst diamonds hosted by kimberlite rock. In contrast, the second part of the reaction path in Fig. 1b($\log\xi > -0.30$), corresponding to much larger extents of fluid–rock interaction, could correspond to the formation of diamond, clinopyroxene and garnet-metasomatized eclogite. Under these circumstances, diamonds may well be preserved in their eclogitic host–rock during the kimberlitic eruption process. If such large extents of fluid–rock interaction are uncommon, the overall model inFig. 1b may help to explain the comparative rarity of diamonds set in a matrix of eclogite[27] rock.

Changes in the aqueous phase during fluid–rock interaction

It can be seen in Fig. 1a that the $\log f_{O2}$ is almost unchanged (<0.01 log units) during the entire reaction path. However, in Fig. 2a, it can be seen that the pH decreases during most of the reaction path. It should be noted here that neutral pH under these conditions is about 2.7 because of the drastic increase in the dissociation constant of water at this temperature and pressure. The reason for the pH decrease of the reacting fluids can be seen in the irreversible reaction path traced on the diagram in Fig. 1a. In contrast to traditional COH models, the equilibrium between a given activity of the magnesite component of a carbonate mineral and diamond represented by the solid line in Fig. 1a is univariant at constant temperature and pressure according to the reaction

$$MgCO_3 + 2H^+ = C + Mg^{2+} + O_{2,g} + H_2O \qquad (1)$$

The univariance of the equilibrium in equation (1) arises, because HCl is a chemical component that provides ions[28].

It can be seen in Fig. 1b that reaction of the fluid with eclogitic minerals in the earliest stages of reaction progress ($-2.0 < \log\xi < -0.30$) causes the precipitation of diamond and a secondary Mg-rich carbonate mineral according to the reactions

$$\begin{aligned} &NaAlSi_2O_6 + Mg^{2+} + CO_2^0 + H_2O \rightarrow MgCO_3 + Na^+ \\ &+ AlO_2^- + 2SiO_2^0 + 2H^+ \end{aligned} \qquad (2)$$

$$\begin{aligned} &Mg_3Al_2Si_3O_{12} + 3CO_2^0 + H_2O \rightarrow 3MgCO_3 \\ &+ 2(AlO_2)SiO_2^- + SiO_2^0 + 2H^+ \end{aligned} \qquad (3)$$

These reactions remove the Mg^{2+} ion and add the $(AlO_2)SiO_2^-$ complex and H^+ ions to the fluid, causing the ratio of $a_{Mg^{2+}}/(a_{H^+})^2$ and the pH to decrease, which results in the fluid chemistry moving into the diamond stability field in Fig. 1a.

In the later stages of reaction progress ($-0.30<\log\xi<+0.1$), the fluid is directly altering primary eclogitic clinopyroxene to secondary garnet with an increasing pyrope component (Fig. 1c) according to the reaction

$$2NaAlSi_2O_6 + 3Mg^{2+} + 2H_2O \rightarrow Mg_3Al_2Si_3O_{12} + 2Na^+ + SiO_2^0 + 4H^+ \quad (4)$$

These reactions remove the Mg^{2+} ion and add H^+ ions to the fluid, again causing the ratio of $a_{Mg^{2+}}/(a_{H^+})^2$ and the pH to decrease, which results in the fluid chemistry moving further into the diamond stability field in Fig. 1a.

Aqueous organic carbon species during diamond formation

The overall decrease in the pH of the fluid seen in Fig. 2a also causes changes in the most abundant carbon species in the fluid (Fig. 2b): the concentration of the aqueous species $CO_{2,aq}$ increases by 623 mmolal, which is mainly balanced by decreases in the concentrations of the $Ca(HCO_3)^+$ complex, HCO_3^- and CO_3^{2-} (−300, −75 and −62 mmolal , respectively) plus a decrease of about −160 mmolal of C from organic carbon species (Supplementary Table S1). However, it can be seen in Fig. 2c that the trend of the number of moles of carbon precipitated as diamond is closely mirrored by the changes in the organic carbon species formate. In fact, a total of 529 mmol of carbon is precipitated as diamond, which comes from the destruction of 405 mmolal formate and 264 mmolal carbon from propionate, according to the reactions:

$$HCOO^- + H^+ + H_{2,aq} \rightarrow C_{diamond} + 2H_2O \quad (5)$$

$$CH_3CH_2COO^- + H^+ \rightarrow 3C_{diamond} + H_{2,aq} + 2H_2O \quad (6)$$

(as noted above, the small excess of organic carbon destroyed compared with diamond precipitated was converted to CO_2 and $Ca(HCO_3)^+$). Although each of the reactions in equations (5) and (6) are redox reactions, the sum of these reactions is independent of the dissolved $H_{2,aq}$ concentration and therefore of the f_{O2}. Consequently, the overall reaction precipitating diamond is independent of f_{O2} and the f_{O2} barely changes during the calculated reaction path in Fig. 1a.

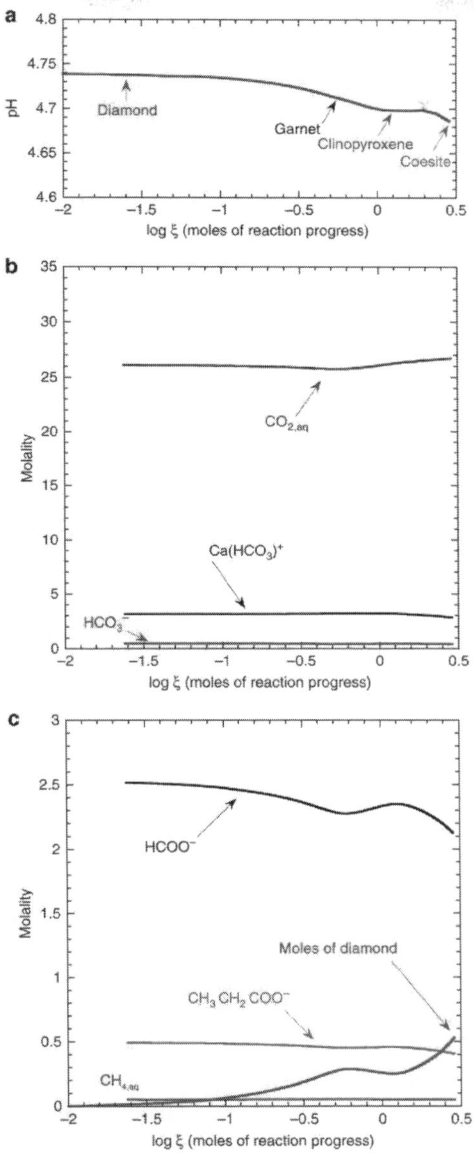

Figure 2 Predicted aqueous phase evolution during fluid–rock interactions forming diamond. (a) Evolution of the pH of the fluid during reaction with eclogite. (b) Evolution of the concentrations of the major inorganic oxidized carbon species. (c) Number of moles of diamond precipitated per kg of water compared with the evolution of the concentrations of the major aqueous organic carbon species. In a,b,c, the x axis represents the logarithm of the reaction progress variable x, which is equal to the number of moles of each reactant mineral destroyed during the reaction progress. All calculations refer to 900 C and 5.0 GPa.

DISCUSSION

Diamond precipitation in Fig. 1a,b is driven by pH drop and not by changes in redox conditions. The formation of diamond is a direct consequence of changes in aqueous fluid chemistry associated with metasomatic reactions involving silicate minerals under upper mantle conditions. Furthermore, in the same reacting chemical system, diamond dissolution can be driven by changes in fluid chemistry at almost constant redox states. For example, it can be seen in Figs 1b and 2cthat diamond is predicted to partially dissolve during the interval of ($-0.25<\log\xi<+0.1$) because of small fluctuations in solution chemistry associated with the disappearance of the Mg-rich carbonate solid solution. This predicted behaviour may mirror the often noted precipitation and dissolution features in natural diamonds[1].

It is also interesting that the precipitation reactions for diamond in Figs 1 and 2 involved aqueous organic carbon species, not just the CO_2 or CH_4 previously pictured as the only possible sources of carbon for diamond formation. The involvement of aqueous organic species in our model arises, because it is predicted that at pressures above ~3.0 GPa it is possible to have a complete equilibrium between all oxidation states of aqueous carbon species[29]. The proposed involvement of these organic carbon species in the formation of diamond has implications for the carbon isotopic composition of diamond[29, 30, 31, 32, 33, 34, 35]. For example, it can be expected that the carbon isotopic compositions of diamonds in equilibrium with an aqueous fluid will be a function of both pH and redox state. A change in either pH or redox state could, in principle, cause changes in the carbon isotopic composition of diamond, depending on the magnitude of the equilibrium fractionation factor between diamond and aqueous C-species. This inference is exactly analogous to that predicted for graphite under crustal hydrothermal conditions[36]. To date, most studies of the carbon isotopic compositions of diamonds invoke redox fluctuations[3]. However, in at least one instance, changes in the carbon isotopic composition of diamond have been suggested to be independent of redox changes[37]. The potential for a dependence on pH changes offers an alternate hypothesis that might be testable with experiments. Similar considerations involving a potential dependence on either redox state or pH variations during metasomatic reactions may apply to the large variability exhibited in the nitrogen isotopic compositions of individual diamonds[38].

The predicted model evolution of the fluid chemistry during the precipitation of diamond is shown inFig. 3 for comparison with worldwide trends in the chemistry of fluid inclusions in diamonds. The initial fluid chemistry is Ca-rich and Si-poor but progressively changes towards a silica-rich composition. After equilibration with the reactant eclogitic minerals, the final fluid chemistry (Supplementary Table 2) is extremely enriched in Al and Si, and depleted in Ca relative to the initial fluid. It can be seen in Fig. 3 that these changes are qualitatively consistent with one of the major trends in the chemistry of fluid inclusions in diamonds that extends from a carbonatitic fluid to a silicic fluid[6]. This trend has previously been suggested to be the result of equilibration with eclogites[39]. As a first step, given our current limited information about element solubility and speciation in high-pressure fluids, this approximation shows a reasonable agreement between predicted and actual fluid compositions.

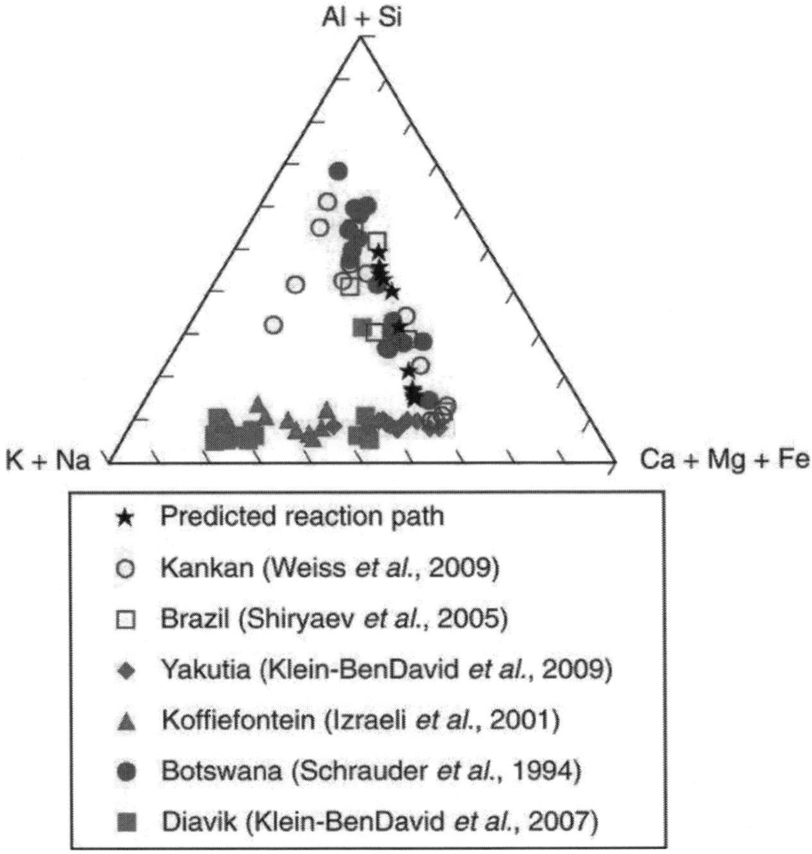

Figure 3: Predicted versus measured fluid chemistry during diamond formation. Theoretical evolution of the fluid chemistry from the model in Figs 1 and 2 (black stars). Measured compositions of fluid inclusions from worldwide diamonds[6, 39, 55, 56, 57, 58] (coloured symbols). The predicted fluid composition evolves from a carbonatitic end-member fluid towards the silicic end member of fluid composition from natural diamonds.

Significant differences, however, are apparent when a more detailed comparison of individual elemental concentrations is made. For example, in Table 1 a more detailed comparison of the final eclogitic fluid composition from the model (Fig. 2) is made with the composition of a silicic fluid inclusion from a diamond from Brazil. To facilitate the comparison of elemental values, the fluid inclusion analysis was normalized to the molality of K in the predicted eclogitic model fluid. Excellent agreement can be seen between the concentrations of Si, Al and Na. However, the predicted model concentrations of Mg, Ca and Fe are either too high or too low by factors of 3.3, 1.9 and 6, respectively. This could be due to inaccurate speciation of these elements in the fluid model, for which we need more experimental data. The most significant difference that can be seen in Table 1 is that the wt.% CO_2 is much higher in the model fluid (47%) than has been derived for the natural fluid inclusions (11%). A possible cause of this discrepancy might be the simplistic assumption of a unit activity coefficient for the the aqueous CO_2 species in the model. Finally, it should be emphasized that the calculations plotted in Fig. 3 refer only to 900 °C and 5.0 GPa, whereas the fluid inclusion compositions of the natural diamonds plotted in Fig. 3 probably refer to a very wide range of temperatures, pressures and geologic environments[13]. Nevertheless, the overall comparison is encouraging. The model fluid speciation presented here is only a first step in modelling the chemistry of the evolution of fluids during diamond formation. Much more experimental and theoretical work needs to be done, to characterize the most important fluid species under such extreme conditions.

Our theoretical calculations provide a model clearly demonstrating that diamond can form by pH decreases during water–rock interaction under constant redox conditions in subcratonic lithospheric mantle environments. The theory implies that diamonds and their solid and fluid inclusions could be the natural, and perhaps even the common result of changes in water chemistry rather than redox changes alone. We wish to emphasize that our proposed pH-change mechanism for diamond formation represents a mechanism in addition to potential changes in redox, temperature and pressure[2]. Indeed, combined pH decreases and f_{O2} increases could result in multiple cycles of diamond precipitation and dissolution, which might explain some observed diamond resorption features[1] as well as the evidence for evolving fluid chemistry during individual diamond formation[40].

The theoretical model developed here could also be used to investigate subcratonic lithospheric diamond formation in peridotites. Furthermore, the model could be expanded to include more sophisticated models of solid solution in the minerals, the partitioning of trace elements between the minerals and fluids, a much wider range of aqueous organic species, and liquid and crystalline hydrocarbons. It would then be possible to integrate quantitative theoretical models of evolving fluid chemistry with detailed studies of natural diamonds including their solid and fluid inclusions, their trends with time during diamond growth and their relationships to host eclogites and peridotites[7, 8,9, 27, 41, 42]. Ultimately, such integrated studies hold the potential for helping to unravel specific reaction histories of fluids in the deep Earth and in deep time. The theoretical model could also be used to predict the cooling behaviour of individual fluid inclusions to room temperature, including the formation of daughter crystals, enabling independent comparisons to be made with the extensive studies of fluid inclusion and their daughter crystals[5, 10].

METHODS

Conceptual model of diamond formation

Diamond formation occurs particularly in the roots of continents known as mantle keels that extend from about 40 to 250 km depth. In this subcratonic mantle lithosphere environment, diamonds have formed at temperatures in the range of 900–1,400 °C and pressures of about 4.0–8.0 GPa (ref.13). Diamonds are also described from ultra-high pressure metamorphic rocks referring to temperatures as low as 600 °C and pressures of >3 GPa (ref. 13). In the present study a model was constructed for subcratonic lithospheric mantle diamond formation in eclogite at 900 °C and 5.0 GPa. Although this temperature is at the low end of the range of recorded temperatures for subcratonic mantle diamond formation, the reactions described here should also apply to higher temperatures. It should be emphasized that the model is a generic one for the purpose of illustrating the role of pH in diamond-forming systems in which the fluid chemistry is coupled to the silicate rock mineralogy of eclogites.

The initial fluid was derived by reaction of fluids assumed to be low in chloride with a carbonated peridotitic mineral assemblage envisioned as passing through the carbonated mafic part of a subducting slab[24]. As a first approximation, the fluid was assumed to have moved through the slab at constant temperature and pressure, to infiltrate and react with a model metasedimentary eclogite consisting of clinopyroxene, garnet and coesite also based on previous model calculations[26]. Ideal site mixing was used for the silicate solid solution model.

Calculation methods

The calculations referred to in Figs 1, 2, 3 were carried out with versions of the Fortran computer codes EQ3NR and EQ6 (refs 22, 23) modified for use at elevated pressures and temperatures. The code EQ3NR is used to set up a speciated initial fluid chemistry in equilibrium with a specified set of minerals that represents a model of a rock as described above. This fluid is then used in the code EQ6, to react with eclogitic minerals as described above. The irreversible mass transfer calculations were carried out assuming relative reaction rates of unity with excess amounts of each reactant mineral. In this mode of calculation, if the fluid reaches equilibrium with a given reactant mineral, it is subsequently assumed that the mineral remains in equilibrium with the fluid, although some of it may dissolve or precipitate during further reaction progress.

The two codes use a database of equilibrium constants calculated in advance using the DEW model[18]. The DEW model is based on an extension of the Helgeson–Kirkham–Flowers aqueous species equation of state, traditionally limited to an upper pressure of 0.5 GPa because of lack of knowledge of the dielectric constant of water at high pressures[43, 44]. We have taken advantage of recent experimental and theoretical advances[19, 20] that enable this limitation to be overcome and we have further shown that the Helgeson–Kirkham–Flowers formalism can be applied to experimental solubility and aqueous speciation data for quartz, forsterite+enstatite, corundum, calcite and aragonite at pressures up to 6.0 GPa, enabling theoretical predictions of equilibrium constants involving aqueous inorganic and organic ions, neutral species and metal complexes up to pressures and temperatures of the upper mantle. Overall uncertainties in the DEW model may be of the order of ±0.3 to 0.5 units in equilibrium constants for aqueous species at high temperatures and pressures. It should be emphasized that all the aqueous species used in the calculations rest on experimental calibrations for as wide a range of temperatures or pressures as is available (see also below). Activity coefficients for neutral aqueous species were set to unity, whereas those for ions were approximated by the Debye–Hückel equation in which the long-range interaction parameters were calculated using dielectric constants and densities of water from the DEW model. Owing to a lack of sufficient data on mineral solubilities at elevated pressures as a function of ionic strength, the short-range interaction parameter in the Debye–Hückel equation was set to zero[21].

Comparison of model calculations and experimental data

Remarkably, little experimental data exist for mineral and silicate-rock solubilities in aqueous fluids at elevated temperatures and at pressures >3.0 GPa, in particular as the melting point of the rocks is approached[45, 46, 47, 48, 49, 50, 51]. Considerable uncertainties surround the nature of the fluid phase in equilibrium with silicate rocks under these conditions. In the present study, we calibrated our high temperature–pressure aqueous model using experimental measurements of a synthetic K-free eclogite at 900 °C and 5.0 GPa (ref. 50). The high solubilities of the elements Si, Mg, Fe and Al in these experiments almost certainly reflect complexes between metals and silicate as the temperatures approach the melting point of the eclogite[51, 52]. Although the exact nature of these complexes remains uncertain and is the topic of intense interest[49, 53, 54], as a first approximation we assumed they were 1:1 metal–bisilicate complexes, for example, $Mg(HSiO_3)^+$, because of the predicted abundance of the bisilicate anion $HSiO_3^-$ at high pressures[29]. Based on equilibrium constants calculated using the DEW model, the experimental eclogite solubility data were used to retrieve new equilibrium constants for the dissociation of the aqueous silica monomer to the bisilicate anion, and the bisilicate complexes of Mg, Fe, Ca and Al (Supplementary Table 1). Comparison between the experimental and calculated fluid composition is shown in Supplementary Fig. 1. Given the overall uncertainties in the experiments and the calculations, the agreement inSupplementary Fig. 1 was deemed to be sufficiently good for the application of the present study to diamond formation. Sensitivity calculations with and without metal–silicate complexes emphasized that the overall conclusion of the present study regarding the importance of a pH drop causing diamond formation did not depend on either the precise values of the equilibrium constants or even the existence of the metal–silicate complexes in the model. This is a consequence of the fundamental importance of reactions such as the one given in Eqn. (4) in controlling the pH of the fluids.

ADDITIONAL INFORMATION

How to cite this article: Sverjensky, D. A. & Huang, F. Diamond formation due to a pH drop during fluid–rock interactions. *Nat. Commun.* 6:8702 doi: 10.1038/ncomms9702 (2015).

ACKNOWLEDGEMENTS

This research was supported by grants from the Sloan Foundation through the Deep Carbon Observatory (Reservoirs and Fluxes and Extreme Physics and Chemistry programmes) and grant DOE DE-FG-02-96ER-14616. Graham Pearson provided generous guidance on the petrology of diamonds and their inclusions. The scientific content of the manuscript was greatly improved by reviews by Oded Navon and two other reviewers. We are grateful for the help and support of the Johns Hopkins University and the Geophysical Laboratory of the Carnegie Institution of Washington and acknowledge helpful discussions with R.E. Cohen, I. Daniel, Y. Fei, M.S. Ghiorso, R.M. Hazen, R.J. Hemley, A. Jones, R. Kessel, S. Kohn, S. Lobanov, C.E. Manning, S. Mikhail, S.B. Shirey, C.M. Schiffries, E. Shock, V. Stagno, E. Thomassot and Y. Weiss. The senior author acknowledges W. Link for his encouragement and invaluable advice on science and life.

CONTRIBUTIONS

D.A.S. initiated the project, carried out the calculations for Fig. 1 and wrote the manuscript; F.H. prepared Fig. 2a, contributed to the characterization of the thermodynamic data underlying the modelling and participated in discussion of the modelling.

COMPETING FINANCIAL INTERESTS

The authors declare no competing financial interests.

REFERENCES

1. Shirey, S. B. *et al.* Diamonds and the geology of mantle carbon. *Rev. Mineral. Geochem.* **75**,355–421 (2013).
2. Stachel, T. & Luth, R. Diamond formation: where, when and how? *Lithos* **220**, 200–220(2015).
3. Cartigny, P., Palot, M., Thomassot, E. & Harris, J. W. Diamond formation: a stable isotope perspective. *Annu. Rev. Earth Planet. Sci.* **42**, 699–732 (2014).
4. Frezzotti, M.-L., Huizenga, J.-M., Compagnoni, R. & Selverstone, J. Diamond formation by carbon saturation in C-O-H fluids during cold subduction of oceanic lithosphere. *Geochim. Cosmochim. Acta* **143**, 68–86 (2014).
5. Frezzotti, M. L., Selverstone, J., Sharp, Z. D. & Compagnoni, R. Carbonate dissolution during subduction revealed by diamond-bearing rocks from the Alps. *Nat. Geosci.* **4**,703–706 (2011).
6. Klein-BenDavid, O., Izraeli, E. S., Hauri, E. & Navon, O. Fluid inclusions in diamonds from the Diavik mine, Canada, and the evolution of diamond-forming fluids. *Geochim. Cosmochim. Acta* **71**, 723–744 (2007).
7. Tomlinson, E. & Müller, W. A snapshot of mantle metasomatism: trace element analysis of coexisting fluid (LA-ICP-MS) and silicate (SIMS) inclusions in fibrous diamonds. *Earth Planet. Sci. Lett.* **279**, 362–372 (2009).
8. Tomlinson, E. L., Jones, A. P. & Harris, J. W. Co-existing fluid and silicate inclusions in mantle diamond. *Earth Planet. Sci. Lett.* **250**, 581–595 (2006).

9. Miller, C. E., Kopylova, M. & Smith, E. Mineral inclusions in fibrous diamonds: constraints on cratonic mantle refertilization and diamond formation. *Miner. Petrol.* **108**, 317–331 (2014).

10. Weiss, Y., Kiflawi, I. & Navon, O. IR spectroscopy: quantitative determination of the mineralogy and bulk composition of fluid microinclusions in diamonds. *Chem. Geol.* **275**,26–34 (2010).

11. Bureau, H. *et al.* The growth of fibrous, cloudy and polycrystalline diamonds. *Geochim. Cosmochim. Acta* **77**, 202–214 (2012).

12. Weiss, Y., McNeill, J., Pearson, D. G., Nowell, G. M. & Ottley, C. J. Highly saline fluids from a subducting slab as the source for fluid-rich diamonds. *Nature* **524**, 339–342 (2015).

13. Stachel, T. & Harris, J. The origin of cratonic diamonds- constraints from mineral inclusions.*Ore Geol. Rev.* **34**, 5–32 (2008).

14. Palyanov, Y. N. *et al.* Mantle-slab interaction and redox mechanism of diamond formation.*Proc. Natl Acad. Sci.* **110**, 20408–20413 (2013).

15. Sokol, A. G., Palyanova, G. A., Palyanov, Y. N., Tomilenko, A. A. & Melenevsky, V. N. Fluid regime and diamond formation in the reduced mantle: Experimental constraints. *Geochim. Cosmochim. Acta* **73**, 5820–5834 (2009).

16. Zhang, C. & Duan, Z. A model for C-O-H fluid in the Earth's mantle. *Geochim. Cosmochim. Acta* **73**, 2089–2102 (2009).

17. Luth, R. W. Diamonds, eclogites, and the oxidation state of the Earth's mantle. *Science* **261**,66–68 (1993).

18. Sverjensky, D. A., Harrison, B. & Azzolini, D. Water in the deep Earth: the dielectric constant and the solubilities of quartz and corundum to 60 kb and 1,200 °C. *Geochim. Cosmochim. Acta* **129**, 125–145 (2014).

19. Pan, D., Spanu, L., Harrison, B., Sverjensky, D. A. & Galli, G. The dielectric constant of water under extreme conditions and transport of carbonates in the deep Earth. *Proc. Natl Acad. Sci.* **110**, 6646–6650 (2013).

20. Facq, S., Daniel, I. & Sverjensky, D. A. *In situ* Raman study and thermodynamic model of aqueous carbonate speciation in equilibrium with aragonite under subduction zone conditions. *Geochim. Cosmochim. Acta* **132**, 375–390 (2014).

21. Manning, C. E. Thermodynamic modeling of fluid-rock interaction at mid-crustal to upper-mantle conditions. *Rev. Mineral. Geochem.* **76**, 135–164 (2013).

22. Wolery, T. J. *EQ3NR: A Computer Program for Geochemical Aqueous Speciation-Solubility Calculations, User's Guide and Documentation: UCRL-53414* Lawrence Livermore Lab., Univ. Calif. (1983).

23. Wolery, T. J. *EQ6---A computer Program For Reaction-Path Modelling of Aqueous Geochemical Systems: User's Guide and Documentation: UCRL-51* Lawrence Livermore Nat. Lab., Univ. Calif. (1984).

24. Kerrick, D. M. & Connolly, J. A. D. Metamorphic devolatilization of subducted oceanic metabasalts: implications for seismicity, arc magmatism and volatile recycling. *Earth Planet. Sci. Lett.* **189**, 19–29 (2001).

25. Hacker, B. R. H2O subduction beyond arcs. *Geochem. Geophys. Geosyst.* **9**, Q03001(2008).

26. Kerrick, D. M. & Connolly, J. A. D. Metamorphic devolatilization of subducted marine sediments and the transport of volatiles into the Earth's mantle. *Nature* **411**, 293–296 (2001).

27. Howarth, G. H. *et al.* The secondary origin of diamonds: multi-modal radiation tomography of diamondiferous mantle eclogites. *Int. Geol. Rev.* **56**, 1172–1180 (2014).

28. Bowers, T. S., Jackson, K. J. & Helgeson, H. C. *Equilibrium Activity Diagrams* Springer-Verlag (1984).

29. Sverjensky, D. A., Stagno, V. & Huang, F. Important role for organic carbon in subduction-zone fluids in the deep carbon cycle. *Nat. Geosci.* **7**, 909–913 (2014).

30. Navon, O. in *Proceedings of the 7th International Kimberlite Conference* **2**, 584–604Cape Town: Red Roof Design.

31. Kirkley, M. B., Gurney, J. J. & Levinson, A. A. Age, origin, and emplacement of diamonds; scientific advances in the last decade. *Gems Gemol.* **27**, 2–25 (1991).

32. Nisbet, E. G., Mattey, D. P. & Lowry, D. Can diamonds be dead bacteria? *Nature* **367**, 694(1994).

33. Cartigny, P. Stable isotopes and the origin of diamond. *Elements* **1**, 79–84 (2005).

34. Mikhail, S. *et al.* Empirical evidence for the fractionation of carbon isotopes between diamond and iron carbide from the Earth's mantle. *Geochem. Geophys. Geosyst.* **15**,855–866 (2014).

35. Thomassot, E., Cartigny, P., Harris, J. & Viljoen, K. F. Methane-related diamond crystallization in the Earth's mantle: stable isotope evidences

from a single diamond-bearing xenolith. *Earth Planet. Sci. Lett.* **257**, 362–371 (2007).

36. Ohmoto, H. Systematics of sulfur and carbon isotopes in hydrothermal ore deposits. *Econ. Geol.* **67**, 551–578 (1972).

37. Shiryaev, A. A., Zubavichus, Y. V., Veligzhanin, A. A. & McCammon, C. Local environment and valence state of iron in microinclusions in fibrous diamonds: X-ray absorption and Mössbauer data. *Russ. Geol. Geophys.* **51**, 1262–1266 (2010).

38. Mikhail, S. *et al.* Constraining the internal variability of the stable isotopes of carbon and nitrogen within mantle diamonds. *Chem. Geol.* **366**, 14–23 (2014).

39. Weiss, Y. *et al.* A new model for the evolution of diamond-forming fluids: Evidence from microinclusion-bearing diamonds from Kankan, Guinea. *Lithos* **112**, 660–674 (2009).

40. Palot, M., Pearson, D., Stern, R., Stachel, T. & Harris, J. Multiple growth events, processes and fluid sources involved in diamond genesis: a micro-analytical study of sulphide-bearing diamonds from Finsch mine, RSA. *Geochim. Cosmochim. Acta* **106**, 51–70 (2013).

41. Weiss, Y., Kiflawi, I., Davies, N. & Navon, O. High-density fluids and th growth of monocrystalline diamonds. *Geochim. Cosmochim. Acta* **141**, 145–159 (2014).

42. Klein-BenDavid, O. *et al.* The sources and time-integrated evolution of diamond-forming fluids: trace elements and isotopic evidence. *Geochim. Cosmochim. Acta* **125**, 146–169(2014).

43. Shock, E. L., Oelkers, E. H., Johnson, J. W., Sverjensky, D. A. & Helgeson, H. C. Calculation of the thermodynamic and transport properties of aqueous species at high pressures and temperatures: effective electrostatic radii to 1000 °C and 5 kb. *Faraday Soc. Trans.* **88**,803–826 (1992).

44. Shock, E. L., Sassani, D. C., Willis, M. & Sverjensky, D. A. Inorganic species in geologic fluids: correlations among standard molal thermodynamic properties of aqueous cations, oxyanions, acid oxyanions, oxyacids and hydroxide complexes. *Geochim. Cosmochim. Acta***61**, 907–950 (1997).

45. Kessel, R., Pettke, T. & Fumagalli, P. Melting of metasomatized peridotite at 4–6 GPa and up to 1200 °C: an experimental approach. *Contr. Mineral. Petrol.* **169**, 1–19 (2015).

46. Kessel, R., Schmidt, M. W., Ulmer, P. & Pettke, T. Trace element signature of subduction-zone fluids, melts and supercritical liquids at 120–180 km depth. *Nature* **437**, 724–727(2005).

47. Kessel, R., Ulmer, P., Pettke, T., Schmidt, M. W. & Thompson, A. B. A novel approach to determine high-pressure high-temperature fluid and melt compositions using diamond-trap experiments. *Am. Miner* **89**, 1078–1086 (2004).

48. Hermann, J. & Spandler, C. J. Sediment melts at sub-arc depths: an experimental study. *J. Petrol.* **49**, 717–740 (2008).

49. Zheng, Y.-F. & Hermann, J. Geochemistry of continental subduction-zone fluids. *Earth Planets Space* **66**, 1–16 (2014).

50. Kessel, R., Ulmer, P., Pettke, T., Schmidt, M. W. & Thompson, A. B. The water-basalt system at 4 to 6 GPa: Phase relations and second critical endpoint in a K-free eclogite at 700 to 1400 °C. *Earth Planet. Sci. Lett.* **237**, 873–892 (2005).

51. Wohlers, A., Manning, C. E. & Thompson, A. B. Experimental investigation of the solubility of albite and jadeite in H2O, with paragonite+quartz at 500 and 600 °C, and 1–2.25 GPa.*Geochim. Cosmochim. Acta* **75**, 2924–2939 (2011).

52. Manning, C. E., Antignano, A. & Lin, H. A. Premelting polymerization of crustal and mantle fluids, as indicated by the solubility of albite+paragonite+quartz in H2O at 1 GPa and 350–620 °C. *Earth Planet. Sci. Lett.* **292**, 325–336 (2010).

53. Hermann, J., Zheng, Y.-F. & Rubatto, D. Deep fluids in subducted continental crust.*Elements* **9**, 281–287 (2013).

54. Newton, R. C. & Manning, C. E. Evidence for SiO2-NaCl complexing in H2O-NaCl solutions at high pressure and temperature. *Geofluids* doi: 10.1111/gfl.12153 (2015).

55. Shiryaev, A., Izraeli, E., Hauri, E., Zakharchenko, O. & Navon, O. Chemical, optical and isotopic investigation of fibrous diamonds from Brazil. *Russ. Geol. Geophys.* **46**, 1185–1201(2005).

56. Klein-BenDavid, O. *et al.* High-Mg carbonatitic microinclusions in some Yakutian diamonds' a new type of diamond-forming fluid. *Lithos* **112**, 648–659 (2009).

57. Izraeli, E. S., Harris, J. W. & Navon, O. Brine inclusions in diamonds: a new upper mantle fluid. *Earth Planet. Sci. Lett.* **187**, 323–332 (2001).

58. Schrauder, M. & Navon, O. Hydrous and carbonatitic mantle fluids in fibrous diamonds from Jwaneng, Botswana. *Geochim. Cosmochim. Acta* **58**, 761–771 (1994).

Chapter 2

Subduction Model for the Origin of Some Diamonds in the Phanerozoic of Eastern New South Wales

L. M. Barron,1 S. R. Lishmund,[1] G. M. Oakes,[1] B. J Barron[2] And F. L. Sutherland[2]

[1]New South Wales Department of Mineral Resources, St Leonards, NSW 2065, Australia.

[2]Division of Earth and Environmental Sciences, Australian Museum, Sydney NSW 2000, Australia

ABSTRACT

Eastern New South Wales has produced over 500 000 carats of diamonds, mostly from the Copeton-Bingara area in the north. A local derivation is indicated by their distinct tribal character and lack of surface damage, while carbon isotopic values and composition of mineral inclusions are unlike those in diamonds from conventional diamond-producing areas, for example Yakutia (Siberia), Kapvaal (South Africa), or Argyle (Western Australia). The eastern New South Wales tectonic setting is also unlike that for 'conventional' diamonds, involving a Phanerozoic sequence of accreted subduction terranes, with relatively thin hot crust. A subduction diamond model is developed to explain the origin and geology of eastern New South Wales diamonds. This model involves prolonged subduction of mature oceanic crust, resulting in the development of a low-temperature metamorphic window into the diamond stability field within the downgoing slab at half the depth required by conventional models. The diamonds are preserved at depth by termination of subduction, and brought to the surface by obduction or by entrainment in suitable magmas. The type of diamond formed depends on the original source rock (diamond affinity is shown in brackets): carbonaceous-rich and carbonaceous-poor trench sediments (CaFeMgMnsilicate, eclogitic), graphite-poor basalt and gabbro (eclogitic) and lherzolite (peridotitic). The model predicts that the New South Wales diamonds are young (Phanerozoic) and that their features, age, associated minerals and geographic distribution relate to New South Wales tectonic provinces. The subduction diamond model extends the range of indicator minerals to include macro-diamond, corundum, Na-garnet, with a new series of carrier magmas (basanite, nephelinite, leucitite). Unsuccessful previous exploration in New South Wales, based on conventional models, has produced some positive results when reinterpreted in terms of the subduction diamond model. A reevaluation of diamond potential of eastern New South Wales is warranted.

Keywords: basanite, blueschist, calc-silicate, diamond, eclogite, nephelinite, peridotite, Phanerozoic, subduction.

INTRODUCTION

Alluvial diamond is known from numerous localities in New South Wales, and elsewhere in the eastern Australian highlands (MacNevin 1977; Sutherland et al. 1994). The most significant production of alluvial diamond was from deposits in and around the Copeton and Bingara areas (New England), the Cudgegong area and the Macquarie River. Minor production occurred at Mt Airly (western Sydney Basin), near Mittagong (southern Sydney Basin), Crookwell (Lachlan Fold Belt) and other parts of the eastern highlands. Since the mid- 1960s there has been intensive exploration for the source of these diamonds, but this has been unsuccessful in locating either kimberlite or lamproite, although minor diamonds have been found in basanite and nephelinite diatremes. In terms of conventional models, which require ancient thick lithosphere for diamond formation, the occurrence of significant diamonds (perhaps 500 000 carats) in the mid-Phanerozoic fold belt of eastern New South Wales is anomalous. Most previous workers therefore assumed a remote ancient cratonic source for the New South Wales alluvial diamonds, but the nearest cratonic block is about 1000 km away, west of Broken Hill. The Copeton diamonds exhibit many features which indicate a local high-grade source so Barron et al. (1994) developed a new conceptual model for forming diamond rich eclogite at approximately half the conventional diamond-formation depth, involving prolonged subduction of mature oceanic crust under continental lithosphere. This subduction diamond model is developed further here. There are implications for diamond exploration within all Phanerozoic terranes that have alluvial diamonds.

CONVENTIONAL MODELS OF DIAMOND FORMATION

Conventional constraints on diamond formation

The most important constraint on the formation of diamond within the earth, for a typical geothermal gradient (oceanic), is that a depth of at least 250-300 km is required for diamond to be stable. This is a mantle depth well within the asthenosphere. Although the geothermal gradient for most continental lithosphere is considerably lower than oceanic geotherms, such continental lithosphere is1 not thick enough to reach the diamond stability field. For diamonds to form within continental lithosphere, the geothermal gradient must be much lower again; for instance the lower lithospheric keel portions of Archaean/Proterozoic cratons are several hundreds of degrees cooler than the average continental geothermal gradient, enough for the base of the cratonic keel to lie within the diamond stability field (150-250 km, 1050°C and higher, and 4.5-5.5 GPa). These keel conditions were probably established by the mid-Proterozoic, because most diamond crystallisation ages are older than 1000 Ma (S. E. Haggerty pers. comm. 1993). Although the diamonds are ancient they are delivered to the surface as xenocrysts and in xenoliths by kimberlite/lamproite carrier intrusives, some of which are as young as Mesozoic. This time-delay constraint requires continuous preservation of the ancient diamonds at depth, which can only happen if the entire column of lithosphere, from the surface down, is tectonically stable over the same period of time. Hence the diamonds brought up in diatremes cannot have a crystallisation age older than the cratonised rocks at the surface.

Diamonds are generally considered to be of two main types: P-diamonds having peridotitic affinity with primitive Archaean origin; and E-diamonds of eclogitic affinity, inferred to have formed from surface-derived carbon that was subducted and then trapped at the required depths before 1000 Ma. Such diamonds should grow in a mechanically static environment at relatively constant high pressure and elevated temperature, but many diamonds are strongly zoned with both growth and resorption structures, so some sort of fluctuating growth mechanism is involved.

The traditional constraints force exploration companies to search for kimberlite/lamproite because these are the only magmas known to originate from the required depths. They rely heavily on drainage sampling for kimberliteassociated minerals (the so-called indicator minerals), together with mapping and airborne geophysical techniques, to locate target intrusives, and subsequently on petrographic and mineralogical analysis to determine prospectivity. The aim of this approach is to identify intrusives which may have entrained lithospheric diamonds. Some exploration companies also rely on searching for detrital micro-diamonds (by convention those diamonds less than 0.4 mm in diameter) to locate prospective intrusives. Experimentally, kimberlite magma acts as a significant catalyst in the liquidus crystallisation of well formed octahedral micro-diamonds (Arima et al. 1993) so the naturally occurring micro-diamonds probably form as crystals in the magma while it is rising through the asthenospheric diamond stability field.

Conventional models applied to eastern New South Wales

The conventional models imply that eastern New South Wales and much of eastern Australia is not prospective for diamonds because the lithosphere is relatively thin, only reaching to about 85 km (E. Scheibner pers. comm. 1994), a little more than half the depth to the conventional diamond stability field. Kimberlite and lamproite are not known in the region, while the maximum age of the basement is about 550 Ma. Notwithstanding these difficulties, the significant alluvial deposits (about two million diamonds) of the region have attracted exploration based on conventional concepts and techniques since the late 1960s. This exploration located many diatremes of nephelinitic, basanitic and basaltic chemistry (Ferguson & Sheraton 1979; L. M. Barron et al. unpubl. data 1995). These magma types originate at 2.0-3.0 GPa, 80-100 km depth (Green 1991), well above the conventional diamond stability field. Nonetheless, small numbers of diamonds have been recovered from such intrusives in New South Wales; exploration companies have dismissed these as freak occurrences or the result of contamination during sample processing.

Characteristic features of eastern New South Wales diamonds which are difficult to explain by conventional models include:

1) They have a distinct but local deposit tribality in terms of colours, shapes, mineral inclusions and surface features, including a lack of abrasion on most stones (historical observations noted in MacNevin [1977], and supported by our own observations on 350 Copeton diamonds recovered in the late 1970s, and 150 diamonds from other New South Wales localities). This tribality is a clear indication of a local diamond-rich source.

2) They have mineral inclusions such as grossular garnet and coesite (Sobolev 1984; Meyer et al. 1994) which are unknown from the conventional diamond paragenesis.

3) They may be composed of anomalously heavy carbon (Sobolev 1984; Sutherland et al. 1994; Meyer et al. 1994), different from other analysed diamonds which are composed of light carbon (Galimov 1991).

4) They commonly have laminar features (called seams or knots by diamond polishers).

5) They commonly have many deep euhedral dodecahedral-like indentations to 1 mm in size.

6) Many have a yellow colour related to high nitrogen contents (> 550 ppm, Taylor et al. 1990; 3000 ppm, Meyer et al. 1994), applicable to about half

of the diamonds (MacNevin 1977). They appear to be mostly unzoned (Meyer etal. 1994).

7) Many are white with variable but lower nitrogen contents (as low as 10 ppm) and may be complex zoned (Meyer etal. 1994).

8) Some have an extreme toughness and durability, in terms of gem processing and industrial use, which has been noted for a large fraction of the New South Wales diamonds (Chalmers 1956; MacNevin 1977; Joris 1986; Sutherland et al. 1994). This toughness is unmatched in the much larger pool of diamonds processed from conventional sources (A. Joris pers. comm. 1994; F. Kaminsky pers. comm. 1994).

Other areas where alluvial diamonds occur in a setting of relatively young lithosphere include western North America, Kalimantan, southern China, Kamchatka and the Urals. Some of these localities have been mined for diamond, and all are difficult to explain in terms of conventional models.

THE SUBDUCTION DIAMOND MODEL

The subduction diamond model involves forming diamond within a subducting slab of oceanic lithosphere. Non-carbonate carbon is present in the upper portions of such a slab in the form of organic matter and inorganic precipitated carbonaceous material in trench sediments, and as graphite dispersed in altered basalts/gabbros/ serpentinised ultramafics. If the downgoing slab is thin,

its internal temperature would increase rapidly as it descended and graphite would remain stable until the diamond stability field is reached at 170 km or more (Kesson & Ringwood 1989). However, if the downgoing slab is thick and mature (i.e. 100-200 million years has passed since formation at a spreading centre), and descends at an appropriate rate, it will remain many hundreds of degrees cooler than the asthenosphere during subduction (Peacock 1990; Wiens et al. 1993; Frohlich 1994). If this process is aided by conductive cooling on the side of the continental block due to prolonged subduction, geophysical modelling indicates that the temperature in the slab can be up to 1100°C cooler than the asthenosphere at the same depth. This anomalously low temperature in the descending slab creates an unusual window into the diamond stability field at pressures as low as 2.0-2.5 GPa at about 80-100 km depth (see Figures 1, 2). During this cold subduction, the upper portions of the subducting slab, comprising trench sediments and basalt and rodingite, undergo near-isothermal prograde metamorphism to lawsonite blueschist, then eclogite and eventually diamond eclogite facies. The middle and lower portions of the slab comprise carbonated serpentinised lherzolite which is metamorphosed to ultramafic granulite ' and metaharzburgite, a variation on Schultze's (1986) concept. The subduction diamond window can thus generate both eclogitic and peridotitic diamonds.

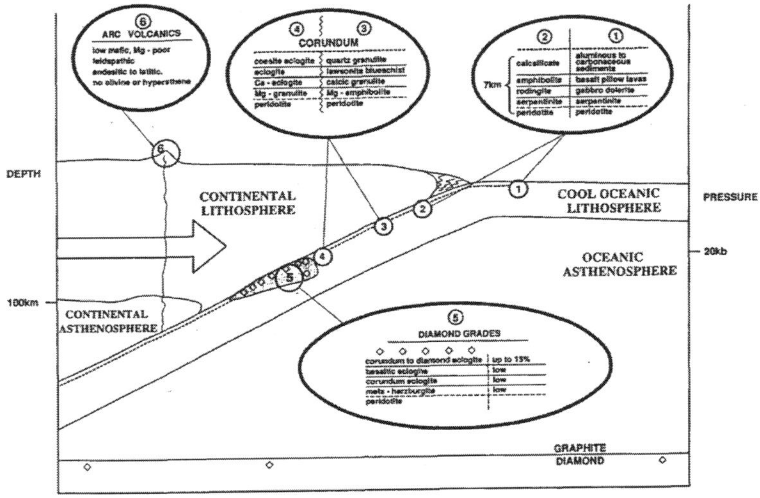

Figure 1. Diagram showing the prograde formation of diamonds during Phanerozoic subduction within a low-pressure dynamic window into the diamond stability field. The window is based on calculated thermal profiles (Peacock 1990; Wiens et al. 1993; Frohlich 1994) for prolonged subduction of mature oceanic crust which indicate a relict minimum slab temperature of 200-400°C within the window. The diamond window closes with increasing depth below about 110 km because there is asthenospheric heating from both below and above the subducting slab.

Under optimal subduction conditions, diamonds can form continuously as the cold slab descends. However, these diamonds are ephemeral, being destroyed as subduction carries them below the subduction diamond window. Termination of subduction is necessary to preserve the subduction diamonds for transport to the surface, and the diamonds so preserved must have an age marking the end of subduction. In New South Wales, the ages (± 10 million years) of 520, 435, 380, 315, and 254 Ma represent the known major terminations of Phanerozoic subduction (E. Scheibner pers. comm. 1994) so crystallisation ages of subduction diamonds found at the surface may match one of these dates. Mineral inclusions in five Copeton (New South Wales) diamonds have been Ar-Ar dated with an average age of 320 Ma with a range of ±31 million years (P. Kennewell pers. comm., Guff Resources Pacific 1993), close to one of these inferred dates. This diamond age also falls within a Phanerozoic period with a significantly isotopically heavy atmospheric carbon pool (230-330 Ma, Berner 1991) so it can explain the heavy isotopic composition of some Copeton diamonds (Sobolev 1984). Prolonged insertion of the subducting slab at the base of continental lithosphere would be a mechanism for generating a range of ages (1-10 million years) for preserved diamond, but age dating is probably not accurate enough to measure this sort of variation. Blueschist and phengite eclogite are ephemeral in essentially the same way as subduction diamonds, so when exposed at the surface their ages also represent possible ages for the formation of subduction diamonds. Note that the age of crystallisation of subduction diamonds erupted on the surface must be equal to or less than the age of metamorphosed and folded basement rocks at the surface, analogous to the age constraint on diamond for conventional models.

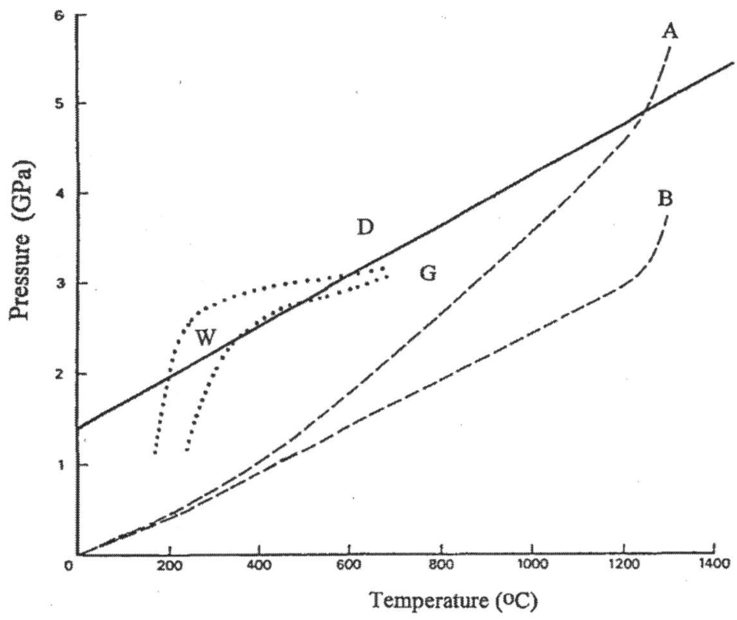

Figure 2. Pressure-temperature diagram showing the stability field of diamond relative to graphite, several geotherms, and the subduction paths producing the subduction window into the diamond stability field. A, average continental geotherm based on a surface heat flow of 44 mW/m2 and B, average ocean basin geotherm based on a surface heat flow of 46 mW/m2 (both after Harte 1978); D, diamond and G, graphite with the boundary taken from Kennedy and Kennedy (1976); W, subduction diamond window with subduction profiles affecting the top (left) and bottom (right) portions of the window.

Mechanisms for bringing subduction diamonds to the surface range from obduction of portions of the subducted slab by closure of an ocean basin to piercing of a static detached slab by carrier intrusives. For the former, the age of presentation at the surface may be nearly the same as the age of crystallisation of the diamonds. The relatively shallow depth of formation of subduction diamonds (80-100 km) allows them to be sampled in a detached static slab by magmas such as leucitite, melilitite, nephelinite, basanite, ultramafic lamprophyre and possibly alkali basalt. Diamonds have been found in several diatremes and dykes with these compositions in eastern Australia (see below, and MacNevin 1977; Ferguson & Sheraton 1979; Barren et al. 1994; Sutherland et al. 1994).

Static detached slabs are common (Pe-Piper & Piper 1994) at lithospheric depths, but can they stay cool enough for subduction diamonds to persist? Tertiary diatremes in Colorado carry xenoliths with surviving blueschist eclogite mineralogy formed during the Franciscan period confirming that a cold slab stored at the base of the crust can retain its original low temperature for more than 50 million years (Helmstaedt & Doig 1975). In New South Wales the presence of diamonds in Jurassic to possibly Cainozoic intrusives within the Palaeozoic Lachlan Fold Belt implies that subduction diamonds can be held in 'cold storage' at depth for over 100 million years.

Palaeozoic tectonic history in New South Wales has been dominated by accretion with tectonic units occurring in elongate north-south belts, predominantly younger to the east (see Figure 4). The subduction diamond model indicates that diamond provinces should be bounded by these major tectonic units. Thus subduction diamonds found in intrusives at the surface should follow the same basic age constraint, so that New South Wales diamonds are expected to be older in the western provinces and younger in the eastern. There may be local reversals in this simple age pattern, caused by overlapping of subduction zones, and tectonic rearrangements. Subduction diamond crystallisation requires an optimal combination of conditions, so diamonds presented at the surface are expected to be confined to much smaller regions within these tectonic provinces.

DISCUSSION

Many authors have invoked subduction in the formation of diamond, but the role of subduction has been restricted mostly to that of supply of carbon, with diamond formation taking place later at conventional pressures and temperatures (Helmstaedt & Doig 1975; Schulze 1986; Schulze & Helmstaedt 1988; Schulze 1992; Pearson etal. 1993). In contrast, Robinson (1978), Sobolev (1984) and O'Reilly (1989) offered brief suggestions that diamonds can form during actual subduction but with the depth unspecified. Also favouring the formation of diamond during actual subduction, Kesson and Ringwood (1989) concluded that there is partial melting within the subducting slab at 170-450 km, and that eclogitic diamonds crystallise from these melts within the subducting slab, while peridotitic diamonds crystallise from these melts after they have migrated into the overlying asthenosphere and become heavily hybridised. Sutherland et al. (1994) provisionally adopted this model to explain the Copeton diamonds. However, these previous subduction diamond models cannot explain the presence of diamond in eastern Australia because their diamond crystallisation depths mean that only kimberlite/lamproite could carry such diamonds to the surface, yet such rocks are absent.

The critical feature of our subduction diamond model is the dynamic window into the stability field of diamond at pressures about half of that required for conventional diamonds to form. This enables diamonds to be generated below regions such as eastern New South Wales. Although this model is based.on the Copeton region (a Phanerozoic terrane), there seems no essential requirement that it be restricted to the Phanerozoic. However, in other than cratonic keels there must be a limit to the survival of diamonds at depth, requiring a loose correlation between the age of crystallisation of diamonds and the age of emplacement of the diamonds at the surface.

Characteristics of subduction diamonds

The affinity of subduction diamonds will depend on the source material in the descending slab. Each different source rock chemistry will impose a characteristic carbon isotopic composition and growth mechanism on diamond, as well as a characteristic mineralogy and element geochemistry for included minerals and gases and associated minerals, as summarised in Table 1. This table has been assembled with the specific aim of conceptualising the inferred characteristics of diamonds formed by the subduction diamond model. Note that diamond colours on Table 1 refer only to the two most common colours. The most important of the predicted characteristics are discussed below.

Table 1. Some significant characteristics of subduction diamonds.

Original rock	Diamond affinity	Nitrogen content; source	Carbon isotopic composition	Crystallinity; surface features; durability	Diamond grade; colour	Potentially anomalous components
Trench sediments	calc-silicate, eclogitic	extreme, variable; sediment, sea	air, sea, at time of subduction	unzoned or zoned; growth twins, indentations, shiny; tough or soft	< 30%; yellow, white	Na, K, Ca, Mg, Fe, Mn, Si, Ti, Al, P, S, Zr, N, B, halogens
Basalt	eclogitic	high; SM*	primitive, SM	zoned growth/ ablation; stepped or etched; soft	low; white	Na, Ca, Al, Ti, S, Fe, Mn, halogens
Gabbro	eclogitic	moderate; primitive but SM	primitive, SM	zoned growth/ ablation; stepped or etched; soft	trace; white	Na, Ca, Al, Ti, S
Ultramafics	peridotitic	low; primitive but SM	primitive, SM	zoned growth/ ablation; stepped or etched soft	trace; white	Na, Mg, Cr, Mn, Co, Ni

*SM, may be contaminated by mixing with seawater

NITROGEN CONTENTS

Because oceanic crust is involved in the subduction diamond model, nitrogen levels in the source rocks should fall into categories dictated by contributions from the obvious natural sources, namely detrital nitrogenous material, atmospheric nitrogen dissolved in sea water and primitive nitrogen. Trench sediments have complete access to the first two, while pillow basalts, gabbros and ultramafics have progressively less access to the first two sources. There may be departures from this simple relation, for instance through contamination/hydration of the upper part of the ultramafic lithosphere. Nitrogen levels in subduction diamonds are expected to follow these trends, namely potentially very high but strongly variable (trench sediments), moderate but variable (pillow basalts), low uniform (gabbro) and very low uniform (ultramafics). Nitrogen determinations by Taylor et al. (1990) and Meyer et al. (1995) are consistent with this simple pattern.

A short summary of nitrogen aggregation in synthetic diamonds is presented in Kesson and Ringwood (1989) involving dispersed substitution by N atoms for single carbon atoms and conversion of this substitution to N-N substitution for neighbouring pairs of carbon atoms, which form quartets, minor triplets and platelets. Nitrogen aggregation in diamonds is a diffusion-limited process, promoted by high temperatures and prolonged but suitable confinement times. Kesson and Ringwood (1989) estimated that at 1050^, it takes about a billion years to convert 20% of the nitrogen pairs into groups. Calibration errors, 80% conversion, and 50°C change in temperature each change the confinement period by a factor of ten, so the estimation is crude. In addition, the interpretation of measured aggregation states is based on the assumption that the initial state of nitrogen in diamond is as completely dispersed single N atoms substituted for carbon atoms within the diamond lattice. This assumption is undoubtedly valid for natural and synthetic diamonds formed at high temperature, but need not be valid for diamonds formed at low temperature from low-temperature source materials. For diamonds formed by our subduction model, the dynamics of nitrogen aggregation suggest that present aggregation states may represent the entrapment states, which may be highly anomalous. For subduction diamonds derived from trench sediments, the aggregation states of nitrogen may be affected also by the loss of hydrogen from ammonia/ amide-based complexes trapped within the diamond.

GROWTH MECHANISMS

The subduction diamond model predicts that diamond growth from graphite is caused by rapidly increasing pressure at near-constant low temperature. Graphite has a flat hexagonal geometry for C-C bonding which is very different from the three dimensional tetrahedral C-C bonding in diamond. This difference in carbon bonding makes conversion of one form of carbon to the other difficult. For instance, diamonds at the Argyle deposit have remained metastable relative to graphite for more than 1100 million years. In the manufacture of diamond from graphite, a metal catalyst which becomes liquid under the P/T conditions of synthesis, is used to aid the conversion from hexagonal to tetragonal carbon bonding. This raises the question whether there are catalysts available for the formation of subduction diamonds. Fullerenes are inorganic hollow cage-like molecules of carbon and experiments (Regueiro et al. 1992) show that fullerenes convert to diamond at room temperature under high-pressure shearing conditions, the conversion being aided by the tetrahedral C-C bonding in fullerenes. Synthetic fullerenes can be produced under atmospheric conditions by incomplete combustion of hydrocarbons, and natural fullerenes are known to occur (Buseck et al. 1992). If traces of fullerene were present in trench sediments they would catalyse the formation of diamonds under the shear-stressed conditions of the subduction diamond window.

Geologically rapid solid-state growth is envisaged for subduction diamonds formed in trench sediments with abundant graphite. Diamonds that crystallise in this way will be single-stage growth crystals (unzoned), while closely spaced growth centres may result in growthtwinned crystals. These crystal features are consistent with a relatively slow rate of growth (on a human time scale) with a low to moderate driving force, using Sunagawa's (1990) work on diamond growth and morphology. The growth mechanism has important implications for the interaction of diamond with other minerals in the model source rock. Prior to diamond forming, the hard prograde minerals such as garnet and omphacitic pyroxene would present euhedral faces to the soft graphite-rich matrix. Rapid growth of diamond, to the extent of partially enveloping these other minerals, would result in diamond carrying euhedral 'indentations' which possess the crystallography of these other minerals. This is in complete contrast to the way diamond imposes its crystallography on the shapes of mineral inclusions (Sobolev 1977). This model growth mechanism explains the common and large euhedral indentations characteristic of eastern Australian diamonds (see Figure 5) typical of the 'tough' Copeton diamonds. Since Sobolev (1984) reported that most of the Copeton diamonds are of eclogite affinity, garnet is the most probable indenting mineral. The common occurrence of these euhedral indentations at apices of the diamond suggests that they represent a restriction to diamond growth. Furthermore, the indentation faces are sometimes curved (Figure 5), indicating that there was some limited coeval growth. Eastern New South Wales diamonds with these indentations are rounded tetrahexahedroid/rhombododecahedral crystals.

The implication is that this shape may represent a growth-limited shape rather than a resorption or ablation shape for the case of subduction diamonds. The features result in the following interpretation: the growth environment resulted in close packing of diamonds which were forced to grow and partially envelope pre-existing euhedral minerals trapped between them.

Due to the low levels of highly dispersed graphite in graphite-poor trench sediments and in basaltic to gabbroic and ultramafic source rocks, diamonds within these rocks would grow from a vapour stream (methane?) and so exhibit a complex sequence of asymmetric growth/ablation features.

TOUGH AND SOFT DIAMONDS

A significant fraction (> 60%) of Copeton diamonds are known for their toughness, a feature established more than sixty years ago by gem cutters through polishing, and widely quoted (Chalmers 1956; MacNevin 1977; Joris 1986; Sutherland et al. 1994). On the basis of a prolonged involvement with diamond polishing and industrial use of diamonds, Joris (pers. comm. 1994) has indicated that this toughness is absent in diamonds from conventional diamond provinces. Joris (1986) proved that this toughness is related to two features, namely a simple low-temperature mechanical toughness due to the presence of the abundant 'Naats' or 'Knots' (growth twinplanes) compounded by an extreme resistance to hightemperature ablation. This ablation resistance was established not only by extensive comparative hightemperature industrial use of diamonds from many countries, but also by comparative experimentation in a high-temperature furnace (Joris 1986). In our preliminary model we predicted that this toughness would correlate with diamonds formed within the graphite-rich trench sediments; they would be unzoned single-growth stage crystals with high nitrogen levels. Meyer et al. (1994) have established that diamonds with these characteristic comprise a dominant fraction of the Copeton diamonds. Other diamonds noted for their extreme ablation resistance are the northern Siberian polycrystalline micro diamonds formed from graphite by meteorite impact (F. Kaminsky pers. comm. 1994). The common element for these two types of tough diamond is that rapid solid-state growth results in the formation of unzoned diamonds.

The converse to this prediction also appears to be true, namely that strongly zoned diamonds will be soft when compared with the tough Copeton diamonds. The complex growth/resorption structures must present points of weakness to ablation and wear. This concept applies not only to about 40% of the eastern New South Wales diamonds (soft diamond, Sutherland et al. 1994; complex zoned diamond, Meyer et al. 1994, 1995), but also to the conventional diamond sources (soft diamond, A. Joris pers. comm. 1994) and to diamonds from the western Urals (soft diamond, F. Kaminsky pers. comm. 1994). The subduction diamond model identifies the most likely growth mechanism and source rocks for these soft diamonds, namely vapour growth within graphite-poor trench sediments, basalts, gabbros and ultramafics.

The isotopic composition of carbon in subduction diamonds will reflect the relative contributions of primordial. versus surface pool carbon, the latter being influenced by the ratio of inorganic to organic carbon in the source rock. Certain source conditions will influence the isotopic composition of carbon so strongly that the source conditions should be revealed by cross-element correlations. For instance, subduction diamonds derived from trench sediments are predicted to show a negative correlation between 5I3C and nitrogen content, this correlation being related to the fraction of organically fixed nitrogenous low-8I3C carbon. Diamonds from the other host rocks are expected to be composed of primordial carbon with weak to strong contamination due to the redistribution implied by a vapour growth mechanism.

The sulfur isotopic composition of sulfide inclusions in subduction diamonds should also be an indication of source rock and geological age. Holser's (1977) work showed that there has been a Z-shaped variation in 834S of oceanic sulfate over the last 650 million years, so it might be feasible to qualitatively date sulfide inclusions in Phanerozoic diamonds.

ASSOCIATED MINERALS

The subduction diamond model predicts that unusually high to extreme levels of many elements may be present in mineral inclusions in subduction diamonds. The elements (see Table 1) would be aligned to the affinity of the host rock, with the subduction model predicting three types of host rock/anomalous element packages, namely trench sediments, basalts/gabbros and ultrabasics. These three groups appear to match the diamond affinities reported by Sobolev (1984) and Meyer et al. (1994) for Copeton diamonds, namely calc-silicate, eclogite and peridotite. New South Wales eclogitic diamonds with elevated levels of some elements (N) are associated with orange-brown to pink or pale garnets with elevated levels of Na, Ca (Sobolev 1984) and Fe, Mn, Ti (see Figure 3). Elevated levels (> 0.05 wt%) of Na (in garnet, Sobolev 1984) and K (in clinopyroxene, Meyer et al. 1995) probably reflect the pressures necessary to make diamond rather than unusual source rocks (McCandless & Gurney 1989; Schulze 1992). The strongest variation in included mineral types is expected within diamonds derived from trench sediments, because of large natural variations in these source rocks. Chemical anomalies in the included minerals arise because the extreme compositional variations established in original and low-grade alteration phases may carry through into the high-grade metamorphic minerals. For the Copeton diamonds, examples of such compositional variations include grossular garnet, coesite and anomalous Mg-rich and Fe-rich clinopyroxene as inclusions in calc-silicate and eclogitic diamonds (Sobolev 1984; Sutherland et al. 1994; Meyer et al. 1995). The work of Evans (1990) on theoretical blueschist facies implies that lawsonite may be stable for pressure/temperature conditions within the lower temperature portion of the diamond stability window, indicating that lawsonite might be found as a mineral inclusion in subduction diamonds.

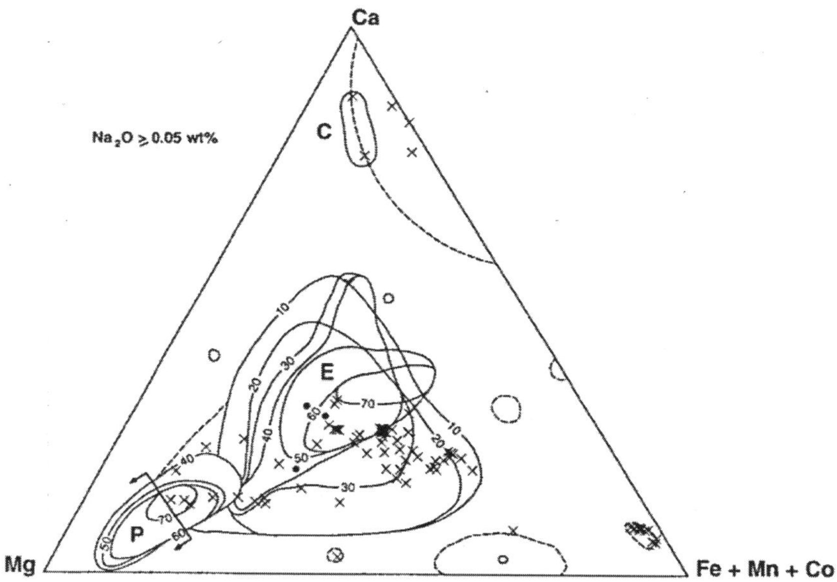

Figure 3 Major element chemistry for garnet associated with diamond, contoured for N^O > 0.05 wt% (1300 published microprobe analyses from Fipke et al. 1989; Sobolev et al. 1994; Boyd & Meyer 1979; Glover & Harris 1985; Ross 1989; Schulze 1992). —, defines the maximum Na2O content (expressed as 100*wt%) found in garnet coexisting with diamond, representing a topographic high in Na2O chemical space. , contains highNa garnets with an uncertain relation to diamond (e.g. garnets in diamondiferous diatremes and alluvials). •, three experimental ultra-high pressure > 17.5 GPa garnets (Na2O>2 wt%, Irifune et al. 1994). X, High-sodium garnets from New South Wales. The related diamond affinity is indicated as: P, peridotitic; E, eclogitic; and C, calcsilicate (two analyses from garnet in a Copeton diamond [NajO>0.22 wt%, Sobolev 1984]). The location of G10 garnets (after Sobolev 1977) is shown by that portion of the P field between the arrowed line and the Mg apex.

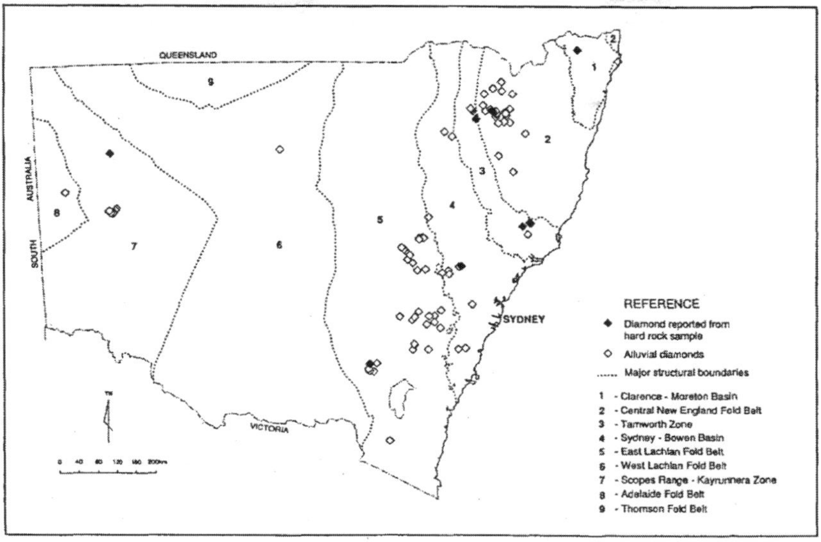

Figure 4 New South Wales diamond provinces based on the major tectonic units of Scheibner (1993)

GRAPHITE-RICH TRENCH SEDIMENTS

Eclogitic/calc-silicate subduction diamonds generated from graphite-rich trench sediments should have 'tribal' characteristics in surface deposits because a small xenolith of diamond eclogite can release a large number of fraternal. diamonds. Figure 5 shows an example of a diamond which has important characteristics identified as tribal through examination of large populations of eastern Australian diamonds. Such diamonds may also exhibit unusual levels of trace and volatile components (see Table 1). Their 8l3C should be characteristic of the subduction age, allowing for organic fractionation. Based on critical similarities between the geological history of northern New South Wales and the Urals (Zonenshain et al. 1990; Dobretsov & Dobretsova 1988) during the Late Carboniferous, we predict that southeastern Urals eclogitic diamonds may exhibit some of the unique characteristics of New South Wales diamonds, such as toughness, growth twins, large euhedral indentations, high nitrogen contents and heavy isotopic composition.

Additional aspects of the subduction diamond model

Other aspects inherent in this conceptual model include the following points:

1) Within favourable structural provinces, subduction diamonds should occur in intrusives situated between correlative mafic-poor latitic arc volcanism

and lawsonite blueschist/eclogite, these being indicators of the termination of prolonged subduction of mature oceanic crust.

2) Micro-diamonds may be present in subduction diamond settings, but cannot crystallise directly from potential carrier magmas. Based on evidence from New South Wales, micro-diamonds in subduction diamond terranes are most probably derived from macro-diamonds through ablation or fragmentation, and may be anomalously scarce.

3) Corundum-bearing assemblages can be produced by prograde metamorphism in the descending slab and this could explain the association of certain types of sapphire with alluvial diamond apparent in eastern Australia (Lishmund & Oakes 1983; Oakes et al. 1996) and also observed in northern California (W. Rohtert pers. comm. 1994).

4) Heating of the detached slab during the storage period may cause thermal destruction of some subduction diamonds and this would change the relative abundances of diamonds of different affinity. Heating from below the slab would decrease the fraction of P-diamonds, while heating from above would decrease the eclogite plus calcsilicate fraction. For the same degree of heating, the 'softer' zoned diamonds show experimentally almost catastrophic ablation compared with the tougher Copetontype diamonds (Joris 1986, p. 19), so the response of dominant diamond affinity to heating would not be uniform.

IMPLICATIONS FOR EXPLORATION

Diamond exploration in eastern New South Wales has been based historically on conventional diamond models, despite the disparities with observed tectonics, lithologies and mineral chemistries. The results obtained have been disappointing in the context of conventional models, in that diamonds have been found only in source rocks that have originated from too shallow a depth to be considered prospective for diamonds. Other target intrusions have been abandoned without evaluation because of this. However, the diamond-in-diatreme results are highly encouraging in the context of the subduction diamond model presented here. All known diatremes should be reassessed, particularly given that relatively few targets have been properly sampled or evaluated, while numerous potentially prospective targets remain untested. Elsewhere in the world, experience from exploration based on conventional diamond genesis models indicates that a large number of targets may need to be tested before an economic deposit is found.

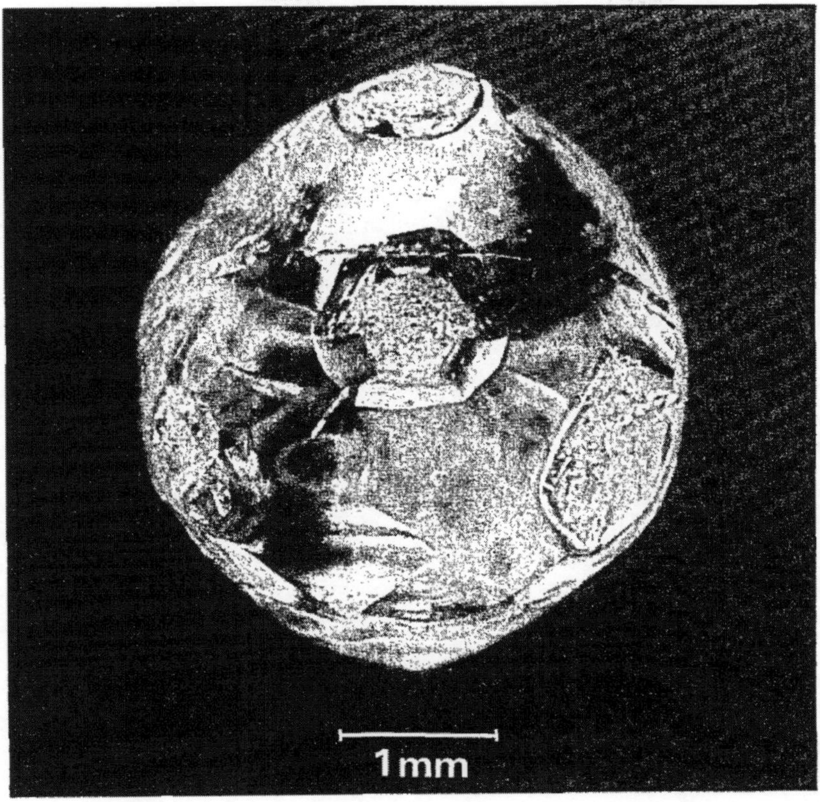

Figure 5. Copeton diamond (white 0.43 ct diamond #96.10, courtesy G. White) rather rounded tetrahexahedroid with four euhedral dodecahedral-like indentations. This diamond shows many important characteristics of Copeton diamonds, including the numerous large indentations, and the growth twin creating a'Knot'.

The subduction diamond model provides a sound basis for reassessing the results of previous exploration and for locating and evaluating new targets. Considering the outstanding success exploration companies have had in locating intrusives with para-kimberlitic indicator mineral searches, such techniques remain an important first step in eastern Australia, requiring only the inclusion of the new indicator minerals defined by the subduction diamond model. Prospective regions can be refined by considering the new additional factors such as the occurrence and distribution of mafic-poor arc volcanics, lawsonite blueschists and eclogite

It should also be noted that G10 garnets, regarded as highly significant in the context of conventional models for diamond formation, occupy less than 7% of the overall compositional fields for garnet coexisting with diamond world-wide (see Figure 3). A focus on G10 garnets alone would significantly weaken exploration strategies for locating diamond-bearing intrusives.

High-sodium garnets of any composition should be treated as anomalous. Several New South Wales localities are confirmed with such garnets (see Figure 3). These analyses mostly overlap the P, E and C fields, but a new field of uncertain relation to diamonds is created at high Fe + Mn, with elevated levels of Mn. High-sodium garnet and diamond occur at Copeton, Beryl/Cudgegong, Isabella Bridge/Oberon (basanite), Jugiong* (nephelinite monchiquite), Kayrunnera/White Cliffs* (leucite basanite), Lee Creek/Rylstone (basanite), Mt Brown/Kyogle* (leucite nepheline basanite), Miowera/Bingara (basanite), Prince Charlie Creek/Gloucester*, Ruby Hill/Bingara* (basanite), The Ovens/Rylstone* (nepheline leucite basanite), and Umbiella/Rylstone (leucite nepheline basanite).

The subduction diamond model also has implications for the minimum size of target which should be sought. Many exploration programmes based on conventional models focus on targets larger than a certain size, which can be as much as 5 ha. However, because of the potentially very high diamond content of diamond eclogites formed from trench sediments in the subduction model, much smaller targets could be economic. There is also the enhanced potential for small but rich alluvial deposits such as those at Copeton.

CONCLUSIONS

More than two million diamonds were produced in eastern New South Wales in spite of a lack of conventional target rocks and a Phanerozoic setting more than 1000 km distant from the nearest craton. Conventional models of diamond genesis do not apply to these diamonds which are characterised by many unique features. Diamonds can be produced by continental collision through rapid doubling of lithosphere with deep formation of diamond (Wang et al. 1990). This process is an unlikely occurrence in the Copeton region because the indicative molasse is absent in New England (E. Scheibner pers. comm. 1995).

We have developed a subduction diamond model to explain the nature and distribution of the eastern New South Wales diamonds. Critical to this model is a dynamically maintained 2.0 GPa 'shallow' window into the diamond stability field, which appears because of an optimal coincidence of many features, including slab age, subduction rate and termination of subduction. Previous authors (see Discussion) have invoked subduction in the formation of diamonds. For the conversion of graphite to diamond, none of these authors considered a depth shallower than that involved in conventional diamond models. Without this shallow diamond window, deeply sourced carrier magmas of kimberlite or lamproite composition are required if diamonds are being sought in diatremes.

Our subduction diamond model predicts that the diamond growth environment (rapid diamond growth as a result of rapidly increasing pressure at almost constant low temperature) is very different from conventional diamond genesis models which assume a rather static environment with a slow fluctuating diamond growth/ resorption rate. Based on an earlier version of the subduction diamond model, Barron et al. (1994) made predictions concerning major to cryptic features about the eastern New South Wales diamonds. These predictions are consistent with recent measurements on Copeton diamonds (Meyer et al. 1994, 1995) including diamond affinity, nitrogen content and internal zonation. The extended subduction diamond model explains the occurrence of diamond in Phanerozoic mobile belts, obducted terranes and other settings outside ancient cratons. For such diamonds in eastern New South Wales, their strongly tribal characteristics, age of crystallisation and anomalous aspects are natural consequences of the subduction diamond model. The model provides regional and local optimal indicators for formation of subduction diamonds.

It explains the occurrence of diamond-bearing intrusives which are not kimberlitic or lamproitic. The subduction diamond model provides a basis for a new approach to exploration in eastern Australia and in similar settings elsewhere and the results of previous diamond exploration in this region now can be seen as highly encouraging instead of enigmatic or negative.

ACKNOWLEDGEMENTS

It is a pleasure to acknowledge the loan of New South Wales diamonds from the Museum of Victoria (W. Birch), Australian Museum (R. Pogson), A. Joris, C. and B. Ribaux, T. Smallwood, J. L. Tottenham, M. Walburn and G. White. The photography was performed by D. Barnes. Constructive comments by P. Temby and another reviewer are appreciated. This paper is published with the permission of the Director General of the New South Wales Department of Mineral Resources.

REFERENCES

1. ARIMA M., NAKAYAMA K., AKAISHI M., YAMAOKA S. & KANDA H. 1993. Crystallisation of diamond from a silicate melt of kimberlite composition in high-pressure and high-temperature experiments. Geology 21, 968-970.

2. BARRON L. M., LISHMUND S. R., OAKES G. M. & BARRON B. J. 1994. Subduction diamonds in New South Wales: Implications for exploration in Eastern Australia. Quarterly Notes Geological Survey of New South Wales 94, 1-23.

3. BERNER R. A. 1991. The long term carbon cycle and atmospheric CO2 over Phanerozoic time. American Geophysical Union Transactions 72, 165.

4. BOYD F. R. & MEYER H. O. A. 1979. eds. Kimberlites, diatremes and diamonds: their geology, petrology and geochemistry. 2nd International Kimberlite Conference Proceedings. American Geophysical Union, Washington DC.

5. BUSECK P., TSIPURSKY S. J., WANG S. & HETTICH R. 1992. Geological occurrence of fullerenes. Science 257, 215-217.

6. CHALMERS R. O. 1956. Diamond. In Some aspects of New South Wales gemstones, pp. 91-94.

7. Journal of the Proceedings of the Royal Society of New South Wales 89. DOBRETSOV N. L. & DOBRETSOVA L. V. 1988. New mineralogic data on the Maksyutovo eclogite glaucophane schist complex, Southern Urals. Transactions of the USSR Academy of Sciences, Earth Sciences Section 300, 111-116.

8. EVANS B. W. 1990. Phase relations of epidote-blueschists. Lithos 25, 3-23.

9. FERGUSON J. & SHERATON J. W. 1979. Petrogenesis of kimberlitic rocks and associated xenoliths of southeastern Australia. In Boyd F. R. and Meyer H. O. A. eds. Kimberlites, diatremes and diamonds: their geology, petrology and geochemistry, pp. 140-160. 2nd International Kimberlite Conference Proceedings. American Geophysical Union, Washington DC.

10. FIPKE C. E., GURNEY J. J, MOORE R. O. & NASSICHUCK W. W. 1989. The development of advanced technology to distinguish between diamondiferous and barren diatremes. Geological Survey of Canada Open File Report 2124.

11. FROHLICH C. 1994. A break in the deep. Nature 368, 100-101.

12. GALIMOV E. M. 1991. Isotope fractionation related to kimberlite magmatism and diamond formation. Geochimica et Cosmochimica Acta 55, 1697-1708.

13. GLOVER J. E. & HARRIS P. G. 1985. eds. Kimberlite occurrence and origin: a basis for conceptual models in exploration. Geology

Department and University Extension, University of Western Australia, Publications 8.

14. GREEN D. H. 1991. The earth's lithosphere and asthenosphere—concepts and constraints derived from petrology and high pressure experiments. In Drummond B. J. ed. The Australian Lithosphere, pp. 1-22. Geological Society of Australia Special Publication 17.

15. HARTE B. 1978. Kimberlite nodules, upper mantle petrology and geotherms. Philosophical Transactions of the Royal Society of London A288, 487-500.

16. HELMSTAEDT H. & DOIG R. 1975. Eclogitic nodules from kimberlite pipes of the Colorado Plateau: samples of subducted Franciscan-type oceanic lithosphere. Physics and Chemistry of the Earth 9, 95-111.

17. HOLSER W. T. 1977. Catastrophic chemical events in the history of the ocean. Nature 267, 403.

18. IRIFUNE T., RINGWOOD A. E. & HIBBERSON W. O. 1994. Subduction of continental crust and terrigenous and pelagic sediments: an experimental study. Earth and Planetary Science Utters 126, 351-368.

19. JORIS A. 1986 A Destiny in Diamonds. Boolarong Publications, Brisbane.

20. KENNEDY C. S. & KENNEDY G. C. 1976. The equilibrium boundary between graphite and diamond. Journal of Geophysical Research 81, 2467-2470.

21. KESSON S. E. & RINGWOOD A. E. 1989. Slab-mantle interaction: 2. The formation of diamonds. Chemical Geology 78, 97-118.

22. LISHMUND S. R. & OAKES G. M. 1983. Diamonds, sapphires and Cretaceous/Tertiary diatremes in New South Wales. Quarterly Notes Geological Survey of New South Wales 53, 23-27.

23. MACNEVIN A. A. 1977. Diamonds in New South Wales. Geological Survey of New South Wales Mineral Resources 42.

24. MCCANDLESS T. E. & GURNEY J. J. 1989. Sodium in garnet and potassium in clinopyroxene: criteria for classifying mantle eclogites. In Ross J. ed. Kimberlites and related rocks: Volume 2. Their mantle/crust setting, diamond and diamond exploration, pp. 827-832. Geological Society of Australia Special Publication 14.

25. MEYER H. O. A., MILLEDGE H. J. & SUTHERLAND F. L. 1994. Unusual diamonds and inclusions from S. E. Australia. Abstracts Geological Society of America, Seattle Meeting, 417.

26. MEYER H. O. A., MILLEDGE H. J. & SUTHERLAND F. L. 1995. Unusual diamonds and unique inclusions from New South Wales, Australia. Abstracts 6th International Kimberlite Conference, Novo Sibirsk, 379-381.

27. OAKES G. M., BARRON L. M. & LISHMUND S. R. 1996. Alkali basalts and associated volcaniclastic rocks as a source of sapphire in eastern Australia. Australian Journal of Earth Sciences 43, 289-298.

28. O'REILLY S. 1989. Nature of the east Australian lithosphere. In Johnson R. W. ed. Intraplate Volcanism in Eastern Australia and New Zealand, pp. 290-297. Cambridge University Press, Cambridge.

29. PEACOCK S. M. 1990. Numerical simulation of metamorphic pressure temperature time paths and fluid production in subducting slabs. Tectonics 9, 1197-1211.

30. PEARSON D. G., DAVIES G. R. & NIXON P. H. 1993. Geochemical constraints on the petrogenesis of diamond facies pyroxenites from the Beni Bousera Peridotite Massif, North Morocco. Journal of Petrology 34, 125-172.

31. PE-PIPER G. & PIPER D. J. W. 1994. Miocene magnesian andesites and dacites, Evia, Greece: adakites associated with subducting slab detachment and extension. Lithos 31, 125-140.

32. REGUEIRO M. N., MONCEAU P. & HODEAU J. L. 1992. Crushing C60 to diamond at room temperature. Nature 355, 237-239.

33. ROBINSON D. N. 1978. The characteristics of natural diamond and their interpretation. Minerals Science Engineering 10, 55-72.

34. Ross J. 1989. ed. Kimberlites and related rocks: Volume 2. Their mantle/crust setting, diamonds and diamond exploration. Geological Society of Australia Special Publication 14.

35. SCHEIBNER E. 1993. Structural Framework of New South Wales. Quarterly Notes Geological Survey of New South Wales 93, 1-35.

36. SCHULZE D. J. 1986. Calcium anomalies in the mantle and a subducted metaserpentinite origin for diamonds. Nature 319, 483-185.

37. SCHULZE D. J. 1992. Diamond eclogite from Sloan Ranch, Colorado and its bearing on the diamond grade of the Sloan Kimberlite. Economic Geology 87, 2175-2179.

38. SCHULZE D. J. & HELMSTAEDT H. 1988. Coesite sanidine eclogites from kimberlite: products of mantle fractionation or subduction. Journal of Geology 96, 435-443.

39. SOBOLEV N. V. 1977. Deep-seated inclusions in Kimberlites and the problem of the composition of the Upper Mantle. (D. A. Brown transl.; F. R. Boyd ed.). American Geophysical Union, Washington.

40. SOBOLEV N. V. 1984. Crystalline inclusions in diamonds from New South Wales, Australia. In Glover J. E. and Harris P. G. eds. Kimberlite occurrences and origin: A basis for conceptual models in exploration, pp. 213-226. Geology Department and University Extension, University of Western Australia, Publication 8.

41. SOBOLEV V. N., TAYLOR L. A., SNYDER G. A. & SOBOLEV N. V. 1994. Diamondiferous eclogites from the Udachnaya Kimberlite Pipe, Yakutia. International Geology Review 36, 42-64.

42. SUNAGAWA I. 1990. Growth and morphology of diamond crystals under stable and metastable conditions. Journal of Crystal Growth 99, 1156-1161.

43. SUTHERLAND F. L., TEMBY P., RAYNOR L. R. & HOLLIS J. D. 1994. A review of the East Australian Diamond Province. In Meyer H. O. A. and Leonardos O. H. eds. Diamonds: Characterisation genesis and exploration, pp. 170-184. Compahia Pesquisa Recursos Minerais Special Publication, Brasilia 1B/92.

44. TAYLOR W. R., JAQUES A. L. & RIDD M. 1990. Nitrogen-defect aggregation characteristics of some Australasian diamonds: time-temperature constraints on the source regions of pipe and alluvial diamonds. American Mineralogist 75, 1290-1310.

45. WANG X., JING Y., LIOU J. G., PAN G., LIANG W., XIA M. & MARUYAMA S. 1990. Field occurrences and petrology of eclogites from the Dabie Mountains, Anhui, central China. Lithos 25, 119-131.

46. WIENS D. A., MCGUIRE J. J. & SHORE P. J. 1993. Evidence for transformational faulting from a deep double seismic zone in Tonga. Nature 364, 790-793.

47. ZONENSHAIN L. P., KUZMIN M. I. & NATAPOV L. M. 1990. Uralian Fold Belt. In Page B. M. ed. Geology of the USSR: a plate tectonic synthesis, pp. 27-54. American Geophysical Union, Geodynamic Series 21.

Chapter 3

Defects and Impurities in Graphene-Like Materials

Paulo T. Araujo[a], Mauricio Terrones[b,c,] and Mildre[d] S. Dresselhaus[a,d,*]

[a]*Department of Electrical Engineering and Computer Sciences, Massachusetts Institute of Technology, Cambridge, MA 02139, USA*

[b]*Department of Physics, Department of Materials Science and Engineering and Materials Research Institute, The Pennsylvania State University, University Park, PA 16802, USA*

[c]*Research Center for Exotic Nanocarbons (JST), Shinshu University, Wakasato 4-17-1, Nagano-city 380-8553, Japan*

[d]*Department of Physics, Massachusetts Institute of Technology, Cambridge, MA 02139, USA*

Defects in materials are normally considered to be detrimental to the properties of materials and to devices based on such materials. However, defects can also be beneficial to materials in providing dopants to control both their carrier concentration and whether the carriers are n-type or p-type1-16. Isotopic defects can also be used to probe materials' properties by separating their electronic properties from behaviors associated with their phonon dynamics. Isotopic defects consist of defects that do not alter the electronic properties of the material. Because of the significant percentage difference between the mass of the 12C and 13C isotopic species, the graphene phonon frequencies for 13C isotopes differ sufficiently from those of 12C so that they can be used to probe related phonon phenomena in a layered compound such as graphene, without changing the electronic properties17-25.

In this review we also consider defect-induced modifications to few-layer graphene, one of the simplest systems under current investigation in materials science, and yet informative about the effects of defects and their relevance to the functional performance of materials26’ 27’ 28’ 29’ 30 and 31. In this context we consider in-plane defects, which occur ubiquitously in few-layer graphene, whether we are considering a single layer or a thin film of ten layers. In-plane defects are symmetry-breaking and can include point defects, such as vacancies32’ 33’ 34’ 35’ 36’ 37’ 38’ 39’ 40’ 41’ 42’ 43’ 44’ 45’ 46 and 47, substitutional impurities1’ 2’ 3’ 4’ 5’ 6’ 7’ 8’ 9’ 10’ 11’ 12’ 13’ 14’ 15 and 16, or interstitial impurities1 ’ 2’ 3’ 4’ 5’ 6 ’ 7’ 8’ 9’ 10’11’ 12’ 13’ 14’ 15 and 16, and the impurity can be a chemical impurity or an isotopic impurity17’18’ 19’ 20’ 21’ 22’ 23’ 24 and 25. Grain-boundary defects have been observed in graphene prepared by chemical vapor deposition (CVD) but they also occur in mechanically exfoliated graphene, though some differences in properties are found when comparing these two types of graphene48’ 49’ 50 and 51. Sample boundaries or edges are also symmetry-breaking defects, but are interesting in their own right, especially for the case of narrow ribbons where electron and phonon mean free paths are longer than the ribbon widths, so that these boundaries serve as scattering centers for electrons and phonons1’2’ 52’ 53’ 54’ 55’ 56’ 57 and 58. Also of interest are interplanar defects, such as stacking faults within interlayer stackings, different interplanar isotopic arrangements, or different chemical arrangements that might be synthesized for the study of specific graphene-related phenomena[39]. In order to understand the effects of the various interesting defects, in graphene-like systems that occur or are intentionally introduced, a variety of experimental probes are used, and in this review special emphasis will be given to atomic-level structural information derived from aberration corrected high-resolution transmission electron microscopy1’ 35’ 36’ 59’ 60’ 61’ 62 and 63, scanning tunneling microscopy1’ 64’ 65 and 66, and spectroscopic information derived from Raman1’ 2’ 58’ 67’ 68’69’ 70 and 71 and scanning tunneling spectroscopies1’ 64’ 65 and 66.

TYPES OF DEFECT

In this review we consider structural defects associated with both naturally occurring imperfections and growth-induced defects1’ 2’ 3’ 4’ 5’ 6’ 7’ 8’ 9’ 10’ 11’ 12’ 13’ 14’ 15’ 16’ 17’ 18’ 19’ 20’21’ 22’ 23’ 24 and 25, on one hand, and defects introduced for enhancing the understanding of graphene-related phenomena, on the other hand. The defects include point defects32’ 33’34’ 35’ 36’ 37’ 38’ 39’ 40’ 41’ 42’ 43’ 44’ 45’ 46 and 47, cluster defects32’ 33’ 34’ 35’ 36’ 37’ 38’ 39’ 40’ 41’ 42’ 43’44’ 45’ 46 and 47 and boundaries or edges48’ 49’ 50’ 51’ 52’ 53’ 54’ 55’ 56’ 57 and 58, which also behave like defects in lowering the symmetry of the infinite crystal.

Point defects occur predominantly within the graphene plane in the form of lattice vacancies and impurity atoms1’ 2’ 3’ 4’ 5’ 6’ 7’ 8’ 9 ’ 10’ 11’ 12’ 13’ 14’ 15’ 16’ 17’ 18’ 19’ 20’ 21’ 22’ 23’24 and 25. The occurrence of impurities can be either in substitutional or interstitial sites and can be in the form of isotopic impurities17′ 18′ 19′ 20′ 21′ 22′ 23′ 24 and 25, which predominantly perturb the phonon spectra. Impurities in the form of foreign species (dopants) tend to mainly introduce local electric fields and strain fields associated with their local charge because of the small size of carbon atoms relative to all other species1′2′ 3′ 4′ 5′ 6′ 7′ 8′ 9′ 10′ 11′ 12′ 13′ 14′ 15′ 16′ 17′ 18′ 19′ 20′ 21′ 22′ 23′ 24 and 25. Boron, which is even smaller in size than carbon, is a special impurity atom that constitutes a common *p*-type dopant for graphene. Nitrogen, being larger than carbon, does not dope graphene as readily as boron but would be more likely to enter the lattice of narrow few layer graphene ribbons than bulk graphite. However, the solubility of both, boron and nitrogen in graphene is relatively low (< 2 at.%) 1′ 2′ 3′ 4′ 5′ 6′ 7′ 8′ 9′ 10′ 11′ 12′ 13′ 14′ 15 and 16.

Hydrogen or fluorine tend to either physisorb in molecular form on specific graphene defect sites or to form graphane (hydrogenated graphene)[72], and fluorinated graphene when added covalently and in high concentrations73 and 74. Generally speaking, the graphene structure forms well-ordered lattices, especially on epitaxial substrates, such as copper (111) or BN, and is relatively inert. When left in an ambient environment, oxygen and nitrogen molecules could be adsorbed on the graphene basal plane and relatively high annealing temperatures (800 °C) are needed to desorb such molecules[1]. The naturally occurring vacancy concentration is determined by the growth temperature, which in turn is related to the growth method. Increasing the growth temperature reduces the naturally occurring vacancy concentration. The introduction of impurities is further discussed in connection with the chemical or electrochemical doping of graphene2′ 18′ 26′27′ 30′ 31′ 75′ 76′ 77 and 78. Ion implantation has been another common technique used for the introduction of point defects under controlled conditions by varying the ion species dose and ion energy33′ 34′ 58′ 79 and 80.

Graphene also accommodates two common types of carbon cluster defects with environments locally perturbed from the perfect sp^2 hybridized carbon 32′ 33′ 34′ 35′ 36′ 37′38′ 39′ 40′ 41′ 42′ 43′ 44′ 45′ 46 and 47. The most common cluster defect is the 5–7-7–5 Thrower-Stone-Wales defect (Fig 1a), but also frequently occurring are the 5–8-5 defects (Fig 1b), both of which maintain an average of the six-carbon planar rings of graphene, and therefore graphene can be annealed at sufficiently high temperatures to lower its defect concentration. Since the 5–7-7–5 and 5–8-5 defects are symmetry-breaking, they can be observed spectroscopically in the Raman D-band (~1350 cm^{-1} at 2.41 eV laser excitation), but the defects invoke other symmetry-breaking spectroscopic features, such as the D′-band at 1620 cm^{-1} and combination modes such as the Raman D+G band 2930 cm^{-1}, each with a different dispersion with laser excitation energy E_L 1′ 2′ 32′ 33′ 34′ 35′36′ 37′ 38′ 39′ 40′ 41′ 42′ 43′ 44′ 45′ 46 and 47.

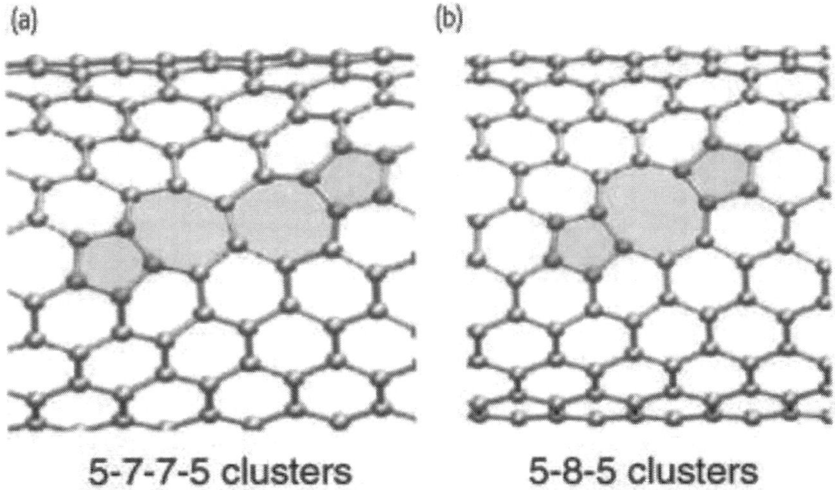

Figure1. Molecular models representing: (a) the 5–7-7–5 Thrower-Stone-Wales defect and (b) the 5–8-5 vacancy-like cluster defect. Because these defects maintain an average of the six carbon atoms per polygon in the graphene lattice, they can be annealed at sufficiently high temperature.*Reprinted with permission from* [32]. *© 2006 by the American Physical Society.*

At present there is no clear information available on the use of the D-band lineshape, linewidth, laser energy dependence to quantitatively discriminate between the density of 5–7-7–5 and 5–8-5 defects/μm^2 in actual graphene samples, or between defects occurring in graphene samples prepared by different growth techniques. For the structural observation of individual isolated 5–7-7–5 and 5–8-5 defects, the TEM operated under high-resolution conditions, including aberration correction is especially powerful, especially for its ability to observe such individual defect clusters, or as either aggregated clusters or defect clusters in grain boundaries. *In situ* annealing studies of such defects in a dedicated environmental chamber under high-resolution conditions would be rewarding to study the effect of each symmetry-breaking mechanism. The ability to do combined tip-enhanced Raman and TEM studies on the same physical cluster defect could provide unique information about the dynamics of such defects and the changes of the different vibrational modes. Recently, highly informative combined Raman and TEM studies on the same sample have been proposed and discussed [4].

CHARACTERIZATION OF DEFECTS

As explained above, defects in graphene-like materials including carbon nanotubes can naturally arise from imperfections and growth-induced defects. These systems can even show defects that are artificially introduced to fulfill some needs for technological applications. In this section, we discuss promising techniques used to characterize defects in graphene-like systems.

MICRO-RAMAN SPECTROSCOPY

Characterization of in-plane defects

Among several characterization techniques, micro-Raman spectroscopy (MRS) has been a powerful and widely used technique to study defects in carbon-based materials. Indeed, MRS is readily accessible, fast, non-destructive and gives a wealth of information about both the electronic and phonon structures of materials[1,2 and 58]. Particularly special, carbon-based materials show a very characteristic Raman feature called the D-band (the D denoting disorder-induced), which is a symmetry-breaking Raman peak that has no intensity in the absence of defects. Every time a given impurity breaks the translational symmetry of the carbon material's lattice, D-band intensity will appear in the Raman spectrum and its Raman cross-section (σ) will be proportional to the defect concentration[1,2,33,34,53,54 and 58]. It is worth noting that not only can the D-band feature be used to understand defects in carbon materials, but also the symmetry allowed G-band and G'-(2D) band in the Raman spectra provide valuable information about defects especially when the impurity in question dopes the material by changing the bonding strength of the atomic species to the host carbon atoms.

Back in 1970, Tuinstra and Koenig[81] showed that the ratio between the D- and G-band Raman intensities (I_D/I_G) is directly related to the crystallite size (L_a) of 3D graphite. At that time, they explained their findings for only one excitation laser energy (E_L), namely, 514 nm (2.41 eV). More than 30 years later, in 2006, Cançado *et al.*[82] successfully extended Tuinstra and Koenig's findings to several E_L values and now the ratio I_D/I_G is more fully described by the equation:

$$\frac{I_D}{I_G} = L_a \frac{E_L^4}{560} \quad (1)$$

where L_a is given in nm, E_L is given in eV and the constant 560 is given in units of eV^4/nm. Although a general equation appeared through this scenario, a basic question still remained: whether the crystallite size L_a is a very special characterization parameter in graphite or would other types of symmetry-breaking features also follow a similar equation? The answer is, not exactly.

In 2010, Lucchese *et al.* [33] used Raman scattering to study disorder in graphene caused by low energy (90 eV) Ar^+ ion bombardment. By varying the ion dose, the authors studied different densities of defects induced by the ion bombardment and were able to understand the evolution of the ratio I_D/I_G with ion dose (see Fig 2). With this experiment the authors provided a Raman spectroscopy-based method to quantify the density of defects in graphene, using HOPG (highly oriented pyrolytic graphite) for calibration. InFig 2 the density of defects due to different ion doses is probed by L_D, which denotes the average distance between defects. This study revealed important information about the defect evolution by observing that I_D/I_G could be fitted by a phenomenological model forI_D/I_G*vs.* L_D. The model considers that the impact of a single ion in the graphene sheet causes modifications on two length scales, here denoted by r_A and r_S (with $r_A > r_S$), which are the radii of two circular areas measured from the impact point (the subscript A stands for "activated" whereas the subscript S stands for "structurally-defective"). Qualitatively speaking, only if the Raman scattering process occurs at distances smaller than $|r_A\text{-}r_S|$, will the corresponding "damaged" region contribute strongly to the D-band feature. Considering these assumptions in conjunction with statistical arguments, the I_D/I_G*vs.* $L_D$relation is given by[33]:

$$\frac{I_D}{I_G} = C_A \frac{r_A^2 - r_S^2}{r_A^2 - 2r_S^2}\left[e^{\left(\frac{-\pi r_S^2}{L_D^2}\right)} - e^{\left(\frac{-\pi(r_A^2 - r_S^2)}{L_D^2}\right)}\right] + C_S\left[1 - e^{\left(\frac{-\pi r_S^2}{L_D^2}\right)}\right] \quad (2)$$

where C_A, C_S, r_A*and* r_S are adjustable parameters. The model is in conceptual agreement with a well-established amorphization trajectory for graphitic materials 1' 2'33 [and] 58. Additionally, the results show that the broadly used Tuinstra and Koenig relation between I_D/I_G and L_a should be limited to the measurement of 3D crystallite sizes. Recently, in 2011, Cançado *et al.* extended the proposed phenomenological model to be applicable to several laser lines [34].

Other types of defects are those that dope the carbon material positively (*p*-type defects) or negatively (*n*-type defects) 1' 2' 3' 4' 5' 6' 7' 8' 9' 10' 11' 12' 13' 14' 15 [and] 16, and those that introduce isotopic defects, the latter being neutral impurities that have only their atomic mass different from that of the pristine sample 17' 18' 19' 20' 21' 22' 23' 24 [and] 25. All these impurities can be naturally introduced into the carbon lattice by their substitution in the place of some original carbon atoms or by functionalization

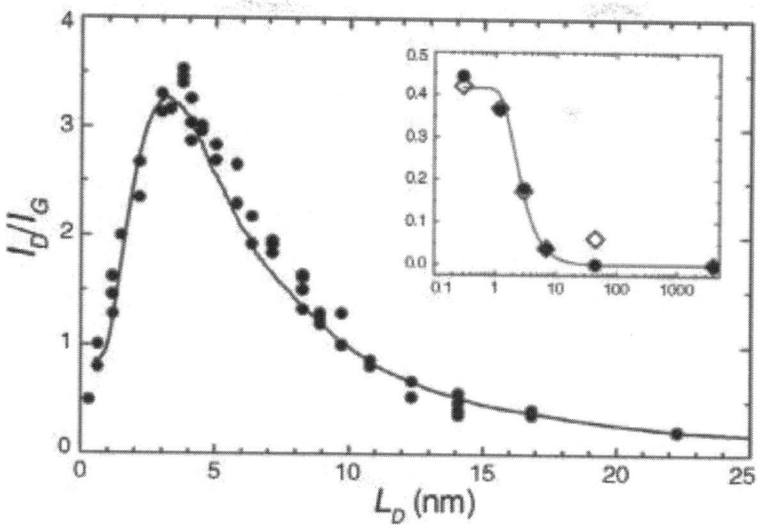

Figure 2. The I_D/I_G data points from three different mono-layer graphene samples as a function of the average distance L_D between defects that are induced by the ion bombardment procedure. The inset shows the Raman intensity ratio I_D/I_G vs. L_D on a semilog scale for two graphite samples: (i) a ~ 50-layer graphene sample found near one of the three mono-layer graphene samples (bullets); (ii) a bulk HOPG sample used for calibration (diamonds). The bulk HOPG values in the inset are scaled by $(I_D/I_G) \times 3.5$. The solid line is obtained from Eq. 2, where $C_A = (4.2 \pm 0.1)$, $C_S = (0.87 \pm 0.05)$, $r_A = (3.00 \pm 0.03)$ nm and $r_S = (1.00 \pm 0.04)$ nm.

of the sample 1' 2' 3' 4' 5' 6' 7' 8' 9'10' 11' 12' 13' 14' 15' 16' 17' 18' 19' 20' 21' 22' 23' 24 [and] 25. In the context of the MRS characterization of these functionalized impurities, a new analysis for the I_D/I_G ratio as a function of dopant concentration and E_L is still required in order to understand in depth the differences in connection with the use of Eq. 1. Experiments of MRS combined with x-ray photoelectron spectroscopy (XPS) to determine the amount of dopants could be carried out in order to correlate the D-band peak with the dopant concentration. Therefore, additional combined measurement techniques are awaiting development.

However, it is interesting to note that, while for graphene, these chemical doping and functionalization procedures are still under development, for single-walled carbon nanotubes (SWNTs) such chemically modified samples have been available for a much longer time so that much is already known about how to understand some of the effects of such defects on the electronic and vibrational properties of SWNTs[2]. For substitutional*n*-type and *p*-type doping, it turns out that the G' (2D) band is extremely sensitive in discriminating between these two different types of doping impurities. More recently, Maciel *et al.* [67] showed that these impurities renormalize both the electronic and the vibrational structures of carbon nanotubes, and by analyzing the particular behavior of the G'(2D)- band in some doped carbon nanotubes, they could both distinguish between*n*-type or *p*-type doping as well as give information about the dopant concentration, as exhibited in Fig 3.

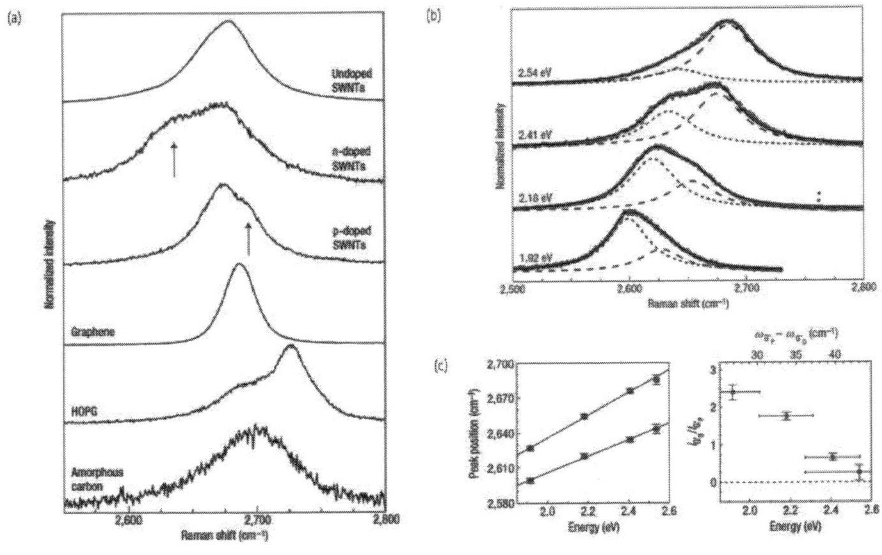

Figure 3. (a) The arrows point to defect-induced peaks in the G'-band for doped SWNTs. The p/n doping comes from substitutional boron/nitrogen atoms, the nearest neighbors of carbon in the periodic table. The spectra of graphene, HOPG and amorphous carbon are shown for comparison. The spectra were measured at room temperature with E_L = 2.41 eV (514 nm). (b) The G'(2D)-band spectra with different E_L for the n-doped SWNT sample. (c) The G' peak position as a function of E_L, obtained by fitting the spectra in (b) with two Lorentizians. The line width γ is determined by the full-width at half-maximum intensity and the values of each Lorentizian linewidth increases linearly with E_L, from $\gamma = 43$ to 53 cm^{-1}. The solid lines are linear fits giving $\omega_{G'P} = (97 \pm 5)E_L + (2,442 \pm 11)$ cm^{-1} and $\omega_{G'D} = (71 \pm 4)E_L + (2,465 \pm 11)$ cm^{-1}. (d) The G'-band intensity ratio ($I_{G'D}/I_{G'P}$) as a function of E_L. The upper scale gives the difference between the G'_Pfrequency ($\omega_{G'P}$) and the G'_D frequency ($\omega_{G'D}$). Here, G'_P stands for the undoped pristine G'-band, while G'_D stands for the doped G'-band.

Regarding defects due to the introduction of isotopes into SWNTs, Costa *et al.* [17] have shown how MRS also reveals special phenomena related to isotopes, as shown in Fig 4. The study reveals linear dependences of the Raman D-band, G-band and G'(2D)-band frequencies with an increase in the relative ^{13}C concentration (x) for each of these nanotube Raman modes, and shows that the effect of the mass variation of the isotope mixture on the phonon frequencies is described through a simple harmonic oscillator model involving ^{13}C and ^{12}C species. The solid lines in Figs. 4a, b, and c are fittings of the experimental data to such a harmonic oscillator model for the frequency and linewidth. Finally, measurements with different E_L were performed and the frequency dispersions of the D and G'(2D)-bands with E_L were observed to be the same for ^{12}C and ^{13}C nanotubes, suggesting no changes in the electronic structure after isotope enrichment, in contrast to what happens for *n*-type and *p*-type impurities [17].

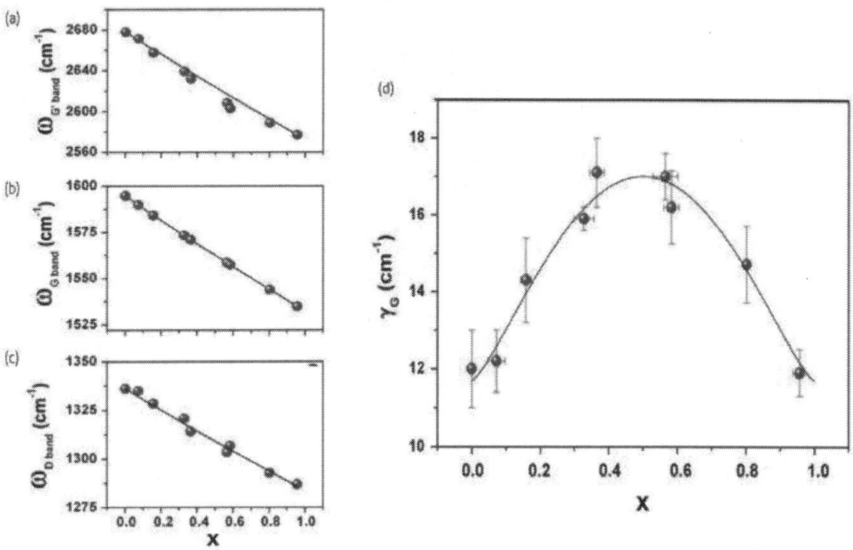

Figure 4. (a) G'(2D)-band, (b) G-band, and (c) D-band frequencies as a function of the 13C content in the sample containing a 13C:12C ratio x. The circles represent experimental points and the solid lines are from model calculations. (d) The data shows a parabolic trend with its maximum at around x = 0.5, and its minima at x = 0 and x = 1.

It is worth mentioning that recent advances in the Raman spectroscopy of *n*-doped and *p*-doped graphene have been reported by Panchakarla *et al.* [9]. The impurities in this case are naturally introduced during the growth process and are, therefore, mainly substitutional. Their results can be summarized in Fig 5. Basically these authors report an asymmetric phonon renormalization of the G-band feature in which defects due to *n*-type and *p*-type both result in a frequency upshift (note that the impurity concentration only changes the renormalization factor). This hardening of the G-band is explained by means of the phonon self-energy in graphene within the non-adiabatic formalism, and its broadening is due to the absence or blockage of the decay channels of the phonons into electron–hole pairs [83]. However the upshift rate is larger for *p*-doped samples in comparison to the *n*-doped ones. From an I_D/I_G analysis, these authors were able to find that the crystallite size L_a for doped graphene is smaller than L_a for undoped graphene, which is expected because of the extra strain introduced by the dopants compared to the strain in a pristine sample.

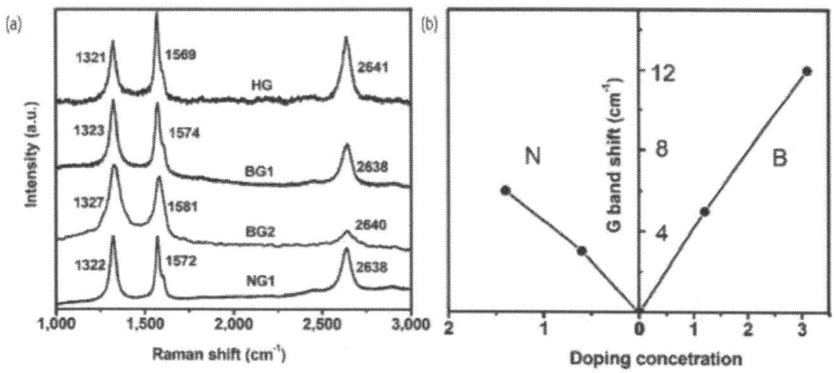

Figure 5. (a) Raman spectra of undoped (HG), boron-doped (BG) and nitrogen-doped (NG) graphene samples. (b) G-band frequency shifts caused by electron (N) doping (with pyridine) and hole (B) doping. From an I_D/I_Ganalysis, it is concluded that the crystallite size is smaller for doped samples than for undoped ones.

Characterization of armchair and zigzag edges

For graphene nanoribbons, edges can be considered to be symmetry-breaking discontinuities. Here, AC-HRTEM (aberration-corrected high resolution transmission electron microscopy) can be used directly to identify the edge type. Since the zigzag edges have a high density of electronic states relative to other edge types[57], polarized Raman spectroscopy can be used to distinguish zigzag edged ribbons from ribbons with armchair and chiral edges, as shown in Fig 6[53].

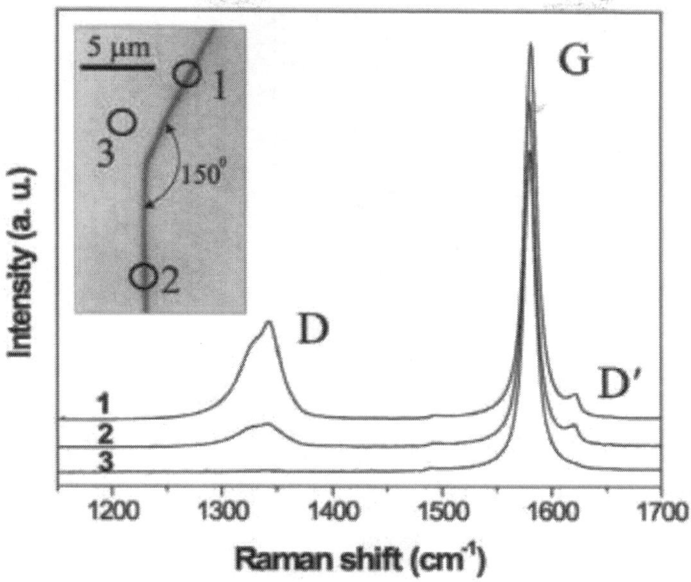

Figure 6. Raman spectra obtained in three different regions of a HOPG graphite sample (open circles in the inset). Namely, the region 1 is an armchair edge, the region 2 is a zigzag edge, while the region 3 is bulk graphite. The spectra were taken at room temperature and the laser energy E_L was 1.96 eV. The inset shows an optical image of the edge regions.

Since the Raman tensor for the zigzag ribbons causes the Raman cross-section to vanish for the optical electric field polarization of light parallel to a zigzag edge, Raman spectroscopy can also be used to distinguish zigzag nanoribbons for ribbons from other edge types, as shown in Fig 7, in which the Raman spectra were obtained for light incident with different polarization angles (θ) with respect to the nanoribbon direction for a narrow graphene ribbon sitting on a bulk graphite substrate[54].

After this brief discussion about MRS and its applications for the characterization of defects, it is important to note that MRS provides us with information about defect concentrations, crystallite sizes and property changes due to *n*-type and *p*-type doping, including similarities and differences. However, MRS cannot tell us by itself the structural differences of, for example, the type of defect which breaks the translational symmetry (vacancies or dopants). MRS cannot tell us if the vacancies are point vacancies or 5–7-7–5 or 5–8-5 defect clusters. Moreover, MRS cannot provide precise information if a given impurity atom is *n*-type, *p*-type, substitutional, or interstitial. For these purposes other observation techniques such as TEM, STM, or near-field spectroscopy would also be required.

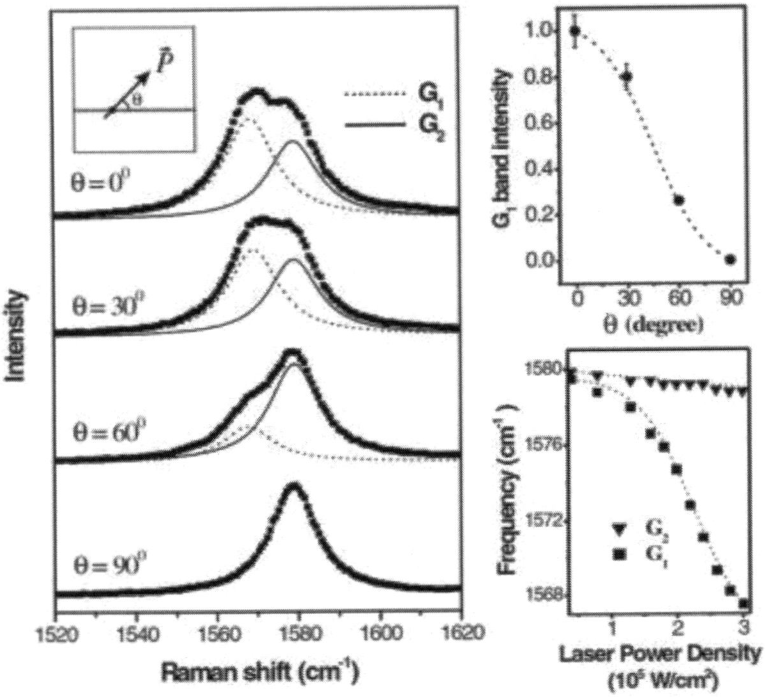

Figure 7. (a) Raman spectra obtained for light incident with different polarization angles (θ) with respect to the ribbon direction. The inset shows a schematic figure of the sample (horizontal gray line) showing the direction between the ribbon axis and the light polarization vector. (b) Intensity of the ribbon's G-band (G_1) peak versus θ (note that G_2 stands for the G-band of the bulk graphite substrate). The dotted line is a $\cos^2\theta$ theoretical curve for the expected θ dependences. The error bars are associated with baseline corrections. (c) Raman frequencies of the graphite substrates G_2 (triangles) and the graphene ribbon G_1 (squares) peaks as a function of the laser power density[54]. Thus, we observe that Raman spectroscopy can distinguish between graphene edges and bulk graphite, and that this technique can also distinguish zigzag edges from armchair edges through the dependence of the Raman intensity on the light polarization angle with respect to the ribbon axis.

TIP-ENHANCEMENT NEAR-FIELD SPECTROSCOPY

Tip-enhancement Near-field spectroscopy (TES) is a technique that combines both: atomic force microscopy (AFM) and confocal optical microscopy. This special combination of techniques provides, simultaneously, sub-wavelength and high spatial resolution of samples (determined exclusively by the diameter of the metallic tip used in the experiment) resulting in a wealth of information at the atomic scale68 [and] 69. Similar to MRS, TES cannot provide information regarding the type of defects that are connected to a given sample, or if the impurity atoms are substitutional or not. However, while MRS probes clusters of defects, TES allows us to determine individual impurities at near atomic resolution, which in most cases reveals the different local interactions of impurities with the pristine material.

Maciel *et al.* [67] also showed that TES can selectively excite regions with and without defects along individual carbon nanotube structures, as shown in Fig 8. It is shown that TES experiments can be used to understand the local differences (in relation to the pristine material) generated by *n*-type and *p*-type dopants. As explained in the last section, the G'(2D) feature measured by MRS consists of two peaks: one is related to the dopants and another, related to the pristine contributions, is shifted in frequency from the first peak. Similarly to Qian *et al.* [69], in the work by Maciel *et al.* [67] the authors could discriminate the optical responses of regions with and without impurities, and by comparing the observed Raman and photoluminescence signals, they clearly could see a localized photoluminescence emission coming from the defective site. However, the work of Maciel *et al.* [67] showed that every time the tip was standing over a defective region, the Raman signal from both the defect and from the pristine nanotube were observed, while only the pristine nanotube signal was observed in regions without a defect. This is an expected result because the Raman signal is strongly connected with the polarizability of the material and therefore the observed spectra will reflect both the electronic and vibrational structure of the material, while photoluminescence is sensitive basically to changes in the electronic structure.

Another application of TES is the length scale study of defect-induced Raman modes in carbon materials. Recently, Georgi and Hartschuh[70], successfully applied TES to investigate with high spatial resolution the D-band scattering near defects in metallic SWNTs. In their work, Georgi and Hartschuh found that the length scale of the D-band scattering process is about 2 nm in SWNTs, similar to this length in graphene[84]. It is worth emphasizing that this 2 nm length is also a value for L_D obtained from ion implantation studies[33].

High resolution transmission electron microscopy

Aberration-corrected AC-HRTEM is highly effective for studying the morphology of carbon based materials1′ 35′ 36′ 59′ 60′ 61′ 62 [and] 63. Of course, AC-HRTEM has the advantage of high (atomic scale) resolution so that it provides valuable information about the atomic structure of pristine materials as well as the impurities in defective materials. This technique has the disadvantage of requiring expensive equipment, is time-consuming in operation and, most of time, AC-HRTEM is destructive in the sense that one cannot be assured that the sample will keep its same characteristics before and after the HRTEM analysis.

Another drawback of this technique is that it relies on additional complex and dedicated equipment that is needed to extract useful information from the images1′ 35′ 36′ 59′ 60′ 61′ 62[and] 63. AC-HRTEM would be ideal to fully characterize, for example, 5–7-7–5 and 5–8-5 defects in nanocarbons1′ 35′ 36′ 59′ 60′ 61′ 62[and] 63. Because this is currently a widely discussed topic in the literature, we will restrict our attention to recent results involving HRTEM techniques35[and] 36.

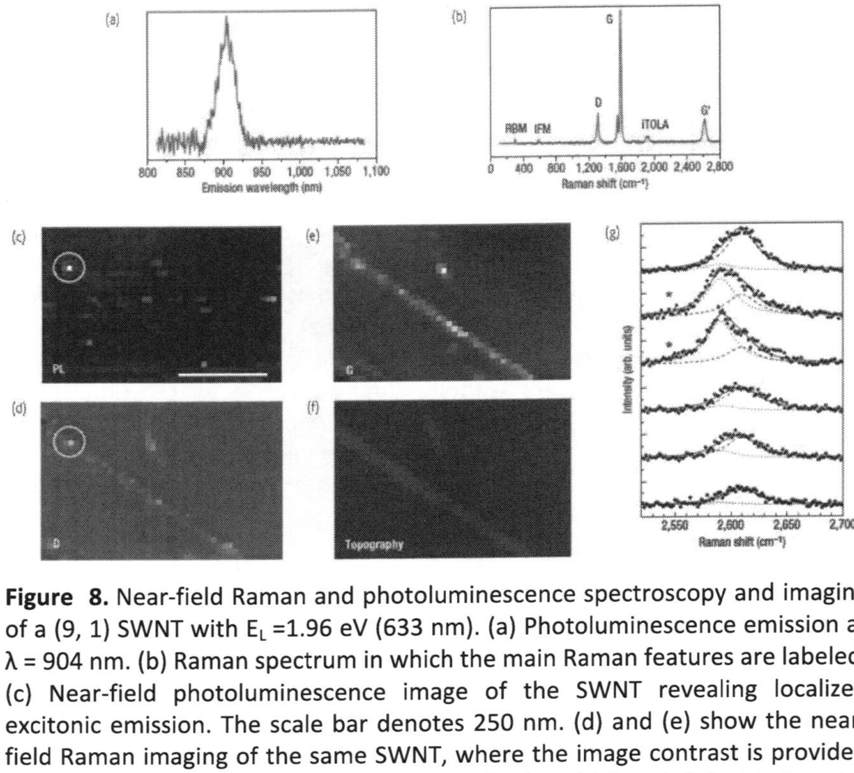

Figure 8. Near-field Raman and photoluminescence spectroscopy and imaging of a (9, 1) SWNT with E_L =1.96 eV (633 nm). (a) Photoluminescence emission at λ = 904 nm. (b) Raman spectrum in which the main Raman features are labeled. (c) Near-field photoluminescence image of the SWNT revealing localized excitonic emission. The scale bar denotes 250 nm. (d) and (e) show the near-field Raman imaging of the same SWNT, where the image contrast is provided by spectrally integrating over the tangential G-band (d) and defect-induced D band (e). (f) Corresponding topography image. (g) Evolution of the G'-band spectra near the defective segment of the (9, 1) SWNT. The spectra were taken in steps of 25 nm along the nanotube, showing the defect-induced G'_D peak (dotted Lorentizian). The asterisks (*) denote the spatial locations where localized photoluminescence and defect-induced D band scattering were measured (see yellow circles in (c) and (e), respectively). The $\omega_{G'P} - \omega_{G'D}$ splitting observed in (g) is smaller than the splitting observed in Fig. 3(b) because the spectra were measured with different excitation laser energies, as discussed in reference[67].

Recently, Kotakoski *et al.* [35] used electron irradiation to create a sp^2-hybridized one-atom-thick flat carbon membrane with a random arrangement of polygons (in other words: a graphene sheet containing heptagons, hexagons and pentagons that resembles the pentaheptite[85] and Haeckelite-like structures[86]). The authors demonstrate, by means of first principles calculations and AC-HRTEM experiments how the transformation occurs step by step by nucleation and growth of low-energy multivacancy structures constructed from rotated hexagons and other polygons (see Fig 9). The observations in this work[35] provide new insights into the bonding behavior of carbon and the dynamics of defects in graphene. It would therefore be ideal to perform *in situ* Raman spectroscopy measurements on the different types of defective graphenes.

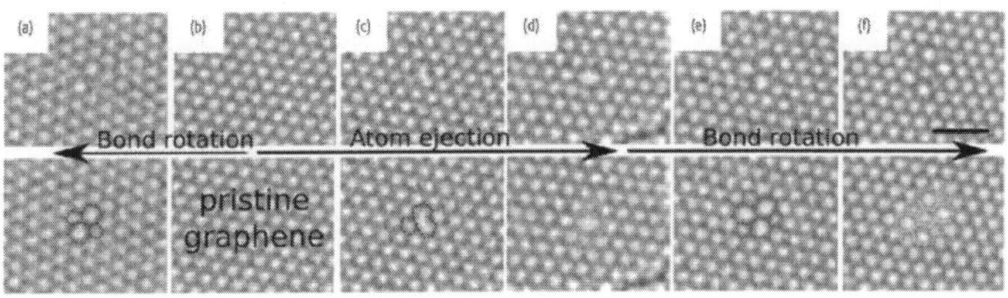

Figure 9. Elementary defects and frequently observed defect transformations under irradiation. Atomic bonds are superimposed on the defective areas in the bottom row. The creation of the defects can be explained by atom ejection and reorganization of bonds via bond rotations. (a) The Thrower-Stone-Wales defect, (b) defect-free graphene, (c) (5–9) vacancy, (d) (5–8-5) divacancy cluster, (e) (555–777) divacancy cluster, (f) (5555–6-7777) divacancy cluster. Scale bar is 1nm.

Another interesting application also involving AC-HRTEM is presented by Meyer *et al.* [36]. They demonstrate an experimental analysis of charge redistribution due to chemical bonding in nitrogen-doped graphene membranes and boron-nitride monolayers. Namely, the electronic charge density distribution of a solid contains information about the atomic structure and also about the electronic properties, such as the nature of the chemical bonds or the degree of ionization of atoms. This work turns out to be important because the redistribution of charge due to chemical bonding is small compared with the total charge density and is difficult to measure. Although the differences are small, a full understanding of how this charge redistribution works is needed. In other words, the electron scattering by a carbon atom next to a nitrogen atom turns out to be significantly different from electron scattering by a carbon atom elsewhere in the graphene sheet.

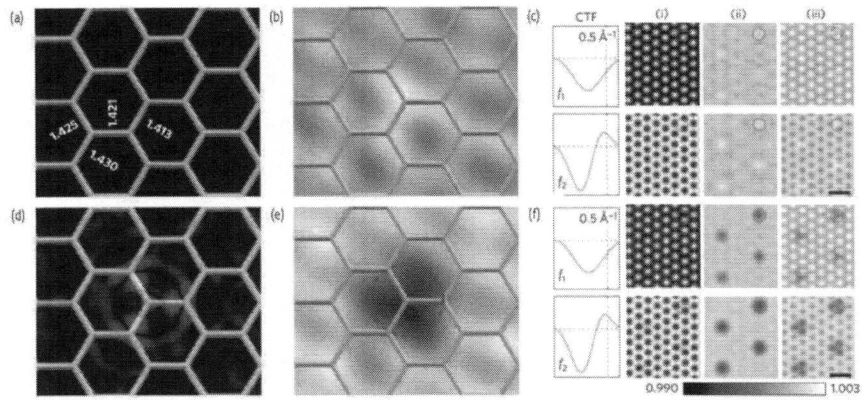

Figure 10. Charge distribution, projected potentials and TEM simulations for nitrogen-doped graphene. (a) The relaxed atomic configuration for a nitrogen impurity substitution in graphene. Bond lengths are given in angstroms. (b) Projected potential based on the IAM (independent atom model), with the periodic components of the graphene lattice removed, and bandwidth-limited to their experimental resolution (about 1:8 Å). Dark contrast corresponds to higher projected potential values, in accordance with conventional TEM imaging conditions. (c) TEM simulation based on the IAM potential, for two different defocus parameters f1 and f2, where the condition f2 stands for a more defocused situation. Filters are: (i) unfiltered, (ii) periodic components removed by a Fourier filter, and (iii) low-pass filtered. (d) Atomic structure (using the same bond lengths), with the changes in the projected electron density due to bonding shown in colour. Blue corresponds to a lower, red to a higher electron density in the DFT result as compared with the neutral-atom (IAM) case. (e) Projected potential, filtered as in (b), based on the all-electron DFT calculation. (f) TEM simulations using the DFT-based potentials. The grey-scale calibration bar applies to columns (ii) and (iii), which are all shown on the same grey-scale range for direct comparison. The scale bars are 5 Å.

The success of Meyer *et al.* relies on AC-HRTEM measurements of defective sites in graphene. In this way, they correct the spherical aberrations of the microscope and do a defocus large enough so that in the contrast transfer function (CTF), features from the nitrogen can be distinguished from the carbon atoms. In order to explain the experiment they use first principles modeling techniques, considering the charge redistribution everywhere the nitrogen atom is located. In this context, they also show that the traditional independent atom model (IAM) analysis is inappropriate to properly interpret HRTEM results in defective samples. Fig 10 shows a comparison between the two methods of analysis and their consequences on the simulated HRTEM images. Note that in Fig 10c it is seen that, no matter what is the defocus status, if one uses the IAM, the nitrogen atom cannot be distinguished. However, in Fig 10f, especially for the defocus status *f*2, it is possible to clearly observe the effects of the hydrogen atom in the graphene lattice.

With this combination of techniques, the authors analyzed the charge transfer on the single-atom level for nitrogen-substitution point defects in graphene, and they confirm the ionicity of single-layer hexagonal boron nitride. Moreover, it is possible to obtain insights into the charge distribution in general nanoscale samples and non-periodic defects can be clearly observed in HRTEM measurements. They have shown that HRTEM experiments when used with first-principles electronic structure calculations opens a new way to investigate electronic configurations of point defects and other non-periodic arrangements of nanoscale objects that cannot be studied by a conventional electron or x-ray diffraction analysis. Moreover, this approach is also very promising for studying vacancy cluster defects, since the existence of vacancies will also require a local redistribution of charge densities. In the future, AC-HRTEM coupled with MRS or TES need to be developed. In this way a clear understanding of the vibrational modes associated with different types of point defects can be achieved. In addition, theoretical calculations indicating the Raman modes associated with the different types of defects still need to be performed. An impressive example of the use of AC-HRTEM to resolve details of grain boundary defects in graphene is shown in Fig 11 where pentagons (blue), heptagons (red) and distorted hexagons (green) can be seen[87]. Here, variants of Thrower-Stone-Wales defects can be directly identified.

Finally, it is worth mentioning that layer stacking defects have also been recently investigated by both AC-HRTEM and optical techniques. In the case of trilayer graphene, the closest packed Bernal ABA stacking is most commonly observed, though ABC rhombohedral stacking has also been identified and investigated88' 89 [and] 90. Since the HRTEM analysis can directly distinguish between ABA and ABC stacking because of the well identified structural stacking differences, this technique is used as a primary standard for the structural identification of graphene samples. Once optical techniques that can also distinguish between these stacking arrangements are established, these more commonly available optical techniques can become secondary laboratory methods for identifying the various structural arrangements that occur commonly in few layer samples88' 89 [and] 90.

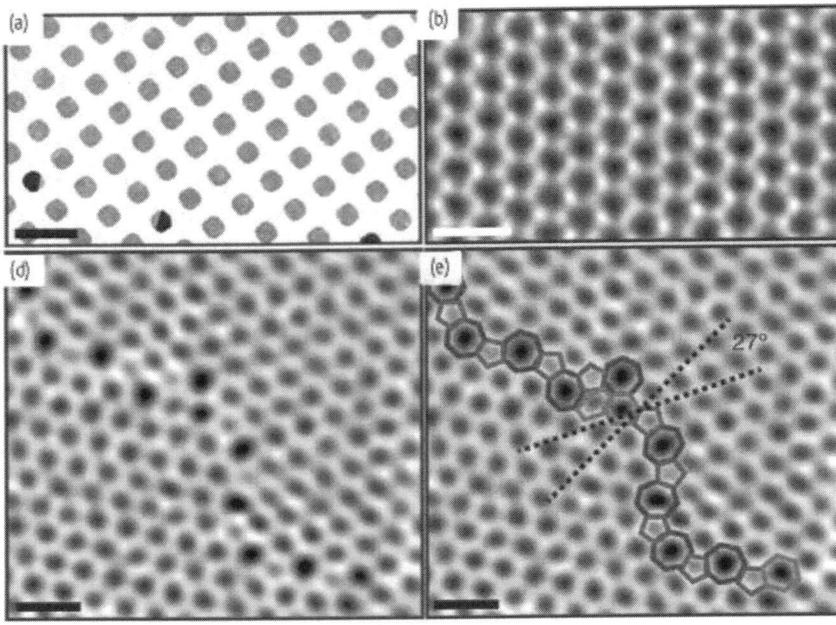

Figure 11. Atomic-resolution AC-HRTEM images of graphene crystals. (a) Scanning electron microscope image of graphene transferred onto a TEM grid with over 90% graphene coverage. Scale bar, 5 mm. (b) AC-HRTEM image showing the defect-free hexagonal lattice inside a graphene grain. (c) Two grains (bottom left, top right) intersect with a 27° relative rotation angle, where an aperiodic line of defects stitches the two grains together. (d) The image from (c) with the pentagons (blue), heptagons (red) and distorted hexagons (green) of the grain boundary outlined. (b)-(d) were low-pass-filtered to remove noise; scale bars, 5Å.

OTHER TECHNIQUES

Besides the techniques presented above, it is worth mentioning that some other approaches (perhaps less well-known or even too new to be known) could be chosen to understand more about the nature of defects. In this context, scanning tunneling microscopy (STM) and spectroscopy (STS) constitute non-destructive techniques that reveal a wealth of information about the electronic structure of carbon materials as well as their impurities1' 64' 65 and 66. An STM in its basic mode of operation will stabilize a sharp metallic tip about 1 nm above the materials surface. In this situation, an electron can tunnel from the tip to the sample and vice versa. Therefore, any tunneling current variation will be related to the density of states in the surface of the material under study, which reveals details about the material's electronic structure. It is worth noting that transport measurements are also powerful for revealing details of the electronic structure of graphene-like materials. Indeed, because the electronic transport is crucial to develop industrial applications, one has to understand how devices (such as transistor and sensors) based graphene-like materials will behave in the presence of dopants75' 76 and 77and/or structural defects79 and 80.

Gate-modulated Raman spectroscopy is another technique that has been extensively used in the field of nanocarbon materials. As an important example, the isotope effect has been exploited to study the effect of the interaction of a graphene layer with a SiO_2substrate[18]. In their work, Kalbac *et al.* [18] prepared two samples containing a single ^{13}C layer (^{13}C 1-LG). In one case, a single ^{13}C layer is in contact with the SiO_2 substrate. The Raman spectrum of this sample is then compared with both a similarly prepared mono-layer ^{12}C sample (^{12}C 1-LG) on a similar SiO_2 substrate and the Raman spectrum from a bilayer sample ($^{13}C/^{12}C$-LG) composed of a single layer of ^{12}C on a SiO_2 substrate on top of which a single layer of ^{13}C is deposited to form the bilayer. In this latter case the two layers of the bilayer graphene are turbostratically arranged. Raman spectra of these three samples taken at a laser energy E_L = 2.33 eV are shown in Fig. 12.

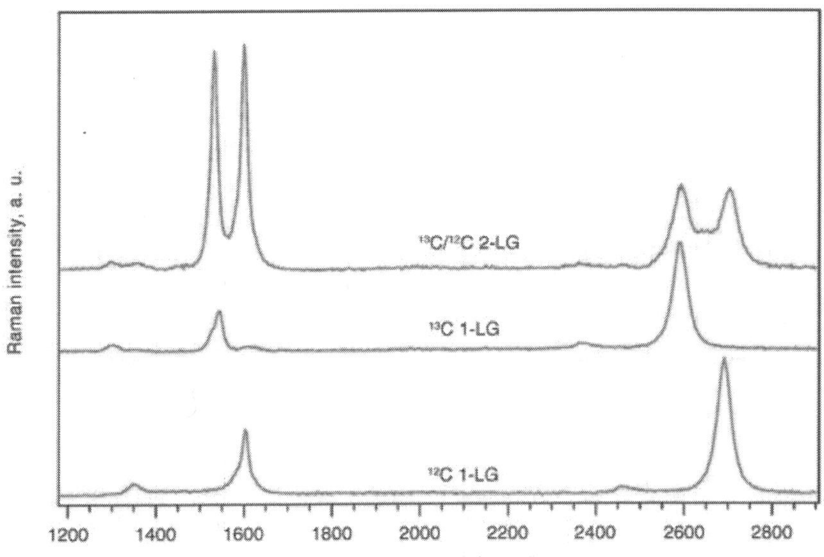

Figure 12. Raman spectra of the ^{12}C 1-LG, ^{13}C 1-LG, and $^{13}C/^{12}C$ 2-LG (^{13}C is on top, ^{12}C at bottom) samples on SiO_2substrates. The spectra are excited using E_L = 2.33 eV laser excitation energy.

Here the frequency ω of the ^{13}C Raman line is down shifted relative to that labeled by $\omega_0$coming from ^{12}C according to the relation $(\omega_0-\omega)/\omega_0 = 1-[(12+c_0)/(12+c)]^{1/2}$ in which c = 0.99 is the ^{13}C concentration in the ^{13}C-enriched sample and $c_0 = 0.0107$ is the natural abundance of ^{13}C in the naturally occurring carbon material. Using electrochemical techniques these authors varied the Fermi level from −1.5 V to +1.5 V in steps of 0.1 V using a calibrated reference potential where the V = 0 potential corresponds to that of the ^{13}C layer which is deposited on the ^{12}C layer of the bilayer graphene sample, and electrochemical Raman results on the G band and on the G' band were obtained[18]. Since gating has become a very important technique for changing the Fermi level in both graphene and SWNTs, the use of electrochemistry for providing calibration standards is important. Electrochemical measurements allow calibration of frequency upshifts for positive potentials and frequency upshifts for negative potentials. That upshifts in frequency occur for both positive and negative potentials is consistent with the chemical doping experiments of Maciel *et al.* [67].

The authors would like to mention that doping techniques to study the electron-phonon coupling in ^{12}C graphene systems have been extensively used in the literature91′92 [and] 93. Finally, transport measurements, optical absorption, photoluminescence, thermo-gravimetric analysis and thermal-transport measurements could also indirectly reveal and quantify defects (impurities) by scanning a structure containing a well-identified defect1′39 [and] 58.

It is noteworthy that the methods to identify, quantify and detect defects discussed here, could be applied to other carbon nanosystems, but some adjustments might be necessary. For example, graphite and graphene are similar systems but their defect detection and quantification, using the techniques above, cannot be simply extrapolated. Raman spectra as well as HRTEM data would be different and defect formation due to ion irradiation needs adjustments and further research along these lines is still needed. A similar scenario is observed for single- and multi-walled carbon nanotubes. Therefore, the present review is also intended to provide carbon researchers with valuable information related to defects and impurities that could be used and adapted to other systems including carbon foams, graphene oxide, nanotubes, etc., but significant further research would be needed to properly characterize heavily irradiated carbons and highly disordered carbons.

CONCLUSIONS AND PERSPECTIVES

In this review article we discussed a variety of defects and impurities in graphene-like materials, including their structure, properties, characterization and application. By envisioning real applications for these structures, one must consider that the material at the same time will somehow become degraded by defects or impurities. As discussed throughout the text, defects such as *n*-type or *p*-type dopants, isotopic substitutions and defect clusters (5–7-7–5 or 5–8-5) could be found in nano-carbons and intermolecular junctions, and are also naturally abundant. In the section *Types of defect* we discussed the nature of such defects and how they occur. Among the many characterization techniques currently available, our discussion in the section *Characterization of defects*focused on the more promising techniques now used to characterize and understand defects. Despite its limitations to distinguish between defects as well as to probe them individually, MRS remains a fast and widely accessible tool which provides a fair understanding of the electronic and vibrational properties of materials. In contrast to MRS are electron microscopy techniques, particularly AC-HRTEM. While, HRTEM is unique in several aspects, its most striking feature is its ability to probe in depth the nature of atomic bonding in crystals. Thus, HRTEM allows for a detailed identification of how specific types of defects occur in materials and how they locally change the material's properties. Other characterization techniques were also briefly discussed, such as scanning probe microscopy (STM, AFM, and others), which greatly amplify our understanding of the electronic properties of materials surfaces.

There had been little attention given to the study of defects in nanostructures and this is because the synthesis of large-scale nanostructures with only a few naturally occurring defects has been difficult. Therefore emphasis has historically been given to promoting the synthesis of defect-free structures with high electronic mobilities and for the scale-up of such structures to the micron and even centimeter length scales for potential practical applications. Now that much progress has been made in materials preparation and characterization, it is time to take stock of where we are now, where we are going in the future, and what scientific barriers remain to be overcome. Among the scientific barriers to future progress is the controlled introduction of defects in the growth process, including point defects, cluster defects and grain boundaries, with grain boundaries arising in the growth process because of the different nucleation sites for graphene grain growth.

Finally, although the topic was not mentioned in the manuscript, the authors believe in the use of defects in applications such as for providing specific sensor sites for molecular adsorption. Enhancing the performance of the different kinds of sensor sites that can now be developed, by increasing the density of such available sites and by expanding the number of ways that such sensor sites could be used reliably and the promotion of new large-scale applications should be emphasized but special niche applications could also be a good strategy for developing technological tools. Carbon, like silicon, is an abundant and cheap element on our planet. Therefore its use for large scale applications is desirable and could perhaps become part of global sustainability planning if graphene and other nanocarbons could by conveniently and cheaply synthesized from CO_2 or waste CH_4.

ACKNOWLEDGEMENTS

PTA acknowledges financial support from CNPq grant No.200616/2010–2. PTA and MSD acknowledge financial support from ONR-MURI-N00014–09-1–1063. MT acknowledges support from the Research Center for Exotic Nanocarbons, Japan regional Innovation Strategy Program by the Excellence, JST.

REFERENCES

1. Jorio, A., Dresselhaus, G., and Dresselhaus, M. S.. Carbon Nanotubes: Advanced Topics in the Synthesis, Structure, Properties and Applications. Springer-Verlag, Berlin, (2008).
2. Panchakarla, L. S., et al., ACS Nano (2007) 1, 494.
3. Koos, A. A., et al., Carbon (2010) 48, 3033.
4. M. Terrones, A. G. Souza Filho, and A. M. Rao, Doped Carbon Nanotubes: Synthesis, Characterization and Applications Vol. 111, 531–566 (Springer Topics in Appl. Phys., Springer, 2008).
5. Rao, A. M., et al., Nature (1997) 388, 257.
6. Villapando-Paez F., et al., Chem Phys Lett (2006) 424, 345.
7. Subrahmanyam, K. S., et al., J Phys Chem C (2009) 113, 4257.
8. Das B., et al., Chem Commun (2008) 5, 155.
9. Panchakarla, L. S., et al., Adv Mater (2009) 21, 4726.
10. Dresselhaus, M. S. and Dresselhaus, G., Adv Phys (1981) 30, 139.
11. Ajayan, P. M., et at., Nature (1993) 362, 522.
12. Endo, M., et at., J Appl Phys (2001) 90, 5670.
13. Terrones, M., et at., Full Sci Tech 6, 787 (1998).
14. Goldberg, D., et at., Carbon (2005) 38, 2017.
15. Reddy, A. L. M., et at., ACS Nano (2010) 4, 6337.
16. Zhang C., et at., Adv Mater (2011) 23, 1020.
17. Costa, S. D., et al., Carbon (2011) 49, 4719.
18. Kalbac, M., et al,. Nano Lett (2011) 11, 1957.

19. Shoushan, F., et at., Nanotechnology (2003) 14, 1118.
20. Rummeli, M. H., et al., J Phys Chem C (2007) 111, 4094.
21. Stuerzl, N., et at., Phys Stat Sol B (2009) 246, 2465.
22. Xiang, R., et al., Jpn J Appl Phys (2008) 47, 1971.
23. Simon, F., et al., Phys Rev Lett (2005) 95, 017401.
24. Simon, F., et al., Phys Rev B (2010) 81, 125434.
25. Miyauchi, Y., and Maruyama, S., Phys Rev B (2006) 74, 035415.
26. Malard, L. M., et al., Phys Rep (2009) 473, 51.
27. Castro Neto, A. H., et al., Rev Mod Phys (2009) 81, 109.
28. Jorio, A., Dresselhaus, M. S., Saito, R., and Dresselhaus, G. Raman Spectroscopy in Graphene Related Systems. WILEY-VCH, Weinheim, (2011).
29. Novoselov, K. S., et al., Nature (2005) 438,197.
30. Malard, L. M., et al., Phys Rev Lett (2008) 101, 257401.
31. Yan, J., et al., Phys Rev Lett (2008) 101, 136804.
32. Kotakoski, J., et al., Phys Rev B (2006) 74, 245420.
33. Lucchese, M. M., et al.,Carbon (2010) 48, 1592.
34. Cancado, L. G., et al., Nano Lett (2011) 11, 3190.
35. Kotakoski, J., et al., Phys Rev Lett (2011) 106, 105505.
36. Meyer, J. C., et al., Nat Mater (2011) 10, 209.
37. Ma, J., et al., Phys Rev B (2009) 80, 033407.
38. Bhowmick, S., et al., Phys Rev B (2010) 81, 155416.
39. Banhart, F., et al., ACS Nano 5, 26 (2011).
40. Stone, A. J., and Wales, D. J., Chem Phys Lett (1986) 128, 501.
41. Li, L., et al., Phys Rev B (2005) 72, 184109.
42. Ma., J., et al., Phys Rev B (2009) 80, 033407.
43. Lee., G. D., et al., Phys Rev Lett (2005) 95, 205501.
44. El-Barbary, A. A., et al., Phys Rev B (2003) 68, 144107.
45. Jeong, B. W., et al., Phys Rev B 78, 165403 (2008).
46. Tsetseris, L., and Pantelides, S. T., Carbon (2009)

47. 47, 901. 47. Lusk, M. T., and Car, L. D., Phys Rev Lett (2008) 100, 175503.
48. Jauregui, L. A., et al., Sol Stat Comm 151, 1100 (2011).
49. Kim, K., et al., ACS Nano (2011) 5, 2142.
50. Yu, Q., et al., Nature Mat (2011) 10, 443.
51. Ruiz-Vargas, C. S., et al., Nano Lett (2011) 11, 2259.
52. Jia, X., et al., Science (2009) 323, 1701.
53. Cançado, L. G., et al., Phys Rev Lett (2004) 93, 047403.
54. Cançado, L. G., et al., Phys Rev Lett (2004) 93, 247401.
55. Campos-Delgado, J., et al., Nano Lett (2008) 8, 2773.
56. Yang, L., et al., Phys Rev Lett (2007) 99, 186801.
57. Nakada, K., et al., Phys Rev B (1996) 54, 17 954.
58. Saito, R., et al., Adv Phys (2011) 60, 413.
59. Hashimoto, A., et al., Nature (2004) 430, 870.
60. Meyer, J. C., et al., Nano Lett (2008) 8, 3582.
61. Gass, M. H., et al., Nature Nanotech (2008) 3, 676.
62. Girit, C. O., et al., Science (2009) 323, 1705.
63. Warner, J. H., et al., Nature Nanotech (2009) 4, 500.
64. Wildoer, J. W. G., et al., Nature (1998) 391, 59.
65. Odom T. W., et al., Nature (1998) 391, 62.
66. Carroll, D. L., et al., Phys Rev Lett (1997) 78, 2811.
67. Maciel, I. O., et al., Nature Mater (2008) 7, 878.
68. Hartschuh, A., et al., Phys Rev Lett (2003) 90, 095503.
69. Qian H., et al., Nano Lett (2008) 8, 2706.
70. Georgi, C., and Hartschuh, A., Appl Phys Lett, (2010) 97, 143117.
71. Doorn, S. K., et al., Phys Rev Lett (2005) 94, 016802.
72. Elias, D. C., et al., Science (2009) 323, 610.
73. Robinson, J. T., et al., Nano Lett (2010) 10, 3001.
74. Cheng, S. H., et al., Phys Rev B (2010) 81, 205435.
75. Guo, B., et al., Insciences J (2011) 1, 80.

76. Hwang, E. H., et al., Phys Rev B (2007) 76, 195421.
77. Farmera, D. B., et al., Appl Phys Lett (2009) 94, 213106.
78. Biel, B., et al., Phys Rev Lett (2009) 102, 096803.
79. Hong, Woong-Ki., et al., Nanotechnology (2006) 17, 5675.
80. Skákalová, V., et al., Phys Stat Sol B (2008) 245, 2280.
81. Tuinstra, F., and Koenig, J. L., J Phys Chem (1970) 53, 1126.
82. Cancado, L. G., et al., Appl Phys Lett (2006) 88, 163106.
83. Lazzeri, M., and Mauri, F., Phys Rev Lett (2006) 97, 266407.
84. Beams, R., et al., Nano Lett (2011) 11, 1177.
85. Crespi, V. H., Phys Rev B (1996) 53, 13303.
86. Terrones, H., M. et al., Phys Rev Lett (2000) 84, 1716.
87. Huang, P. Y., et al., Nature (2011) 469, 389.
88. Lui, C. H., et al., Nano Lett (2011) 11, 164.
89. Cong C., et al., ACS Nano (2011) 5, 8760.
90. Lui, C. H., et al., Nature Phys (2011) 7, 944.
91. Das, A., et al., Nature Nanotech (2008) 3, 210.
92. Basko, D. M., et al., Phys Rev B (2009) 80, 165413.
93. Das, A., et al., Phys Rev B (2009) 79, 155417.

Chapter 4

Diamond Wars? Conflict Diamonds and Geographies of Resource Wars

Philippe Le Billon

Department of Geography, University of British Columbia

ABSTRACT

In the late 1990s, natural resources such as oil, diamonds, and timber came under increased scrutiny by conflict analysts and media outlets for their purported role in many contemporary wars. This article discusses some of the limitations of conventional arguments linking wars and resources. Dominated by econometric approaches and rational choice theory interpretations, arguments pertaining to "resource wars" often oversimplify or overlook the geographical dimensions of resource-related conflicts. By defining spatiality primarily in terms of the location of resource reserves and flows generating revenues for belligerents, these approaches overlook other geographical aspects of resources crucial to conflicts. Focusing on "conflict diamonds" and drawing on recent international relations works and geographical research on the political ecology of violence, commodity chains, and consumption, the article presents an alternative conceptual framework engaging with resource-related spaces of vulnerability, risk, and opportunity for conflicts. This framework, in turn, highlights policy biases resulting from oversimplified readings of "resource war" geographies.

Keywords:conflict, development, diamonds, governance, resource curse, war, **冲突**, 发展, 钻石, **治理**, 资源的诅咒, 战争, conflicto, desarrollo, dia mantes,formas de gobierno, maldición de los recursos, guerra

In the late 1990s, the use of diamonds to fund rebel movements in Angola and Sierra Leone captivated the attention of conflict analysts and media outlets. With the end of the Cold War, more belligerents have come to rely on revenues from commodities such as timber, oil, narcotics, or precious minerals (Jean and Rufin 1996; Kaldor 1999; N. Cooper 2002;Ross 2003, 2006; Fearon 2004; Nordstrom 2004). These "conflict commodities" are not only understood as financing hostilities, but also as shaping the motives of violence and behavior of armed groups (Keen 1998; Weinstein 2007). Among these commodities, conflict diamonds arguably played the largest part in reframing analyses of contemporary armed conflicts and conflict termination initiatives (Reno 1995; Misser and Vallée 1997; Global Witness 1998; Alao 1999;Berdal and Malone 2000; Smillie, Gberie, and Hazleton 2000; Bannon and Collier 2003).

Popularized in the media as "blood diamonds," precious gems are portrayed as the primary motive of greed-driven wars led by rampaging warlords (R. N. Cooper 2001; Campbell 2002; Le Billon 2006). War, it is argued in many of these "strategic narratives" (Shapiro 1997), is largely a criminal project, motivated by the "lust for resources" and facilitated by easily "lootable" resources (Collier and Hoeffler 1998; Keen 1998; United Nations [UN] 2001; Global Witness 2002a; MacGinty 2004).Yet besides "greed-driven" warlords using forced labor, hundreds of thousands of men, women, and children also engaged in diamond mining, hoping, in a context of chronic poverty and widespread abuses, that "luck be stronger than death" (Interview notes Angola, July 2001; De Boeck 2001; Partnership Africa Canada and Global Witness 2004). Numerous diamond companies profiteered from conflict diamonds, initially ducking and deriding attempts to curtail the trade. Even peacekeepers, suggested UN Secretary General Kofi Annan, were not immune to the "poisonous mix" of diamonds and greed fueling these wars (Crossette 2000).

This article examines the geographical dimensions of so-called diamond wars and related regulatory initiatives. Resource spatiality has recently received attention as part of a broader reengagement with the geographical dimensions of wars (Le Billon 2001; Buhaug and Gates 2002; Ward and Gleditsch 2002; Flint 2004; Simmons 2004). Among academics, Michael Klare (2001, 215) pushes the argument the furthest, arguing that a "new geography of war" is emerging in which "resource concentrations rather than political boundaries are the major defining features." Until recently, relatively few contemporary geographers have engaged with the study of resource wars, as a concept relating resources and large-scale organized violence (but see Wescoat 1992; Sidaway 1998; Le Billon 2001; Peluso and Watts 2001; Dalby 2002). Most geographers have focused their attention on specific livelihood conflicts involving struggles over access to resources, a "traditional" area of political ecology (Bryant and Bailey 1997; Robbins 2004; Simmons 2004). Indeed, geographers addressing resource wars have critiqued the neo-Malthusian overtones of this concept and its naturalization of violence (Peluso and Watts 2001; Dalby 2002), as well as the oversimplification of its geographical arguments (Flint 2003, 2004; Le Billon 2005b; O'Lear 2005;Korf and Fünfgeld 2006).

In response to these debates, this article underlines the importance of the materiality, location, flows, and political economy of resources in many armed conflicts, specifically examining the case of diamonds. It also seeks to broaden the scope of geographical contributions beyond locationally probabilistic assumptions about a potential "geographical destiny" in struggles over resources. The study builds on research on diamonds and war economies conducted since 1998, when I was first introduced to this issue through my collaboration with Global Witness, the first advocacy organization to launch a major campaign on conflict diamonds. Related fieldwork included visits of up to five weeks in Angola (1998, 2001), the Democratic Republic of the Congo (2001), Ghana (2003), Sierra Leone (2001, 2006), and South Africa (2001). Interviews were conducted with industry representatives, public officials, and members of nongovernmental organizations (NGOs) in Belgium, Canada, England, and France between 1999 and 2006. The research also included participant observation in four policy meetings related to the Kimberley Process Certification Scheme, the international diamond certification scheme set up to prevent the laundering of conflict diamonds by the industry. Drawing on this research and recent developments in political ecology, resource geography, commodity chain analysis, and geographies of consumption, this article seeks to systematically explore the geographical dimensions of resource wars through a study of conflict diamonds by making three interrelated arguments.

First, the article argues that analyses of resource wars need to incorporate three major dimensions to better relate resource exploitation with conflicts and violence. The first consists of resource-based processes of peripheralization and uneven development, and their spatial dimensions (Roberts and Emel 1992; Auty 2004). The second is the social construction of resource narratives and their intersection with the biophysical aspects of resources and modes of production (Bunker 1985;Swyngedouw 1999; Bridge 2001; Le Billon 2001; Bakker and Bridge 2006). The last is the interplay of resource extraction with accumulation strategies and contestations in resource peripheries (Bridge 2000, 2004; Peluso and Watts 2001; Hayter 2003;Hayter, Barnes, and Bradshaw 2003; Simmons 2004).

Second, the article argues that studies of resource-related conflicts need to broaden their analyses beyond spaces of resource exploitation to include the interrelationship between spaces of production, consumption, representation, and governance (Hartwick 1998; Castree 2001; Neumann 2004; Watts 2004). Geographies of commodity chains connect spaces of production and consumption, and stress the importance of social relations around particular spaces and places of commodity circulation (Bassett 1988; Hartwick

1998; Leslie and Reimer 1999; Castree 2001; Bryant and Goodman 2004; Neumann 2004). Mapping "vertical" connections along the commodity chain between sites of commodity production and consumption, as well as "horizontal" connections between each of these sites and broader social networks, refines resource policies and politics of consumption (Leslie and Reimer 1999; Hartwick 2000). Such critical perspective exposes the limits and ambivalences of "ethical" consumption and reregulation of resource sectors, notably the uncritical acceptance of consumption, prejudiced representations, and biased conceptions of legitimacy (Bryant and Goodman 2004; Barnett et al. 2005; Le Billon 2006).

Third, the article argues that the narrow definition of violence used in much of the literature—usually pertaining to armed conflicts resulting in at least twenty-five battle deaths per year—overlooks multiple forms and scales of violence enacted through resource exploitation and regulation. Recent work on the geography and political ecology of violence reasserts an expanded definition of violence (McIlwaine 1999; Peluso and Watts 2001; Turner 2004), encompassing nonphysical forms of violence but also acknowledging the pervasive presence of (realized or implied) violence in resource (dis)possession regimes (Blomley 2003; Harvey 2003; Harris 2004; G. Hart 2006). War is not the only (or even primary) type of violence associated with resource-extractive industries, and resource war arguments risk essentializing and depoliticizing violence, thereby misreading its causal factors and impacts.

The first section of the article situates conflict diamonds within the broader literature on resources and conflicts, outlines three concepts relating resources and conflicts—resource curse, resource conflicts, and conflict resources—and discusses their main geographical dimensions. In the second section, the article uses the case of diamonds to explore how a constrained understanding of the geography of resources and violence reduces the explanatory value of these resource-related conflicts. This section then scrutinizes the assumption that diamonds are highly facilitative of rebel activity, and highlights the political and tactical ambivalence of diamond exploitation in wars. The final section of the article considers the implications of a geographical perspective linking spaces of exploitation, consumption, and regulation. This viewpoint stresses the impact of ambiguous standards of criminality and legitimacy on artisanal diamond producers, arguably among the most marginalized participants in the global diamond commodity chain. [1]

GEOGRAPHIES OF RESOURCE WARS: AN ANALYTICAL FRAMEWORK

Three distinct yet overlapping theoretical arguments are mobilized to explain resource wars (Le Billon 2001, 2005b). The first,*resource curse*, argues that resource dependence results in economic underperformance and a weakening of governing institutions that makes a society more vulnerable to armed conflict. The second, *resource conflicts*, suggests that grievances, conflicts, and violence associated with resource control and exploitation increase the risk of onset of larger scale armed conflicts. The third, *conflict resources*, recognizes resources as providing financial opportunities motivating belligerents and financially sustaining armed conflicts. Table 1 displays this framework.

Table 1 Analytical framework

Conflict factors	Key concepts	Relations between resource and conflict	Major geographical dimensions	Potential variables and measurements
Vulnerability	Resource curse	Resource sector undermines governance and economic performance, making violent conflict more likely	Peripheralization and uneven development: resource dependence shaping political economy and mode of governance	• Evolution of resource abundance and dependence (annual variations in export and fiscal revenues) • Ownership structure and mode of exploitation (private vs. parastatal, artisanal vs. industrial) • Socioeconomic indicators (spatially disaggregated measures of poverty; capital flight and foreign direct investment)
Risk	Resource conflicts	Resource sector motivates disputed processes of allocation, making violent dispossession and resistance more likely	Territorialization: spatial control of resource central to cost–benefits allocation and identity mobilization	• Conditions of access to resources (legal status and practices) • Instances of conflicts and violence (population displacement, demonstrations, repression, homicides) • Processes of identity formation and mobilization (sectarian institutions and voting patterns)
Opportunity	Conflict resources	Resource sector rewards belligerents, making the escalation and prolongation of violent conflict more likely	Interconnection: ease of access to resource revenues crucial in sustaining conflicts	• Characteristics of the resource (price and weight) • Regulatory environment (level of illicit trade, strength of regulatory institutions) • Proximity to transport and marketing infrastructures (borders, airports, buying offices)

Rethinking the Resource Curse: Conceptualizing Vulnerability

The resource curse argument suggests that resource dependence creates a context for the emergence of armed conflicts through its negative effects on economic performance and the quality of governing institutions (Ross 1999; Le Billon 2001; de Soysa 2002; Auty 2004; Fearon 2005). Rather than simply explaining war through greed-driven belligerents motivated by lootable resources, this argument emphasizes resource dependence as both reflecting and shaping conditions that increase vulnerability to armed conflicts. According to the political economy literature, the characteristics of countries most vulnerable to civil war since 1946 are low per capita income, declining economic growth rate, "weak" state coercive capacity and institutional authority, and political regimes in transition, although the influence of different kinds of inequalities remains debated (Cramer 2003; Fearon and Laitin 2003; Besançon 2005).

Empirical evidence for the resource curse argument is strong, although historically and institutionally contingent, with studies suggesting that resource-dependent countries tend to underperform economically, be more poorly governed, and present lower social indicators compared with other countries with similar income levels (Auty 2001; Ross 2001b; Sachs and Warner 2001; Papyrakis and Gerlagh 2004; Mehlum, Moene, and Torvik 2006; Perala forthcoming). Economic explanations include exposure to high price fluctuations, declining terms of trade, natural capital depreciation, a crowding out of the nonresource sectors through local currency overvaluation and rent seeking, as well as overconsumption and misguided economic and social policies (Auty 2001; Neumayer 2004). Political explanations range from resource rent effects on overoptimistic and shortsighted policies, policy capture by special interests, higher levels of corruption, and state fiscal independence resulting in a lack of democratic bargaining power for the population (Leite and Weidmann 1999; Ross 1999, 2001a; Moore 2001;Sala-i-Martin and Subramanian 2003). Empirical evidence linking resource curse and war, however, is relatively weak in quantitative studies, and little consensus has yet emerged with the exception of oil-dependent countries (for a review, seeRoss 2004b; Fearon 2005).

Theories of uneven development suggest resource dependence has spatial dimensions, expressed as selective processes of modernization and peripheralization combined with the production of hierarchical scales predominantly defined by their relationship with the resource sector (Lanning and Mueller 1979; Harvey 1982; Smith 1984). Resource dependence is also constitutive of, and constituted by core–periphery relations between (and within) producing and consuming countries, which although not fixed in time entail historical legacies with potent political and economic impacts (Bridge 2006). Contemporary low income per capita, for example, is strongly correlated with European extractive institutions set up in colonies that could not be settled (Acemoglu, Robinson, and Johnson 2001). Quantitative studies of the resource curse have not systematically tracked the historical origins and evolution (rather than the level) of resource dependence as variables in relation to conflicts. As the relative economic and political importance of resource production areas increases, other areas are comparatively peripheralized. Within resource production areas themselves, nonresource aspects also become peripheralized. Arguably, resource areas are also peripheralized in the dependency school understanding that they become dependent on an unequal relationship with center or core areas. The uneven development resulting from such processes of relative centralization and peripheralization, in turn, reflects the social construction, spatial distribution, and mode of production of resources (Roberts and Emel 1992).

Peripheralization and uneven development entail tangible spatial effects far beyond the narrow confines of the resource sector, extending into social identities, territorialities, political governance, economic marginalization, and environmental outcomes of relevance to the study of conflicts (Bunker 1985; Watts and Bohle 1993; Bridge 2000;Moore 2001).

Resource Conflicts and Conflict Resources: Conceptualizing Risk and Opportunity

Overlapping with environmental security studies (Dalby 2002), resource conflicts arguments associate resources with specific conflicts (e.g., over resource access) and occurrences of violence (e.g., militarization of resource areas, pollution, or labor abuses). This association has focused on livelihood conflicts pertaining to mostly renewable resources (Homer-Dixon 1999;Peluso and Watts 2001; Simmons 2004; Turner 2004) and national or military resource security pertaining to mostly nonrenewable resources such as oil and "strategic" minerals (Klare 1980; Lesser 1989; Le Billon 2004). Although involving distinct case studies and often guided by very different ideological agendas, these approaches share a central assumption: The diversity of patterns of violence involved in resource control and exploitation can lead to "war," ranging from its discursive deployment to organized physical violence (Klare 2001; Hayter 2003; Neumann 2004; Perreault 2006). The spatial dimensions of resource conflicts are generally conceived in terms of scarcity (i.e., relative distribution of resources), and identity-related endowments (e.g., who controls what and where).

In contrast, the conflict resources argument (the most prominent in the case of conflict diamonds) associates resources with specific opportunities (mostly understood as financial) afforded to belligerents. From this perspective, some resources are more prone to sustaining and motivating war than others, notably highly valuable resources most easily accessible to the weaker party in a war. A narrow definition of conflict resources refers to resources financing rebel groups, as in the case of the official definition of conflict diamonds by the UN General Assembly (2000): "rough diamonds used by rebel movements or their allies to finance conflict aimed at undermining legitimate governments." A broader definition is that of natural resources whose control, exploitation, trade, taxation, or protection contributes to, or benefits from the context of, armed conflict. Broader definitions widen the range of responsibility from illegal armed groups to all those profiteering or maintaining economic relations prolonging hostilities, including companies, governments, and consumers (Global Witness 2002b). As discussed later in the case of diamonds, opportunity is understood to have spatial dimensions reflecting the biophysical aspects and spatiality of this resource and its mining and marketing context (i.e., what is accessible to who and where).

The concepts of resource conflicts and conflict resources are not mutually exclusive. Rather, they are complementary in their respective focus on the risks of engendering armed conflict, and the resource-related economic opportunities available to combatants. Two distinct conceptions of the spatiality of resource-related conflicts underlie these approaches. The geography of risk (resource conflicts) is understood to revolve largely, but not exclusively, around processes of territorialization, predicated on perceptions of resources as spatially fixed but socially constructed "natural" endowments. From this perspective, entitlements are in large part articulated and contested through processes of territorialization (Vandergeest and Peluso 1995; Neumann 2004; Peluso 2005), including a militarization of regulatory responses to resource conflicts. In contrast, the geography of opportunity (conflict resources) reflects processes of interconnection between actors at local, regional, and international scales. Although financing war can rely on the territorialization of resource production areas (possibly entailing a resource conflict), it also involves spatial connections enabling the circulation of commodities (resources, money, arms, labor). Furthermore, belligerents can generate revenues through the control of resource-related flows, such as threats of destruction or obstruction of oil pipelines or the kidnapping and ransoming of resource project staff (Ross 2003). The geography of resource wars is not only defined by frontlines around the production area, but also by spaces along the commodity chain. Regulatory responses affect the behavior of commodity chain actors and thus work in part through strategies of reterritorialization and disconnection.

DIAMOND WARS?

As already presented, the literature on resource wars provides at least three ways of thinking about the relationship between resources and conflicts, each of which has important spatial dimensions: vulnerability and peripheralization or uneven development, risk and territorialization, and opportunity and interconnection. This section of the article explores these concepts with respect to the case of conflict diamonds, and examines how an expanded understanding of geographies of resource wars can contribute to studies of the role of diamonds in armed conflicts. Academic studies of diamond-related conflicts have until recently largely concentrated on the labor, capital, and racial dimensions of the diamond sector (Crush 1994; Zack-Williams 1995), with a focus on late-eighteenth-century Brazil (H. Bernstein 1988; Pinto Vallejos 1985) and late-nineteenth-century struggles over diamond mining in South Africa (Turrell 1981, 1987), the diamond mining and trading activities of De Beers (Newbury 1989; Carstens 2001), and the control of "strategic" industrial diamonds during World War II and the Cold War (Dumett 1985). Given the importance of historical context to political ecology approaches, these studies inform this article, notably with regard to discussions of labor and the political economy of the diamond sector. The article remains largely focused on diamond-producing countries that have experienced wars during the 1990s, which include six of the eight most diamond-dependent countries in the world, all of them in sub-Saharan Africa, with Angola, Sierra Leone, and the Democratic Republic of the Congo (DRC, formerly Zaire) being the countries reported to be most affected by recent diamond wars.

Before considering arguments pertaining to vulnerability, risk, and opportunity, this section briefly reviews evidence presented on the existence and prevalence of diamond wars. One in six countries in the world currently produces diamonds, with a heavy concentration in sub-Saharan Africa (sixteen out of twenty-eight producers and 60 percent of worldwide production value; see Figure 1A, Figure 1B), the most conflict-affected region since the 1980s. Quantitative studies have generally found conditional support for a relationship between diamonds and war (see Table 2). Using a new data set of worldwide diamond deposit locations (see Figure 1A, Figure 1B), Lujala, Gleditsch, and Gilmore (2005) examine the effects of diamond deposits and exploitation on the likelihood of civil war during the 1945 to 2002 period. Although war is more prevalent in diamond-producing countries (74 percent experienced at least one armed conflict compared to 43 percent in general), they find that when controlling for other factors, such as low income, the presence of diamond deposits or diamond production did not increase the risk of war onset. Taking into account the spatiality of resources on conflict likelihood (seeLe Billon 2001), however, they find that the exploitation of secondary or alluvial diamond deposits—a "diffuse" and more "lootable" resource—increases the incidence (or relative frequency) of war, especially in poor countries, and with respect to ethnic conflicts and to the post–Cold War period. In contrast, primary or Kimberlite deposits—a "point" resource less lootable by rebels—had a dampening effect on the incidence of war, and the discovery of diamond deposits alone had no effect. [2]Lujala, Gleditsch, and Gilmore (2005) interpret these findings as confirming the opportunity argument: lootable diamonds (secondary deposits) are more likely to increase the incidence of wars than nonlootable ones (primary deposits). They cast doubt, however, on the vulnerability and risk arguments. Overall, a majority of large-*N* studies assessing general patterns of relations between war and diamonds provide conditional support for the opportunity argument, but less so for vulnerability and even less for risk (see Table 2).

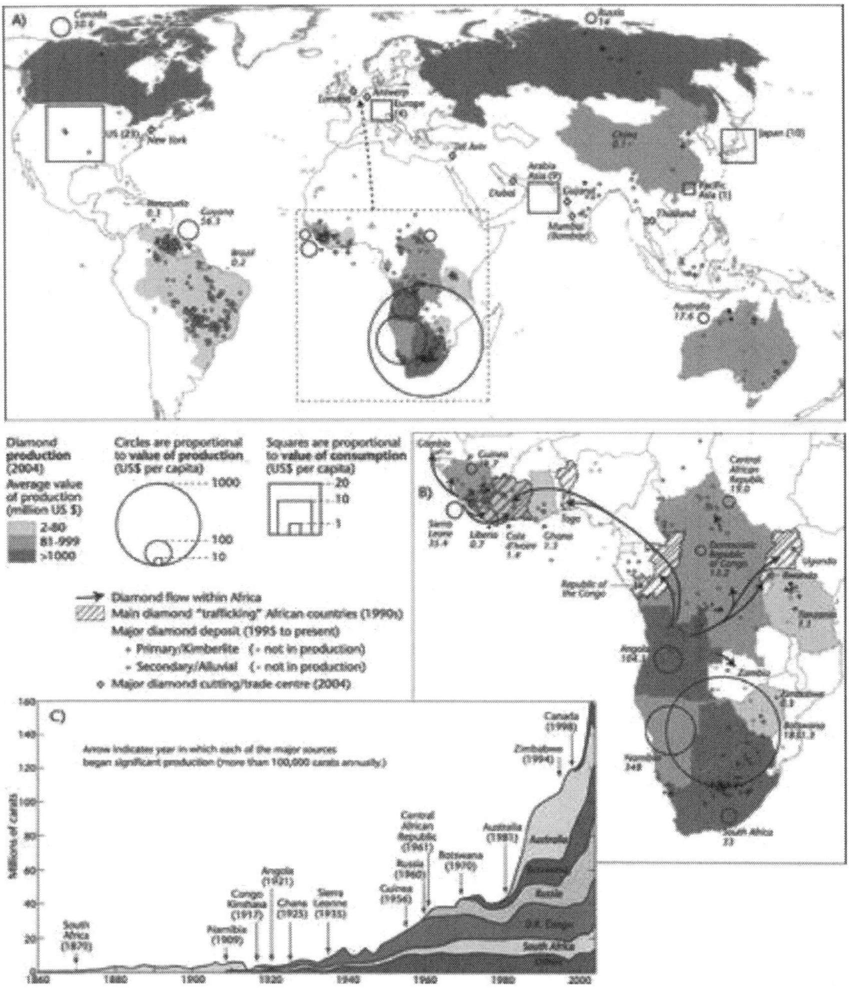

Figure 1 (A) World diamond production and major diamond trading and cutting centers; (B) Sub-Saharan Africa diamond production and "trafficking" flows; (C) Historical world diamond production (1860–2005). *Note:* Prior to the 1720s diamonds came exclusively from India, then mostly from Brazil. In 2004, most diamonds from Africa were still exported to Antwerp and London, with limited direct exports to Tel Aviv, New York, and emerging diamond trading centers. *Source:* (A) and (B): UN (2000a, 2001, 2002), USGS (2005), Gilmore et al. (2005). (C) Levinson (1998) up to 1995, USGS Minerals Yearbook for 1996–2005.

Table 2 Quantitative analyses of diamonds and wars

Studies	Data sets	Resource variables	Findings	Vulnerability	Risk	Opportunity
Collier and Hoeffler (1998)	27 civil wars, 1960–1992	Primary exports/gross domestic product	Increased conflict onset likelihood and conflict duration (curvilinear)	Weak	—	Strong
Buhaug and Gates (2002)	265 civil wars, 1946–2000	Natural resources, including diamonds, in subnational conflict zone	Increased size of conflict zone	—	—	Strong
Fearon (2004)	128 wars, 1945–1999	Contraband goods under rebel control, including diamonds	Prolonged conflict duration	—	—	Strong
Lujala, Gleditsch, and Gilmore (2005)	127 wars, 1945–1999	Diamond type, production, reserves	Secondary diamond production associated with increased onset likelihood of ethnic conflicts but not primary diamonds nor diamond reserves	Weak	—	Strong
Buhaug and Lujala (2005)	252 civil wars, 1946–2001	Presence of diamonds in conflict area	Prolonged conflict duration	—	—	Strong
Humphreys (2005)	122 wars, 1945–1999	Diamond production	Increased conflict onset likelihood, but reduced duration (positively correlated with both military victory and negotiations)	Strong	Medium	Weak
Regan and Norton (2005)	153 civil wars, 1945–1999	Availability of gemstones in country	Reduced conflict onset likelihood and no effect on conflict duration	Weak	Weak	Weak
Ross (2006)	90 wars, 1960–1999	Diamond type, production	Increased conflict onset likelihood for primary diamonds, but not secondary diamonds (reverse for separatist conflicts), no effect on conflict duration	Medium	Weak	Weak
Buhaug and Rød (2006)	Civil wars in Africa, 1970–2001	Diamond type, location	Increased governmental conflict onset likelihood, but reduced separatist onset likelihood for secondary diamonds	Strong	Medium	Strong

The small number of wars in diamond-producing countries should nevertheless caution against any generalizing conclusions on the relationship between war and diamonds based on large-*N* studies (Ross 2006). Moreover, the relative significance of diamonds should also be weighted against other factors (Buhaug and Lujala 2005; Lujala, Gleditsch, and Gilmore 2005). A review of the literature suggests that although an estimated twenty conflicts could plausibly be related to diamonds between 1946 and 2005 (see Table 3), in only four cases were diamonds strongly related to conflict either through vulnerability, risk, or opportunity factors: civil wars in Sierra Leone (Revolutionary United Front of Sierra Leone [RUF], 1992–2001) and Angola (National Union for the Total Independence of Angola [UNITA], 1993–2002), the secession of South Kasai (Independent Mining State of South Kasai, 1960–1962), and the independence struggle of Namibia (South-West Africa People's Organization [SWAPO], 1966–1988). The following subsections examine each argument in more detail.

Table 3 List of diamond-related armed conflicts, 1946–2005

Country	Government (Side A)	Rebel (Side B)	Years	War-related deaths	Vulnerability factor	Risk factor	Opportunity factor (side)	Major rebel funding sources
Sierra Leone	Sierra Leone, ECOWAS, United Kingdom	RUF, AFRC	1991–2000	75,000	Strong	Strong	Strong (A, B)	Diamonds ($25–$75 million per year)
DRC	Congo/Zaire	Independent Mining State of South Kasai	1960–1962	600	Strong	Strong	Strong (A, B)	Diamonds
Angola	Angola	UNITA	1991–2002	700,000	Medium	Medium	Strong (A, B)	Diamonds ($200–$600 million per year)
Namibia	South Africa	SWAPO	1966–1988	25,000	Medium	Strong	Medium (A)	Proxy
DRC	Congo/Zaire	AFDL, Rwanda	1996–1997	230,000	Medium	Medium	Medium (A, B)	Proxy
DRC	Congo/Zaire, Zimbabwe, Angola, Namibia	RCD, MLC, Rwanda, Uganda	1998–2003	2.5 million	Medium	Medium	Medium (A, B)	Gold, coltan, coffee, diamonds
Liberia	Liberia	Military faction	1980	27	Weak	Medium	Medium (A)	N/A
Liberia	Liberia	NPFL	1989–1996	200,000	Weak	Medium	Medium (A, B)	Iron, timber, rubber, diamonds
Liberia	Liberia	LURD, MODEL	2000–2003	2,000	Weak	Medium	Medium (A, B)	Proxy
Central African Republic	Central African Republic, Libya	Military faction	2001–2002	500	Medium	Medium	Weak (A)	Proxy
Ivory Coast	Ivory Coast	MPCI, MJP, MPIGO, FN	2002–2005	850	Weak	Weak	Medium (B)	Cocoa, timber, diamonds
Angola	Angola, Cuba	UNITA, South Africa	1975–1991	500,000	Weak	Medium	Weak (A, B)	Proxy
Guinea	Guinea	Military faction, RFDG	1970, 2000–2001	300; 1,100	Weak	Weak	Weak (A)	N/A
Angola	Portugal	MPLA, FNLA	1961–1974	90,000	Weak	Weak	Weak (A)	Proxy
South Africa	South Africa	ANC, PAC, Azapo	1981–1988	20,000	Weak	Weak	Weak (A)	Proxy
Surinam	Surinam	SLA/Jungle Commando	1986–1988	500	Weak	Weak	Weak (A)	Proxy
Ghana	Ghana	Military factions	1966, 1981, 1983	100	Weak	Weak	Weak (A)	N/A

Note: AFDL = Alliance of Democratic Forces for the Liberation of Congo-Zaire; AFRC = Armed Forces Revolutionary Council; ANC = African National Congress; DRC = Democratic Republic of Congo; ECOWAS = Economic Community of West African States; FN = New Forces; FNLA = National Front for the Liberation of Angola; LURD = Liberians United for Reconciliation and Democracy; MJP = Patriotic Youth Movement; MLC = Movement for the Liberation of Congo; MODEL = Movement for Democracy in Liberia; MPIGO = Ivorian Popular Movement of the Great West; MPCI = Ivory Coast Patriotic Movement; MPLA = Popular Movement for the Liberation of Angola; NPFL = National Patriotic Front of Liberia; PAC = Pan Africanist Congress; RCD = Rally for Congolese Democracy; RUF = Revolutionary United Front of Sierra Leone (also RUF/SL); SLA = Surinamese Liberation Army; SWAPO = South-West Africa People's Organization; UNITA = National Union for the Total Independence of Angola. Battle deaths accounted for 460,000 lives.
Source: Lacina and Gleditsch (2004); Center for the Study of Civil War (2006).

Vulnerability: Is There a Diamond Curse?

What is the impact of diamond wealth on vulnerability to conflict? How should diamond wealth be characterized in this respect? Distinguishing between diamond dependence, diamond abundance, and mode of exploitation (see Figure 2), this subsection discusses each in turn. *Diamond dependence* refers to the importance of the diamond industry within an economy, measured as diamond production in percentage of gross domestic product (GDP). When simply considering GDP growth rates since 1960, the economies of diamond producers have generally outperformed those of non-diamond-producing countries, notably in the late 1960s and during the 1980s.[3] Since the early 1990s, diamond producers and nonproducers are, on average, on par.[4] Not all diamond-producing countries performed equally. Botswana outperformed both other diamond producers and the rest of the subcontinent: Over three decades of diamond production it grew ten times faster than the average for sub-Saharan Africa (see later). Among other diamond producers, a first group of countries including South Africa, Guinea, Tanzania, and to a lesser extent Namibia and Ghana performed on par with the rest of sub-Saharan Africa. A second group, including Central African Republic, Angola, Sierra Leone, Liberia, and the DRC clearly underperformed, and all have been affected by armed conflicts during the past decade. Furthermore, the level of economic dependence on diamonds does not seem to be associated in a general fashion with economic underperformance, at least within sub-Saharan Africa, as both groups include countries with different levels of diamond dependence.

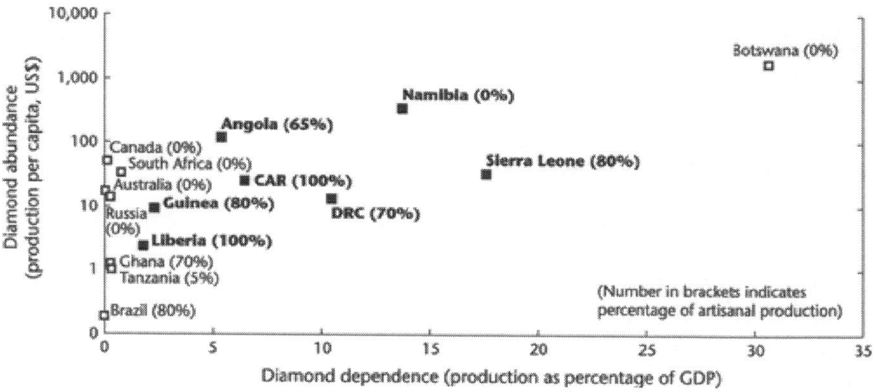

Figure 2 Diamond dependence, abundance, and artisanal production, 2004. CAR = Central African Republic; DRC = Democratic Republic of the Congo. *Source:* Northwest Territories (2005), USGS (2005), Oomes and Vocke (2003).

Diamond abundance refers here to per capita diamond production. The evolution of diamond abundance in terms of production should influence economic performance, especially when measured through GDP that does not capture profit repatriation by foreign mining companies. This relation, however, runs counter to the economic underperformance argument. The level of diamond abundance also varies within these two groups, although to a lesser extent. The underperforming group seems to have a medium level of diamond abundance, with the exception of Liberia's low per capita value of diamonds. The *mode of exploitation* seems to yield support to economic performance, with industrial exploitation being associated with stronger economic growth, notably for Botswana and Namibia. Yet the economic contribution of artisanal production to GDP is generally underestimated when occurring with high levels of diamond (and imported goods) smuggling.

In short, diamond abundance, dependence, and mode of exploitation, on their own, do not seem to predict economic underperformance. This is consistent with views that the diamond sector compares positively with most other primary commodity sectors in terms of macro-economic development, in particular for Africa, where it is one of the few sustained commodity export sectors (Ng and Yeats 2002). Unlike other commodities, and despite a huge increase in production volume (see Figure 1C), diamonds have been largely sheltered from massive price collapse or fluctuation due to the price regulating (and enhancing) monopolistic activities of De Beers and sustained expansion of demand (Khoury 1990; Platt 2006). The diamond sector thus appears relatively privileged by stronger terms of trade and lower volatility compared with most other commodity sectors. This macroeconomic and state-centered reading of the economic side of the diamond curse risks being misleading, however. It is thus important to deal historically with the impact of diamond sectors on economic performance, through more detailed studies disaggregating economic performances and assessing peripheralization and uneven development effects (see later and Table 1).

The second diamond curse argument relates to the quality of governance and political stability, which in turn would influence economic performance (Olsson 2006, 2007). Resource wealth frequently undermines the quality of institutions, resulting in an overextension of the state and policy capture by vested interests (Ross 1999), ultimately weakening state capacity and legitimacy. This institutional effect appears to vary according to the geographic and economic concentration of the resource involved, with more concentrated point resources having a more deleterious effect than diffuse resources (Bulte, Damania, and Deacon 2005; Isham et al. 2005). Is this argument applicable in the case of diamonds? As mentioned earlier, Lujala et al. (2005) find a contrary result to this argument: Diffuse diamonds are associated with war (and institutional breakdown), not point diamonds. Similarly, Snyder and Bhavnani (2005) argue that diffuse diamonds undermine government institutions by making tax collection extremely difficult, whereas point diamonds consolidate them.

In their study of "gemmocracies" in Africa, Misser and Vallée (1997) argue that diamond rents are central to many postindependence African regimes. In the face of structural adjustments and declining terms of trade and market share for most other export sectors, diamond rents have been remarkably resilient, yet with ambivalent political effects. Diamond rents are linked with state overextension, corruption, and capital flight (Reno 1995; Misser and Vallée 1997). Yet rents have also financed patronage politics and sustained the stability of political regimes. Vulnerability to conflicts would respond not only to the political impact of the rent, but more broadly to the governmentality of the diamond sector. [5]

The evolution of diamond rents reflects in part that of the ownership structure, as well as levels of inputs and reserve depletion. Control and access of diamond revenues is also highly dependent on the mode of exploitation, which relates in part to the material specificities of diamond deposits. Alluvial diamonds, retrieved with minimal technological input from vast areas of riverbed gravel, can unsurprisingly prove to be a very uncooperative commodity for the bureaucratic state seeking to capture rents (see Bakker and Bridge 2006). Once retrieved from the soil, diamonds can easily evade taxation (with the complicity of buyers). Control and access to diamonds then relies on the insertion of politicians into informal diamond trading, a mode of governmentality defined by Reno (1995) in his seminal study of Sierra Leone as "shadow state" politics.

The diamond governmentality—and more specifically diamond-related "governable space"(see Watts 2004, 199)—thus relies in part on the spatiality of diamond deposits and modes of exploitation. The exploitation of primary or Kimberlite deposits requires industrialization. In contrast, secondary or alluvial deposits are open to artisanal mining, an option that is widely taken in poorer countries, with or without the official consent of authorities. Because artisanal diamond mining is harder to tax than industrial mining, poorer countries are more vulnerable to weak state capacity. [6] Furthermore, industrial mining is frequently described as minimizing "excessive" and "inefficient" labor inputs, and as a less wasteful mode of exploitation both in terms of reserve recovery and environmental degradation than artisanal mining (Sinding 2005; Davies 2006a). Rising resource prices, deposit discovery, and the legalization of artisanal exploitation are often cited as drawing labor into mining with detrimental effects on the rest of the economy and tax revenues for the state. In this perspective, poorer countries would thus be worse off again, with artisanal mining booms the precursor of economic and political "troubles" leaving a country more vulnerable to conflicts.

This perspective requires nuances. First, artisanal exploitation can be more beneficial to local communities and producing countries, except in terms of fiscal returns for the state (Killick 1973). There is also a need to distinguish between "push" and "pull" factors affecting labor mobility toward the diamond sector. Some artisanal mining booms have also been demonstrated to follow rather than precede economic recession, and to stabilize the economy, thereby reversing the resource curse argument and providing a valuable source of income for the poorest (Heemskerk 2001). The desperation of populations facing failed and oppressive agrarian or urban institutions and economies should thus be acknowledged, notably domestic and colonial legacies of servitude and humiliation in agrarian spaces shaping hopes of emancipation in the mining fields (De Boeck 2001; Richards 2005).

Second, there is a need to refine the link between weak taxation and political vulnerability to conflicts. Lack of fiscal access to diamond revenues due to artisanal exploitation should leave the state less capable of carrying out its role of public service provision, including social services, public infrastructure, and regulatory functions. Low state capacity can in turn reduce political legitimacy and increase vulnerability to conflicts. Some empirical evidence suggests that primary deposits are less frequently associated with civil wars than secondary deposits open to artisanal mining (Lujala, Gleditsch, and Gilmore 2005). Moreover, there is some evidence that private industry expropriation in the diamond sector can aggravate resource curse effects (Acemoglu, Johnson, and Robinson 2003; Olsson 2006; Reno 1995; on the case of oil, see Luong and Weinthal 2006). In this perspective, Botswana's political stability is often presented as the result of its sound (public–private joint venture) industrial exploitation and budgetary allocation of rich primary diamond deposits (Dunning 2005; Olsson 2006), whereas the collapse of formal industrial mining in Sierra Leone since the early 1970s, including through the nationalization and dismantling of the main diamond company, increased vulnerability to civil war in the early 1990s (Reno 1995; Richards, 1996;Snyder and Bhavnani 2005; see Figure 3).

Privileging industrial over artisanal mining has significant political consequences. Compared to artisanal exploitation, industrialization concentrates power in the hands of the state, undermining state accountability to a population made more dependent on state handouts and heightening societal vulnerability to state-level failures. The choice between industrial and artisanal also entails balancing fiscal revenue and direct employment opportunities with consequences for political capacity and domestic political legitimacy. This choice also reflects the preferences of foreign extractive companies and their home governments seeking to foster their industrial ventures. This, in turn, influences foreign relations as investments put a premium on domestic political stability, and thereby on regime survival and the "sovereignty" of ruling elites (see Sidaway 2003), often at the expense of local populations. Industrialization might provide domestic rulers with external allies, but it might also attract competing mining interests seeking to undermine production or gain access to mineral rights through political destabilization, not to mention companies disinvesting and sabotaging their own operations when faced with increased demands by states (Lanning and Mueller 1979).

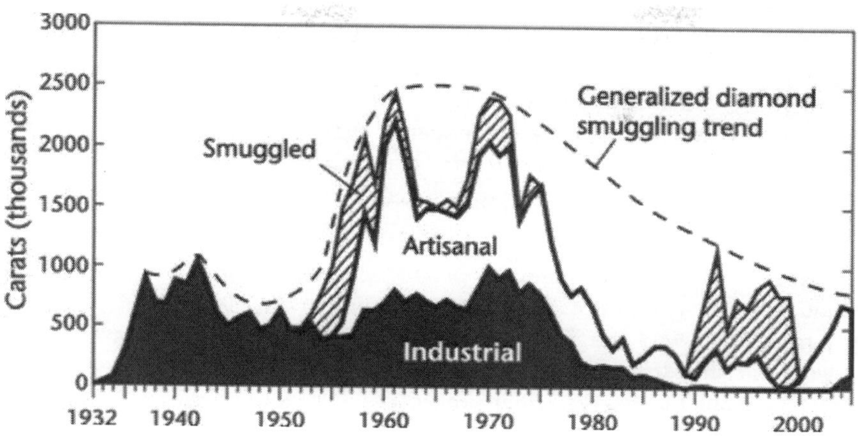

Figure 3 Diamond production in Sierra Leone, 1932–2005. *Note:* Cumulated industrial and artisanal productions and estimated smuggled exports. Smuggled diamonds have generally higher value per volume. Smuggling by value might have reached two-thirds of official production in the mid-1950s, and grew from about three times to nearly the entire official production between the late 1970s and the late 1990s. Periodization and main types of smuggling: (1) 1930s to early 1950s mostly involved "theft" from Sierra Leone Selection Trust (SLST) mining operations, and "value smuggling" through profit repatriation by SLST; (2) early 1950s to early 1970s: increased smuggling from artisanal production; (3) early 1970s to late 1980s: institutionalized smuggling by political cronies; (4) 1992 to 2001: smuggling by belligerents and diamond traders; (5) 2002 to 2005: postconflict smuggling under tightening KPCS regime. *Source:*Saylor (1967), van der Laan (1965), Zack-Williams (1995), Davies (2006a).

To conclude, there is little conclusive evidence for a diamond curse based on a general conception of diamond wealth, or when its three main dimensions—dependence, abundance, and mode of production—are examined in isolation. Yet an assessment of the combined effects of these three dimensions yields more support for the claim of resource curse. A medium level of diamond dependence (5 to 20 percent) and abundance ($12–$150 per capita), as well as predominantly artisanal exploitation (60 to 100 percent) generally characterize diamond-producing countries affected by growth collapse and political instability (Figure 2).[7] This approach posits that countries with such characteristics are more vulnerable to armed conflict.

Comparing Botswana and Sierra Leone

Botswana and Sierra Leone provide illustrations of these arguments and qualifications about forms of violence. Sierra Leone's diamonds were first exploited in the mid 1930s under a British colonial regime, while exploitation in Botswana started in the 1970s under a joint venture between the independent government and De Beers (Hall 1968; Reno 1995; Dunning 2005). Diamonds in Sierra Leone were located in mostly secondary deposits open to artisanal exploitation, rather than exclusively in primary deposits as in Botswana. Whereas Botswana's diamond dependence was higher, labor allocation to mining was minimal compared to Sierra Leone due to industrial exploitation. All three dimensions—abundance, dependence, and mode of exploitation—thus affected state capacity, most notably through fiscal returns and linkages.

The late discovery of diamonds, a postindependence context, and diamond abundance meant that Botswana could bargain harder with foreign investors. In turn, this allowed the government of Botswana to provide greater public services and infrastructures, which in part compensated for the lack of direct access to diamond reserves (that, unlike in Sierra Leone, did not require as much political manipulation and coercive policing due to necessarily industrial exploitation of primary deposits). Having allegedly beaten the resource curse (Sarraf and Moortaza 2001), Botswana is frequently hailed as an African success story. Botswana did experience some resource curse effects as a result of its high diamond dependence (Love 1994), but these aspects have not contributed to economic and political instability due to "sound" macroeconomic management, strong state capacity, and elite cohesion (Dunning 2005). If Botswana has, so far, avoided the scourge of war, it has been ruled by the same party since independence in 1966, and much of its population has endured structural forms of violence leaving the country with the second highest levels of income inequality and second lowest life expectancy in the world (Taylor 2003;Good 2005; UN Development Program 2006).

Sierra Leone's diamond curse has been particularly deleterious (Reno 1995; Silberfein 2004; Keen 2005). Political power in Sierra Leone has long been maintained through predatory institutions, based on a colonial system of indirect rule that increased the power of customary chiefs. This system allowed the British mining company (Sierra Leone Selection Trust [SLST]) to rely on chiefs to minimize illegal mining (although granting them such power occasionally reinforced their ability to protect artisanal mining interests) while also using Sierra Leonean police forces and deploying its own paramilitary forces to stop illegal miners and repress the supporting local populations (Hayward 1972). The chiefs' frequent misuse of delegated powers and privileged access to public revenues also increased the frustration of a neglected population and sidelined local central government officials (Keen 2005; Richards 2005). The 1950s were marked by anti-chief and tax riots, as well as the first large-scale "diamond rush" (see Figure 3). Demands for a radical shift in the control of diamond wealth in the context of active independence movements and loss of effective control on the ground by SLST led to the legalization of artisanal mining in 1956, which greatly increased overall diamond production (Hayward 1972; Reno 1995). By 1961, the artisanal diamond sector was officially exporting twice as many diamonds in volume than SLST, but brought in only half the tax revenue (van der Laan 1965). The diamond rush and renegotiation of the diamond mining regime had several effects (Cartwright 1978). At the national level, the transformations of the 1950s demonstrated the weakness and bias of a colonial regime and "native" government in dealing with a European company, but also the tenuousness of its authority toward local populations. Among mining communities, the diamond rush stirred up a greater sense of ethnic identity (especially among the Kono), emancipation from chiefdom-based rule and identities, and political radicalism (including a greater participation of the provinces under protectorate status, in relation to the colony of Freetown).

By the early 1970s, the political status quo among British corporate interests, national political elites, and chiefs in the diamantiferous districts came again under stress. Siaka Stevens, a former president of the Mines Workers Union, proved a successful political challenger to the ethnic Mende-dominated Sierra Leone People's Party (SLPP) that succeeded British rule at independence in 1961. Elected in 1967 on a populist platform through his All People Congress (APC) party, Stevens nationalized SLST and reoriented diamond production toward clientelized (semi-industrial/artisanal) operations, and ensured that local opponents involved in artisanal mining were either marginalized or joined the APC. As demonstrated byReno (1995), diamond mining and revenue allocation were closely intertwined with corruption and the consolidation of a "shadow state" bringing diamonds into the private sphere of rulers and marginalizing formal institutions and bureaucracies.

When Stevens personally appointed Brigadier-General Joseph Momoh as his successor in 1985, the diamond sector had been thoroughly informalized to the main advantage of Stevens and his cronies. Momoh largely failed to gain the upper hand on diamond revenues, through either formal or informal means. Momoh's attempt at "reindustrializing" (with the help of dubious foreign investors and approval of international creditors anxious to overturn fiscal collapse) and "reformalizing" the sector (including through military crackdown on illegal artisanal mining) in the last years before the war started in 1991 largely failed. The diamond sector thus continued to induce dual processes of peripheralization. The first was continued diamond smuggling (and thereby capital flight) that peripheralized the country's official economy at the international level and the political control of the government at the domestic level (see Figure 4). The second was to draw an increasing number of poor people into the mining sector as the rest of the economic collapse (Zack-Williams 1995; Williams et al. 2002; Partnership Africa Canada 2004). The vulnerability of the government at the time was exposed when the Revolutionary United Front/Sierra Leone (RUF) supported by Liberian warlord Charles Taylor and Libya gained control of eastern parts the country, and when a year later Momoh fled the country as a group of disgruntled frontline soldiers came to the capital to express their grievances (Gberie 2005).

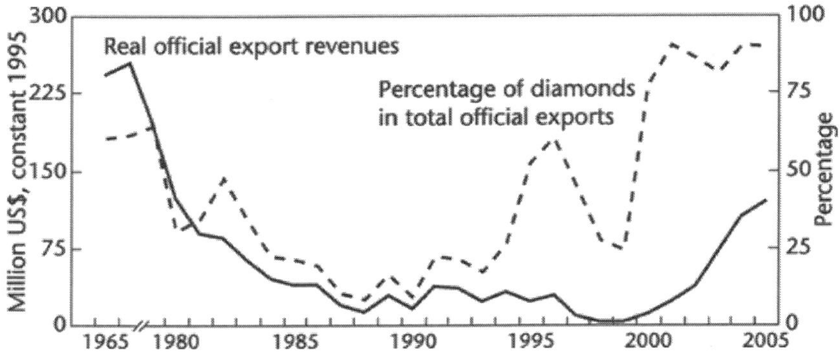

Figure 4 Share of diamond exports in Sierra Leone, 1965–2005. *Source:* Davies (2006b).

From this perspective, vulnerability to conflict in Sierra Leone was not simply the result of the relative economic dependence of Sierra Leone on diamonds, the medium level of diamond abundance, the growth of artisanal diamond mining, or the partial nationalization of the main diamond company. Rather, this historical configuration allowed for the progressive peripheralization of the formal state by ruling elites and their cronies. In turn, this peripheralization left official agencies unable to operationalize legitimate political mandates in a formal way, thereby deepening the informalization of the sector. This peripheralization, which was aggravated in the mid-1970s, was made possible by forty years of prior political status quo favoring exploitative British and local elite private interests and marked by uneven development (Riddell 1985; Keen 2005). From this historical background, the population had initially welcomed a shift in favor of artisanal mining and away from private foreign corporate monopoly. Because of the political character of this informalization, however, artisanal mining and the nationalization of SLST remained essentially beneficial to political networks rather than to mining communities and the government, leaving the country in a political and economic situation vulnerable to armed conflict. Beyond this vulnerability, Sierra Leone's civil war was also related to a higher risk of war resulting from chronic conflict and violence in the diamond fields.

Risk: Are Diamond Fields a Breeding Ground for Wars?

Diamond mining areas are frequently referred to as "breeding grounds" for civil wars. More specifically, artisanal exploitation sites are portrayed as spaces of conflicts and violence, in contrast with industrial mines described as "island[s] of modernity in a sea of civil war" (Harden 2000). In turn, the difficulty of establishing a stabilizing "industrial" order is explained by the diffuse physical geography of alluvial diamond fields. From this defenseless physical geography emerges a threatening human geography of "swarming" and "dangerous" poor young men (Goreux 2001; Partnership Africa Canada and Global Witness 2004). From mining "gangs" would emerge larger rebel movements, miners being thought of as more supportive of anticorruption, propoor, or secessionist agendas, and thereby more likely to start or join a rebellion voluntarily. According to this argument, alluvial diamonds and artisanal mining thus not only represent a weakening factor for the state and a financial opportunity for rebel movements, as discussed later, but also shape a specific geography of risk.

Diamond exploitation is associated with conflicts and various forms of violence (Newbury 1989; Bredeloup 1999a;Leclercq 2001). Like in other extractive sectors, diamond-related conflicts are mostly associated with community relations, land rights and mining access, working conditions, and revenue sharing. The price of a diamond being more subjective than that of gold, for example, disputes over prices can occur more frequently (Interview with Jan Katelaar, Kono, December 2006). Forms of violence include physical ones, such as repression by corporate security forces, or interpersonal and intercommunity violence around artisanal diamond mining, as well as structural and cultural ones. Territorialization constitutes the major geographical dimension of diamond conflicts, in relation to both reserve ownership and production. Whereas diamond wealth should entice secession in producing regions "dispossessed" by central authorities, secessionist wars involving diamonds have been extremely rare since 1946, unlike in the case of oil. Easier access to diamonds by local populations in the case of secondary deposits, according to Ross (2003), would reduce incentives to pursue diamond control through secession. Cases of secessionist wars include South Kasai (1960–1962), and when considering independence struggles, Angola (1961–1974) and Namibia (1966–1988). In each case, diamond reserves included alluvial deposits, but colonial authorities and licensed corporations strictly prohibited access to them by the population.

Conflicts and various forms of violence are certainly involved in the diamond sector, but do chronic conflicts and violence in diamond mining areas increase the risk of war? Conflicts and violence arising from resource exploitation have occasionally acted as precursors to large-scale hostilities, notably through escalating processes of grievances and repression, but such scaling-up processes are rare (Ross 2004b). Artisanal diamond mining is not simply a "violent" space characterized by anarchy as is frequently depicted; on the contrary, studies of artisanal mining document hierarchies providing order and security that are necessary for successful resource extraction (Leclercq 2001). Accommodation of artisanal mining by industrial companies and state officials also often maintains a positive social status quo with communities in mining areas and their customary authorities. Much of the physical violence results from competing claims over diamonds between mining operators backed by central authorities, but challenged by local communities and customary authorities (Misser and Vallée 1997). It is in part for this reason that self-defense groups protecting and regulating artisanal mining areas have a long history in the diamond sector (Leclercq 2001). Although large-scale violent demonstrations by artisanal mining groups are not uncommon, there is little evidence that these groups escalate their activities into full-scale wars. Rather, they frequently merge with wider conflicts, notably through connections between mining groups and competing political authorities.

Relative depravation, discontent, conflicts, and violence relating to the diamond sector have a long history in Sierra Leone, most notably in the Kono district where the SLST had been since the early 1930s. Radical social movements, such as the Kono Progressive Movement, emerged in the 1950s out of frustration with the behavior of the SLST, customary chiefs, and the government in Freetown (Hayward 1972). Its goals converged with those of illicit miners, as resistance to the SLST and reappropriation of Kono diamond wealth, and with those of Freetown-based pro-independence movements seeking "national" emancipation. Mining groups were frequently involved in (violent) political struggles. President Stevens intimidated local political opponents through youth groups also involved in mining, or replaced opponents by political appointees, protecting the interest of his business cronies (Reno 1995; Keen 2005). Appointees who had cracked down on "illegal" mining activities were later the target of artisanal mining groups more likely to have joined the RUF (see later). In contrast, customary chiefs who had protected illegal mining activities maintained greater control of (mostly local) youths, drawing them into Civil Defense Force (CDF) local militias associated with initiation and traditional hunter societies. Conflicts and violence reflected in part the wealth and character of diamond deposits (Reno 2003). The principal economic targets of central authorities were richer upstream alluvial diamond deposits around Kono that could be exploited industrially. In these areas, chiefs not siding with Stevens were more readily intimidated or replaced than chiefs in areas with poorer and shallower deposits around Kenema more suited for artisanal mining. The spatiality of conflicts and violence was also shaped by ethnicized local political affiliations. Mende chiefs around Kenema and Pujehun often supported the opposition party and artisanal mining groups were drawn into violent demonstrations, notably to support local candidates during electoral campaigns such as during the Ndogboyosoi rebellion in 1982. More generally, brutal government "clean-out" operations since the 1950s set a pattern of conflicts and acts of violence in Sierra Leone's diamond fields, but these did not directly initiate the 1991 civil war.

None of the rebel movements in Angola, the DRC, and Sierra Leone started as an insurrection in the diamond mining areas by diamond mining groups (Ross 2004a), although some rebel leaders had a mining background (e.g., Alliance of Democratic Forces for the Liberation of Congo-Zaire (AFDL) with Laurent Kabila, and RUF with Sam Bockarie). All of these movements had broader constituencies, domestic and regional agendas, and geopolitical dimensions (Power 2001). Yet homelessness, injustice, and corruption in the diamond sector did provide a cause, or at least a rationale, for some combatants to join and support the RUF in Sierra Leone (Kandeh 1999; Omasombo Tshonda 2001; Richards 2003; Humphreys and Weinstein 2004). Moreover, labor struggles, negative social and environmental impacts, highly unequal benefits, embezzlement by ruling elites, and abuses by security forces and mercenaries relating to diamonds figure prominently in political discourses of emancipation and social mobilization by radical organizations in Sierra Leone (Hayward 1972; RUF 1995; Abdullah 1998). Beyond discontent relating to diamonds, however, the risk of war in Liberia and Sierra Leone—and individual acts of revenge committed during the conflicts—resulted from widespread and deep-rooted agrarian resentment by nonelite families lacking control over land and their own labor (Richards 2005).

As Scott (1990) suggests, rebel discourses are rarely novel but transpose widely shared "hidden transcripts" within aggrieved groups into the public realm. The risk that diamond exploitation creates is to a significant degree discursively mediated by rebel groups, rather than emerging directly from the social and material practices of the diamond sector. The risk of war would thus result in part from the opening up of political space by risk-taking individuals seeking to achieve a radical, violent transformation. Moreover, given that artisanal mining often provides a space of relative social autonomy outside the traditional confines of chiefdoms and bureaucracies, it is rather its downfall in the face of increased repression, diamond depletion, and growing competition in a context of generalized economic collapse that increases the risk of conflict, as suggested by Keen (2005) in the case of Sierra Leone. The realization of this risk in a context of high vulnerability, however, also requires "radical" violence to be sustained in the face of repression through spaces of opportunity to rebel.

Opportunity: Are Diamonds a Rebel's Best Friend?

The opportunity argument is by far the most prevalent argument in the literature on the diamond wars. Simply put, diamonds are portrayed as the "ultimate loot," financing and rewarding rebel movements. Estimates remain imprecise, but conflict diamonds represented between 4 and 12 percent of the $5-to $8-billion annual international trade in rough diamonds throughout the 1990s. [8] Revenues from diamonds did play a major role in the financing of belligerents, including the provision of arms, local collusion, or foreign support for rebel groups, yet diamonds were rarely the only available source of finance (see Table 2). A look at the territoriality of rebellions in sub-Saharan Africa

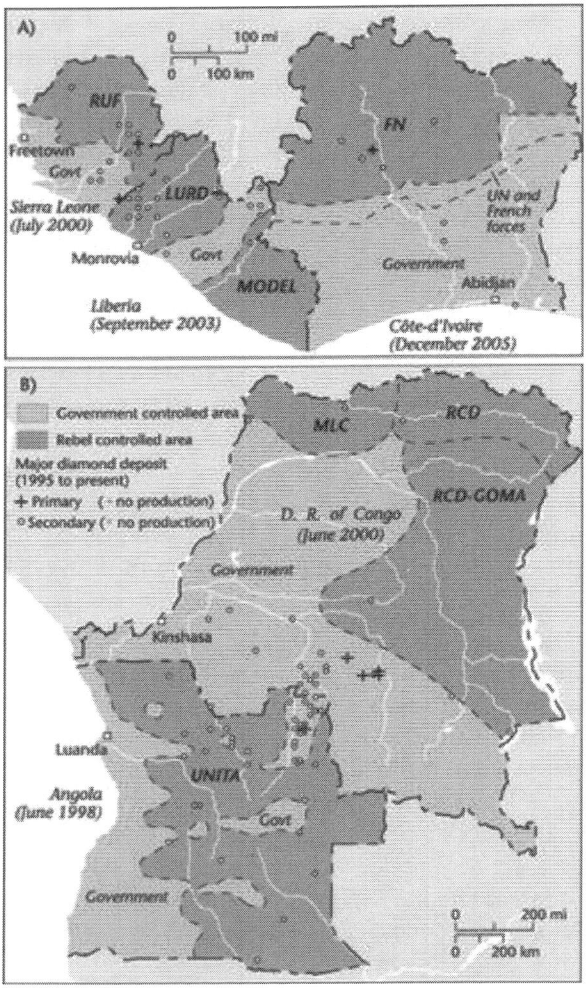

Figure 5 Diamond deposits and rebel-controlled areas. RUF = Revolutionary United Front of Sierra Leone; LURD = Liberians United for Reconciliation and Democracy; FN = New Forces; MODEL = Movement for Democracy in Liberia; MLC = Movement for the Liberation of Congo; RCD = Rally for Congolese Democracy; RCD-GOMA = RCD faction based in Goma; UNITA = National Union for the Total Independence of Angola. *Source:* Gilmore et al. (2005); Silberfein (2004), UN (2004).

suggests a correlation with the location of diamond areas (see Figure 5). In Sierra Leone, diamond areas accounted for the largest number of recorded instances of conflict events (Raleigh and Hegre 2005), confirming that diamond areas were the focus of confrontation over the course of the conflicts (see Figure 6B). Yet diamond areas in Sierra Leone are not significantly related to the level of brutality against civilians (Bellows and Miguel, 2006; Humphreys and Weinstein 2006; see Figure 6C). Although brutalities were perpetrated against civilians in diamond areas (Keen 2005), many sought long-term refuge outside diamond mining areas partly because these were the object of recurring hostilities.

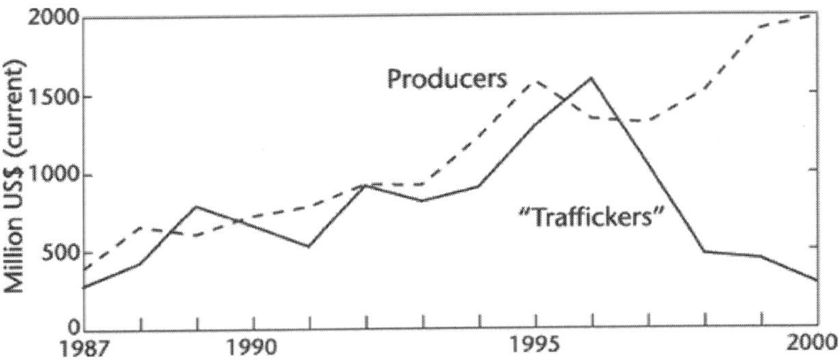

Figure 6 Rough diamond imports from African producing and trafficking countries to Antwerp, Belgium, 1987–2000.*Source:* Statistics provided by the Belgium Ministry of Economic Affairs.

Opportunity stems largely from diamonds' material, spatial, and social characteristics: an easily mined and highly valuable commodity that is spread over vast areas when found in alluvial deposits; that is easy to conceal, transport, store, and trade; and that does not need any transformation before reaching international (rough diamond) markets. As already discussed, rebel access to diamond revenues is in part related to the type of deposits and associated mode of production, and also to the military capacity of the government to secure the deposits and their relative location. In Angola, UNITA's access to diamonds was largely facilitated by the existence of vast alluvial fields, but isolated industrial mining compounds also proved accessible through raids overwhelming army and mercenary protection. In turn, coercion and incentives were used by belligerents to reduce the diffuse character of diamonds once in control of alluvial diamond fields, the impunity of belligerents enabling more drastic forms of disciplining (De Boeck 2001; Interview notes Angola, July 2001). In Namibia, alluvial diamonds buried under the sands of the southern coastline could also have constituted a lootable resource, but the SWAPO, struggling for independence from apartheid South Africa, found it impossible to access the resource. Not only had previous German colonial authorities addressed this lootability problem by defining the area as a strictly enforced *Sperrgebiet*(forbidden zone) in the wake of its genocide of the Herero, but the open terrain of the deserted coast also offered no cover to a guerrilla force. As put by a former SWAPO fighter, now Director of Mines, "We could not have operated there. The South Africans would have simply bombed us." [9] The distinction between point and diffuse resources is thus to a large degree socially produced, with geographies of opportunity reflecting both physical and social conditions of control and access.

Spaces of opportunity are also shaped by the conditions of access to markets. The diamond market often starts in close proximity to the mines, notably through coaxers and "bush" diamond-buying offices staffed by West African and Lebanese traders "roaming" the diamond territory from Guinea to Angola (Bredeloup 1999b). The diamond commodity chain continues through international hubs, bringing together local exporters with (multiple) passport-holding commodity brokers able to enter diamond trading centers (Figure 1A). The long-established culture of clandestine trading and "no questions asked" policy on the part of the diamond industry, as well as the strategic positioning of diamond buying offices in countries neighboring production areas, also allow multiple connections for both smuggled and conflict diamonds to be circulated down the commodity chains and be legally traded in international rough diamonds markets such as that of De Beers or the diamond bourses in Antwerp or Tel Aviv (Misser and Vallée 1997). Import statistics from Antwerp, the world's largest rough diamond market, demonstrate the complacency of the industry toward diamond trafficking. From the late 1980s to the mid-1990s, as many diamonds arrived in Antwerp from trafficking countries as from producing ones in Africa, with no action taken by Belgian authorities to address smuggling issues (see Figure 6). [10] If spaces of opportunity are shaped by connections with savvy companies and wealthy consumers, they also reflect a connection with poverty and coping economies driving civilians to mine in rebel-controlled areas. Although forced labor and extortion were widespread, belligerents in Angola, Sierra Leone, and the DRC used economic incentives to draw workers into mining, including from neighboring countries (UN 2000b, 2002;De Boeck 2001; Kivilu 2001; Keen 2005). As succinctly stated by a Congolese miner in explaining his complicity with the rebellion in Angola, "There is no social welfare in Congo" (Interview notes, Angola, July 2001).

Interconnections at the local and regional levels also reflect broader relationships and geographies that influence arms provision or diplomatic support (on the case of Angola, see Power 2001). Launched in 1991 from Liberian territory with the support of Liberian warlord Charles Taylor as part of a regional project supported by the Libyan and Burkinabe governments, the RUF insurgency continued to find in Liberia its main connections with "diamond markets" (UN 2000a). The Liberian government, and Charles Taylor personally, assisted in most diamonds-for-arms flows among RUF mining sites, camps, and Monrovia, as well as the port of Buchanan (see Figure 7 and Figure 8). The RUF also relied on two other major types of connections. The first was with diamond buyers officially operating from government-controlled areas but using couriers crossing into rebel-held territories. As the diamond sanction regime tightened controls, diamond buyers also set up "paper mines" in frontline areas to launder RUF diamonds (Interview notes, Sierra Leone, April 2001). The second type was with colluding Sierra Leonean troops and individual Guinean officers occasionally trading arms for diamonds (Keen 2005). As illustrated in Figure 7, proximity to borders, transport infrastructures, diamond trading centers, and the collusion of authorities shaped the geography of these interconnections.

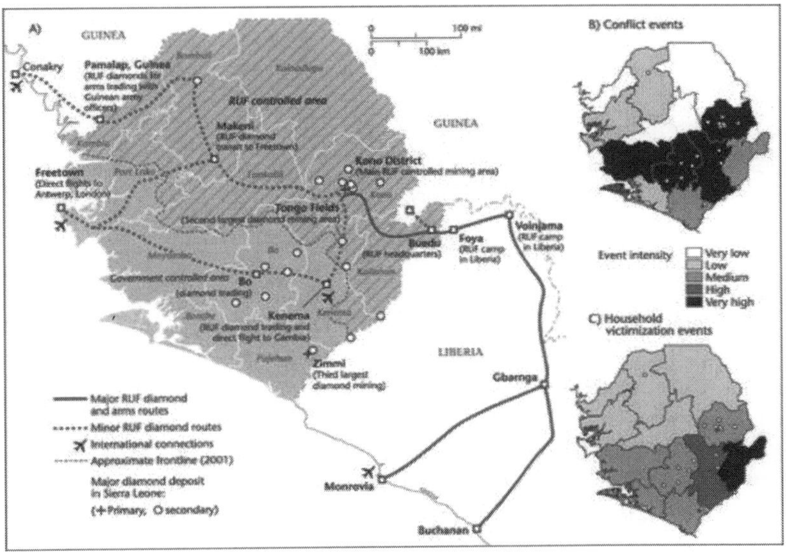

Figure 7 (A) RUF (Revolutionary United Front of Sierra Leone) diamond mining and trading routes. (B) Conflict events intensity (1991–2002). (C) Household victimization events intensity (1991–2002). *Note:* RUF mining areas and trading routes shifted during the decade-long conflict, with the exception of Liberia. Periods of RUF control of major mining areas: 1992–1993 and 1995 (Kono); 1997 (Kono, Tongo Field, Zimmi—jointly with a military junta); and 1999–2002 (Kono, Tongo Field), the period shown in this figure, when RUF mining was most organized and possibly reached a peak. Many other belligerent groups were involved in mining, including Liberian forces, Sierra Leone Army, and Civil Defence Forces (see TRC 2004; Keen 2005). Figure 7B is based on the cumulative number of reported conflict events (e.g., battles) in Sierra Leone between 1991 and 2002. Figure 7C is based on an index of the average of responses to a household survey of war victimization experiences between 1991 and 2002. *Source:* (A) UN (2000a), Keen (2005). (B)Raleigh and Hegre (2005). (C) Bellows and Miguel (2006).

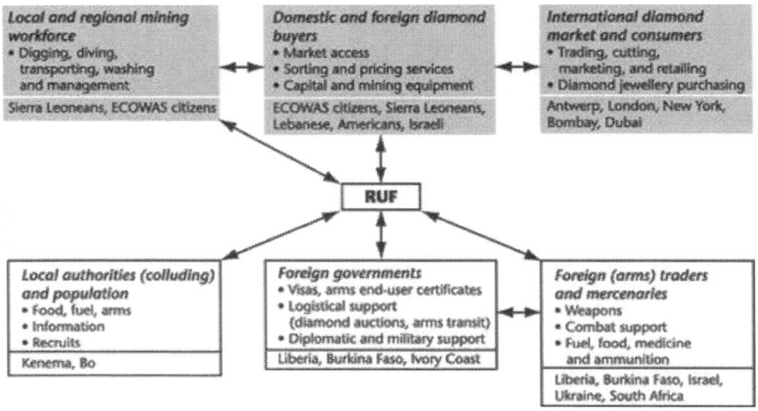

Figure 8 RUF (Revolutionary United Front of Sierra Leone) diamond-funded support network. ECOWAS = Economic Community of West African States. *Source:* UN (2000a); Keen (2005).

Because diamonds can be easily concealed, transported, stocked, and (to some extent) marketed, they have been dubbed a "currency of choice" for individuals and groups faced by volatile currencies, banking scrutiny, or financial sanctions. Al Qaeda operatives, for example, reportedly benefited from the sales of millions of dollars of conflict diamonds mined by rebels in Sierra Leone (Global Witness 2003; Farah 2004). Initial reporting confirmed the picture of West African warlords supporting Islamic terrorists. Deals had been conducted in Liberia under the protection of warlord-turned-elected-president Charles Taylor, an early supporter of the RUF also sponsored by Libyan authorities (TRC 2004). Such interconnections painted West Africa not only as a "wild zone" endangering the locals, but also as a "rogue zone" threatening the West (Le Billon 2006, 783).

The opportunities that diamonds (particularly alluvial) provide for combatants in mining and marketing are conventionally cited as reasons why diamonds are conflict resources, well-suited, given their physical characteristics, to sustain armed conflict. Yet the role of the diamond sector as a rebel's best friend is more ambivalent than is generally portrayed. Diamonds undeniably generated millions of dollars for financing hostilities in Angola, DRC, Sierra Leone, and to some extent in Liberia and the Central African Republic (see Table 2). Although all three major diamond-funded rebel movements—UNITA, RUF, and AFDL—rapidly targeted diamond-mining areas, each originally benefited from the backing of foreign governments and companies, even if some of that support was granted in part with the hope of gaining access to diamonds at a later date, as in the case of Liberian support of the RUF. The financial opportunity factor is thus broadly confirmed, but mostly after these conflicts were already initiated

(Ross 2004a; Arnson and Zartman 2005; Le Billon 2005a).[11] Furthermore, this opportunity had ambivalent effects.

From a political standpoint, diamond financing facilitated the portrayal of rebellions as greedy criminal ventures, thereby undermining their political credentials. Access to diamond revenues also had negative impacts in terms of trust and discipline within the movements, with accusations of private profiteering and widespread abuses against local populations (UN 2000a;Le Billon 2001; Keen 2005). Angolan and Sierra Leonean government troops have abandoned their military duty to search for diamonds, with individual soldiers or officers leaving their posts after a significant finding, or trading with the enemy (Interview with diamond buyer in Angola, July 2001; Keen 2005). UNITA partly avoided this problem by prohibiting its soldiers from participating in mining and employing mostly Congolese diggers and foreign buyers (De Boeck 2001). Defending diamond areas is much more difficult than capturing them, resulting in frequent shifts of control between factions or even local truces enabling both sides to mine.

To sum up, geographies of opportunity relating to diamonds are not only defined by location, understood as vast and remote areas in the case of alluvial diamonds. They are also shaped by interconnections defined by the material and commodity characteristics of diamonds; the domestic, regional, and international regulatory environment of the sector; as well as the transport and marketing infrastructure and institutions. As discussed later, diamonds and wars also mutually shape spaces of opportunity reaching well beyond the confines of diamond-bearing conflict areas.

REGULATION, REPRESENTATION, AND TERRITORIALIZATION

Assumption that a resource curse exists, that artisanal diamond mining breeds war, and that diamonds are a rebel's best friend have contributed to regulatory responses addressing conflict diamonds. Four types of regulatory strategies seek to address the role of commodities in wars (Le Billon 2007). Three of these engage with the opportunity aspect of resource wars: targeted economic sanctions limiting market access for belligerents, military interventions restricting belligerent access to resources, and sharing agreements between belligerents seeking to shift opportunities from war into the realm of peace. The fourth strategy seeks to address the resource curse and to reduce grievances, conflicts, and violence through sectoral reforms, and thus address vulnerability and risk. This section discusses the adequacy of these strategies and their effects on conflict-affected countries and artisanal diamond producers.

From Self-Regulation to International Scrutiny

As legal scholar Lisa Bernstein (1992, 156) notes, "the diamond business has been largely self-regulating, operating outside the law of the state." Reputation, trust, secrecy, and the superiority of private dispute resolution mechanisms have provided the regulatory tenets of the diamond-trading sector. The role of diamonds in motivating or financing violence has long been known within industry circles. Because of the characteristics of the commodity, self-interest within the industry, foreign political support of diamond-funded rebellions, and lack of public recognition of the problem, however, early attempts were derided, such as that of Belgian Members of Parliament (MPs) to curtail the importation of UNITA diamonds into Belgium in 1993. UN sanctions were also slow to materialize (Misser and Vallée 1997). Momentum only grew in the late 1990s when peace processes granting official responsibility over diamond sectors to rebel factions in Angola and Sierra Leone failed, public campaigns by advocacy groups Global Witness and Partnership Africa Canada, and UN sanctions captured the attention of the media and threatened the industry (Global Witness 1998; Smillie, Gberie, and Hazleton 2000; Le Billon 2006).

Three initiatives were successively pursued. First, the UN sanctions on UNITA were drastically strengthened through investigations "naming and shaming" sanction busters, including heads of state (UN 2000b). Similar sanctions and investigations were later imposed on Sierra Leone in 2000, Liberia in 2001, and Ivory Coast in 2005, and investigations were also undertaken in the DRC.

The second initiative consolidated the sanction regimes by mobilizing the industry and key governments in support of diamond trade reforms through public campaigns, international negotiations over diamond certification of origin (the Kimberley Process), and legislative action. The UN Sanctions Committee for Angola lobbied the diamond industry for prompt reforms and warned of a public boycott of diamonds, citing the example of the fur boycott in the 1980s. Campaigning NGOs denounced the complicity or inaction of the diamond industry and governments, yet stopped short of calling for a complete diamond boycott that could affect hundreds of thousands of jobs, and the diamond-dependent economies of Botswana and Namibia.[12] Nor did campaigners systematically launch judicial processes against corporate accomplices of war crimes and sanction-busting. After initially denying any connection with conflict diamonds, major industry actors championed the creation of an international certification agreement to protect the carefully crafted image of its luxury product. By November 2002, after the (official) end of war in Angola and Sierra Leone, thirty-eight countries adopted the international Kimberley Process Certification Scheme (KPCS). The scheme establishes a voluntary system requiring all participants not to trade in rough diamonds with nonparticipating countries, according to principles backed by national legislation, peer review missions, and the possibility of exclusion, as in the case of the DRC in 2004 (Smillie 2005).

The third initiative consisted of a set of "peace building" and "fair trade" programs to improve the social conditions, political economy, and governance of diamond mining in conflict-prone countries. Recognizing that the KPCS did not address the concerns of artisanal mining communities, these programs brought together industry, governments, civil society organizations, and mining communities (Partnership Africa Canada and Global Witness 2004). The Diamond Development Initiative (DDI 2005, 1), for example, sought to "optimize the beneficial development impact of artisanal diamond mining to miners, their communities and their governments" throughout Africa. In Sierra Leone, British and U.S. government-funded projects sought to bring about "peace and prosperity" in mining areas (Maconachie and Binns 2007; Le Billon and Levin 2008).

Regulation as (Punitive) Representation and Territorialization

This set of three initiatives is rightly praised as a success against conflict resources(Tamm 2004; Le Billon 2005a). UN sanctions and the conflict diamond campaign undermined the strength and support for rebel movements in Angola and Sierra Leone, and to a lesser extent in the DRC. With campaigners refocusing on "fair trade" artisanal diamond mining, new ethical dispositions and practices also influenced the industry and consumers. Beyond the diamond sector, the campaign also inspired more ethical practices in other extractive industries, notably for revenue transparency. The role of UN sanctions, expert panels, and the KPCS in ending these conflicts, however, is often overemphasized, compared to the role of controversial military operations that targeted rebel-controlled diamond mining. Hailed for their effectiveness (Shearer 1998), these operations were marked by vested commercial interests and widespread abuses, including the use of helicopter gunships against civilians in diamond mining camps (Dietrich 2000; Richards 2001; Keen 2005).

With the wars in Angola and Sierra Leone over, regulatory attention has focused on sustaining peace. Both the conflict diamond narrative conflating danger with artisanal mining, and the relative effectiveness of military operations to bring an end to the conflicts have informed peace consolidation strategies. In light of the difficulties of directly taxing the artisanal sector, revenue-concentrating modes of exploitation through industrial means and commercialization through trade licensing monopolies were often presented as the best ways for raising taxes and sustaining political stability (Goreux 2001;Snyder and Bhavnani 2005). In turn, violent policing operations sought to secure peace in the diamond mining fields and reinstate the legitimate claims of industrial mining companies (Dietrich 2000; Richards 2001).

Regulatory initiatives reflected a series of discursive representations serving to reposition corporate and political interests in diamond-producing and consuming countries. This resulted in several failures, the first being a lack of historical accountability for past corporate practices. Sierra Leone's government called on British companies to resume mining activities, although companies had, under British rule, pushed for prohibiting the possession of diamonds by natives, underpriced Sierra Leone's diamonds, and obtained from the public purse in 1955 the current equivalent of $55 million in "compensation" for relinquishing, after twenty years, the least valuable parts of a countrywide lease (M. Hart 2001). Compensation to governments for diamonds illegally purchased from rebel groups or to victims of diamond-financed war crimes were not sought.

Second, some companies clearly instrumentalized the conflict diamonds campaign for their own interest. The discursive repositioning of diamonds as conflict-free and ethical reinforced advertising strategies based on corporate integrity and corporate restructuring around branding (e.g., De Beers, Rand diamonds). The "purity" and "whiteness" of Canadian diamonds was also evoked in contrast with tainted "blood diamonds" from Africa, invoking racialized images of Africa as synonymous with violence and primitivism (Le Billon 2006). Linking diamonds to terrorism and feeding the image of "rogue" artisanal mining were also described, ironically, as the best ways to get policy attention and funds from headquarters for "peace-building" diamond programs (Interview with foreign government official, Freetown, December 2006).

Finally, artisanal mining was generally cast as a dangerous nuisance that foreign-driven industrial mining ventures could eliminate. While seeking to improve the developmental and security impact of artisanal mining, peace-building programs frequently bolstered foreign industrial mining investments and overlooked their negative impacts on local livelihoods and community-related conflicts (Interview notes, Sierra Leone, 2006). Stricter international regulations promoted "reputable" industrial companies over artisanal mining and petty trading, with foreign donors and national authorities actively supporting foreign investment. One of the main avenues for the diamond sector to benefit the poorest was thus undermined to the advantage of wealthy corporations.

Sierra Leone's government clearly argued that foreign investment in diamond mining was a "primary focus" to rid the sector of artisanal miners (President Kabbah, cited in Partnership Africa Canada 2005, 1). Merging security and development agendas, President Kabbah awarded a prime diamond mining concession shortly after his election in 1996 to a Canadian mining company related to a British and South African mercenary outfit contracted a year before to oust the RUF (Sheppard 1998; Francis 1999). Many companies have since followed suit, the largest holding mineral licenses covering half of the country. In Angola, peace has failed to bring about a "fundamental change in the pattern of exploitation in the diamond fields," helping to pave the way for violent "clean-up" operations by government security forces and "to legitimize the status quo [of abuses and corruption] in the eyes of international observers and participants in the industry" (Pearce 2004, 15). Although varying in their degree of physical brutality, violent processes of legalization and (re)industrialization have taken place in many postconflict mining areas, most notably through the eviction of artisanal miners whose entitlements relied in part on conditions of chronic insecurity deterring industrial investments and enclosures (Human Rights Watch 2005;Marques and Falcao de Campos 2005; Heaps 2006). Territorializing "peace," in other words, is proving to be a violent and marginalizing process for many artisanal miners and local mining communities.

CONCLUSIONS

As argued in this article through the case of diamonds, resource wars can be conceptualized in relation to three related dimensions: vulnerability, risk, and opportunities. Studies of vulnerability first require a contextualization of the resource curse argument through which resource sectors affect governance and socioeconomic performances. Geographical contributions in this regard can capture processes of peripheralization and uneven development defining social relations around resource exploitation. These processes, in turn, help expose the exclusionary and exploitative histories from which resource wars have emerged.

Analyses of resource wars also require contextualizing processes of territorialization and interconnection shaping spaces of risk and opportunity. Although there is little evidence that so-called diamond wars directly emerged from lived experiences and social relations in diamond mining areas, the diamond sector was frequently rife with conflicts and various forms of violence, especially between (and within) local communities, migrant workers, companies, and authorities. Violent processes of dispossession, including forced displacement of communities and highly coercive "clean-up" operations against "illegal" diamond miners and traders in diamond fields, arguably increased the risk of war by raising grievances, undermining the probity and legitimacy of security forces, and drawing miners into wider conflicts. The diamond sector also contributed to hostilities both financially (diamonds as a source of revenue) and discursively (diamond-related grievances as sources of rebellion justification). Geographical contributions to these two main points include spatialized commodity chain analyses engaging with both the flow of resources and social relations along the chains. This dual engagement allows for both a better understanding of processes of territorialization and interconnection shaping the geography of conflict diamonds, within and beyond the diamond sector itself. It also allows for more nuanced understandings of the physical and social characteristics of conflict diamonds, as well as accounts of the lives and motivations of actors along the chain (see

De Boeck 2001). Although such nuanced accounts might undermine strategic narratives of resource wars that have proven so effective in capturing public attention, these accounts should bring about greater reflexivity on the violence of representation and naturalizing territorialization of resource peripheries.

This article has also engaged the geography of resource wars generally laid out in security studies and the public media. Although their emphasis on the spatial location and flows of resources helps to track diamond flows and reduce revenue access for rebel groups, these perspectives also have ambivalent effects on the interpretation of diamond wars and the policies seeking to address them. Policies deployed to this effect have often focused on military interventions and economic sanctions, with notable successes but also adverse effects in terms of collusion with rebel forces, economic agendas of intervening forces, and human rights abuses against local populations (on the case of Sierra Leone, see Keen 2005). Second, there might be a tacit expectation that once resource locations are militarily "secured" the country will be at peace, yet renewed hostilities have followed the "successful" retaking of diamond areas by government troops and mercenaries, for example in Angola and Sierra Leone (Le Billon and Nicholls 2007).

Finally, this article has pointed out that mainstream perspectives on resource wars rest on narrow definitions of violence. By concentrating on the worst forms of human rights, resource war narratives often succeeded in grabbing the attention of policymakers and consumers, and in promoting significant reforms, as in the case of rough diamond trading. Yet narrow definitions of violence and ensuing policies focused on territorializing peace overlook the violence implicit in regulatory interventions in resource governance, such as the forceful eviction of local populations and artisanal miners from diamond fields. The selective representation of diamonds through their sites and modes of exploitation also contributes to (re)shaping spaces of legitimacy and criminality constitutive of (violent) processes of marginalization. The idealization of industrially mined Canadian diamonds over artisanally mined African diamonds, for example, risks marginalizing a major export commodity for several African economies and aggravating the situation of artisanal miners. Ending diamond-related conflicts in Africa does not entail ending the various forms of violence associated with diamond exploitation. Broadening the definition of violence beyond armed conflicts and giving greater attention to violent processes of uneven development and territorialization can help bring to light exploitative and exclusionary occurrences of "peacetime" diamond exploitation. This, in turn, requires a critical rereading of the selective geographies of exploitation, regulation, and consumption that are shaping postconflict spaces of legitimacy and criminality.

ACKNOWLEDGMENTS

The author would like to thank for their assistance or comments Karen Bakker, Charmian Gooch, Morlai Kamara, Estelle Levin, Nancy Lee Peluso, Ansumana Babar Turay, participants of the University of California Berkeley's Workshop on Environmental Politics, and three anonymous reviewers. Funding was received from the Social Sciences and Humanities Research Council, Canada.

NOTES

1. See Zack-Williams (1995), Bredeloup (1999a), Partnership Africa Canada and Global Witness (2004), and Levin (2006). On industrial diamond mine workers, see Newbury (1989) and Crush (1994). On diamond polishing workers in India, notably children, see Burra (1995). On First Nations and the diamond rush in northern Canada, see Bielawski (2003).
2. The terms *point* and *diffuse* refer to both Euclidian spaces of resource location as well as socially constructed spaces within which political economies and techniques of exploitation are concentrating or diffusing access to resources and their revenues. As such, not only is the physical location of resources important, but also the materiality and modes of production, regulation, and consumption shaping their broader spatiality.
3. Source: GDP in constant 2000 dollars from 1960 to 2004, from the World Bank (2005). A multivariate analysis would be necessary to assess a specific "diamond influence" on GDP, as well as on inequality and fiscal revenues, but that falls beyond the scope of this study.
4. Economist Ola Olsson (2007) found a negative relationship between economic growth and three diamond indicators (i.e., diamond production as share of GDP as a proxy for dependence, as well as diamond volume and value per square kilometer as proxies for abundance) in the 1990s. This relationship could be spurious, however, as all diamond exporters with negative economic growth defining this trend were affected by political instability likely to affect both GDP growth and diamond dependence.
5. *Governmentality* refers to Foucault's notion of *gouvernementalité*, and is applied here to describe the set of institutions, procedures, analyses, and tactics that exercise power over (and derive power from) a resource and populations (dis)associated with it. Forms of governmentality, it is argued here, respond to the socially contextualized materiality and spatiality of resources.

6. Although artisanal diamond mining can be legalized by issuing diamond mining and trading licenses, it is hard to tax directly, largely due to high taxes leading to illegal mining and smuggling and the difficulty of controlling a large number of small operations (Hentschel, Hruschka, and Priester 2002).

7. This pattern is largely confirmed by examining the characteristics of these countries in the year preceding the start of hostilities (see United States Geological Survey [USGS] Minerals Yearbooks for 1974, 1988, 1990, and 1996). Exceptions are Liberia, with lower diamond abundance; Namibia, with industrial exploitation; and Angola, whose diamond sector was still largely dominated by industrial exploitation until the early 1980s.

8. Production of conflict diamonds peaked at about $800 million in 1997 according to estimates based on industry sources, USGS Minerals Yearbook reports, Global Witness (1998) and UN (2000a, 2001, 2002).

9. Interview with Kennedy Hamutenya, Namibian Ministry of Mines and Energy, April 2002.

10. About 80 percent of the world's rough diamonds pass through Antwerp (Global Witness 2000, 40). Data are collected on the basis of "country of provenance" (not "origin") and were obtained for the period 1987 to 2000, after which information on a per country basis is not publicly available. The sharp decline in "trafficked" diamonds from 1996 mostly resulted from a reduction the flow of Angolan (and mostly UNITA-controlled) diamonds through the DRC, although a sharp rise in the value per carat of trafficked diamonds compared to producer diamonds also suggested a shift to a smaller volume of higher quality diamonds.

11. Recruitment of diamond miners also constituted an "opportunity" for rebel groups, notably for the RUF, who initially recruited miners along the Liberian border and among 15,000 "illegal diggers" evicted through "Clean State Operation" in 1990, which aimed at improving the operating conditions of multinationals promoted by the International Monetary Funf as a solution to the fiscal crisis faced by the government of Sierra Leone (Reno 1993; Fithen 1999; Richards 2001, 82). Forced recruitment of miners continued in the mid-1990s (see, e.g., Henry forthcoming). By the end of the war, however, only about 5 percent of combatants were miners when recruited, and 90 percent of combatants had been forcibly recruited (Humphreys and Weinstein 2004; Maclure and Denov 2006). UNITA mostly recruited in Ovimbundu areas with little mining activity, and diamond diggers were mostly Congolese with no military role. Finally in the DRC, diamond diggers were taxed by armed groups but rarely recruited as soldiers.

12. More radical organizations, however, advocated for a total boycott of diamonds, arguing that the industry as a whole was tainted by human rights, environmental, and consumer abuses, including the funding of small arms trafficking, wars fought by child soldiers, slave labor in the cutting and polishing industry, the violation of indigenous land rights (especially the San in Botswana), environmental impacts, miners' exposure to HIV/AIDS, little resale value, overpricing, and constructed desire (Stanton 2000).

REFERENCES

1. Abdullah, I. 1998. Bush path to destruction: The origin and character of the Revolutionary United Front/Sierra Leone. *Journal of Modern African Studies*, 36(2): 203–235.
2. Acemoglu, D., Johnson, S. and Robinson, J. 2003. "An African success story: Botswana". In *In search of prosperity: Analytical narrative on economic growth*, Edited by: Rodrik, D. 80–119. Princeton, NJ: Princeton University Press.
3. Acemoglu, D., Robinson, J. A. and Johnson, S. 2001. The colonial origins of comparative development: An empirical investigation. *American Economic Review*, 91: 1369–1401.
4. Alao, A. 1999. Diamonds are forever ... but so also are controversies: Diamonds and the actors in Sierra Leone's civil war.*Civil Wars*, 2(3): 43–64.
5. Arnson, C. J. and Zartman, W. I., eds. 2005. *Rethinking the economics of war: The intersection of need, creed, and greed*, Baltimore: Johns Hopkins University Press.
6. Auty, R., ed. 2001. *Resource abundance and economic development: Wider studies in development economics*, New York: Oxford University Press.
7. Auty, R. 2004. Natural resources and civil strife: A two-stage process. *Geopolitics*, 9(1): 29–49.
8. Bakker, K. and Bridge, G. 2006. Material worlds: Materiality and the "matter of nature.". *Progress in Human Geography*, 30(1): 5–27.
9. Bannon, I. and Collier, P. 2003. *Natural resources and violent conflict: Options and actions*, Washington, DC: World Bank.
10. Barnett, C., Cloke, P., Clarke, N. and Malpass, A. 2005. Consuming ethics: Articulating subjects and spaces in ethical consumption. *Antipode*, 37(1): 23–45.

11. Bassett, T. J. 1988. The political ecology of peasant-herder conflicts in northern Ivory Coast. *Annals of the Association of American Geographers*, 78(3): 453–72.

12. Bellows, J. and Miguel, E. 2006. *War and local collective action in Sierra Leone*, Berkeley: Department of Economics, University of California. Unpublished paper

13. Berdal, M. and Malone, D. 2000. *Greed and grievance: Economic agendas in civil wars*, Boulder, CO: Lynne Rienner.

14. Bernstein, H. 1988. *The Brazilian diamond in contracts, contraband, and capital*, Lanham, MD: University Press of America.

15. Bernstein, L. 1992. Opting out of the legal system: Extralegal contractual relations in the diamond industry. *Journal of Legal Studies*, 21(1): 115–57.

16. Besançon, M. L. 2005. Relative resources: Inequality in ethnic wars, revolutions, and genocides. *Journal of Peace Research*, 42(4): 393–415.

17. Bielawski, E. 2003. *Rogue diamonds: The rush for northern riches on Dene land*, Vancouver, Canada: Douglas & McIntyre.

18. Blomley, N. 2003. Law, property, and the geography of violence: The frontier, the survey, and the grid. *Annals of the Association of American Geographers*, 93(1): 121–41.

19. Bredeloup, S. 1999a. La fièvre du diamant au temps des colonies (The diamond fever at the time of the colonies).*Autrepart*, 11: 171–90.

20. Bredeloup, S. 1999b. "Territoires du diamant et migrants du fleuve Sénégal (Diamond territories and migrants from the Senegal River)". In *Les territoires de l'identité. Le territoire, lien ou frontière? (The territories of identity. The territory, linkage or frontier?)*, Edited by: Bonnemaison, J., Cambrezy, L. and Quinty-Bourgeois, L. 283–96. Paris: L'Harmattan.

21. Bridge, G. 2000. The social regulation of resource access and environmental impact: Production, nature and contradiction in the US copper industry. *Geoforum*, 31(2): 237–56.

22. Bridge, G. 2001. Resource triumphalism: Postindustrial narratives of primary commodity production. *Environment and Planning A*, 33(12): 2149–73.

23. Bridge, G. 2004. Mapping the bonanza: Geographies of mining investment in an era of neoliberal reform. *The Professional Geographer*, 56(3): 406–21.

24. Bridge, G. 2006. "Exploiting: Power, colonialism and resource economies". In *Companion Encyclopedia of Geography*, Edited by: Douglas, I., Huggett, R. and Robinson, M. 867–86. London: Routledge.

25. Bryant, R. L. and Bailey, S. 1997. *Third world political ecology*, London: Routledge.

26. Bryant, R. L. and Goodman, M. K. 2004. Consuming narratives: The political ecology of "alternative" consumption.*Transactions of the Institute of British Geographers*, 29(3): 344–66.

27. Buhaug, H. and Gates, S. 2002. The geography of civil war. *Journal of Peace Research*, 39(4): 417–33.

28. Buhaug, H. and Lujala, P. 2005. Accounting for scale: Measuring geography in quantitative studies of civil war. *Political Geography*, 24(4): 399–418.

29. Buhaug, H. and Rød, J. K. 2006. Local determinants of African civil wars, 1970–2001. *Political Geography*, 25(3): 315–35.

30. Bulte, E. H., Damania, R. and Deacon, R. T. 2005. Resource intensity, institutions, and development. *World Development*, 33(7): 1029–44.

31. Bunker, S. G. 1985. *Underdeveloping the Amazon: Extraction, unequal exchange, and the failure of the modern state*, Urbana: University of Illinois Press.

32. Burra, N. 1995. *Born to work: Child labour in India*, Oxford, U.K.: Oxford University Press.

33. Campbell, G. 2002. *Blood diamonds: Tracing the deadly path of the world's most precious stones*, Boulder, CO: Westview.

34. Carstens, P. 2001. *In the company of diamonds: De Beers, Kleinzee, and the control of a town*, Athens: Ohio University Press.

35. Cartwright, J. R. 1978. *Political leadership in Sierra Leone*, Toronto: University of Toronto Press.

36. Castree, N. 2001. Commodity fetishism, geographical imaginations and imaginative geographies. *Environment and Planning A*, 33(9): 1519–25.

37. Center for the Study of Civil War. 2006. http://new.prio.no/CSCW-Datasets Armed conflicts Version 4-2006b. Center for the Study of Civil War. Oslo: Peace Research Institute in Oslo (last accessed 2 May 2007)

38. Collier, P. and Hoeffler, A. 1998. On economic causes of civil war. *Oxford Economic Papers*, 50: 563–73.

39. Cooper, N. 2002. State collapse as business: The role of conflict trade and the emerging control agenda. *Development and Change*, 33(5): 935–55.

40. Cooper, R. N. 2001. Conflict goods: The challenges for peacekeeping and conflict prevention. *International Peacekeeping*, 8(3): 21–38. Cramer, C. 2003. Does inequality cause conflict?. *Journal of International Development*, 15(4): 397–412.

41. Crossette, B. 2000. http://globalpolicy.org/secgen/annan/2000/annan3.htm UN chief faults reluctance of US to help in Africa. *New York Times* 13 May 2000 (last accessed 27 February 2008)

42. Crush, J. 1994. Scripting the compound: Power and space in the South African mining industry. *Environment and Planning D: Society and Space*, 12: 301–24.

43. Dalby, S. 2002. *Environmental security*, Minneapolis: University of Minnesota Press.

44. Davies, V. A. B. 2006a. *Alluvial diamonds: A new resource curse theory*, Centre for the Studies of African Economies, Oxford University. Unpublished manuscript

45. Davies, V. A. B. 2006b. "Diamonds, poverty and war in Sierra Leone". In *Small-scale mining, rural subsistence and poverty in West Africa*, Edited by: Hilson, G. 165–80. Rugby, , U.K.: Practical Action.

46. De Boeck, F. 2001. Garimpeiro worlds: Digging, dying and "hunting" for diamonds in Angola. *Review of African Political Economy*, 28(90): 549–62.

47. de Soysa, I. 2002. Ecoviolence: Shrinking pie, or honey pot?. *Global Environmental Politics*, 2(4): 1–27.

48. Diamond Development Initiative. 2005. *Diamond Development Initiative begins: New approach to Africa's diamond problems*, London: Diamond Development Initiative.

49. Dietrich, C. 2000. "Power struggle in the diamond fields". In *Angola's war economy*, Edited by: Cilliers, J. and Dietrich, C. 173–94. Pretoria: Institute for Security Studies.

50. Dumett, R. 1985. Africa's strategic minerals during the Second World War. *Journal of African History*, 26: 381–408.

51. Dunning, T. 2005. Resource dependence, economic performance, and political stability. *Journal of Conflict Resolution*, 49(4): 451–82.

52. Farah, D. 2004. *Blood from stones: The secret financial network of terror*, New York: Broadway Books.

53. Fearon, J. D. 2004. Why do some civil wars last so much longer than others?. *Journal of Peace Research*, 41(3): 275–301.

54. Fearon, J. D. 2005. Primary commodity exports and civil war. *Journal of Conflict Resolution*, 49(4): 483–507.

55. Fearon, J. D. and Laitin, D. D. 2003. Ethnicity, insurgency and civil war. *American Political Science Review*, 97(1): 75–90.

56. Fithen, C. 1999. *Diamonds and war in Sierra Leone: Cultural strategies for commercial adaptation to endemic low-intensity conflicts*, PhD dissertation London: Department of Anthropology, University College London.

57. Flint, C. 2003. Terrorism and counterterrorism: Geographic research questions and agendas. *The Professional Geographer*, 55(2): 161–69.

58. Flint, C., ed. 2004. *The geography of war and peace: From death-camps to diplomats*, Oxford: Oxford University Press.

59. Francis, D. J. 1999. Mercenary intervention in Sierra Leone: Providing national security or international exploitation?. *Third World Quarterly*, 20(2): 319–38.

60. Gberie, L. 2005. *A dirty war in West Africa: The Revolutionary United Front and the destruction of Sierra Leone*, London: Hurst.

61. Gilmore, E., Gleditsch, N. P., Lujala, P. and Rod, J. K. 2005. Conflict diamonds: A new dataset. *Conflict Management and Peace Science*, 22: 257–72.

62. Global Witness. 1998. *A rough trade: The role of diamond companies and governments in the Angolan conflict*, London: Global Witness.

63. Global Witness. 2000. *Conflict diamonds: Possibilities for the identification, certification and control of diamonds*, London: Global Witness.

64. Global Witness. 2002a. *Branching out: Zimbabwe's resource colonialism in the Democratic Republic of Congo*, London: Global Witness.

65. Global Witness. 2002b. *The logs of war: The timber trade and armed conflict*, Oslo, Norway: Fafo Institute for Applied Social Sciences. Fafo Report 379

66. Global Witness. 2003. *For a few dollars more: How Al Qaeda moved into the diamond trade*, London: Global Witness.

67. Good, K. 2005. Resource dependency and its consequences: The costs of Botswana's shining gems. *Journal of Contemporary African Studies*, 23(1): 27–50.

68. Goreux, L. 2001. *Conflict diamonds*, African Region Working Paper Series. Washington, DC: World Bank.

69. Hall, P. K. 1968. *The diamond fields of Sierra Leone*, Freetown: Geological Survey of Sierra Leone.

70. Harden, B. 2000. http://query.nytimes.com/gst/fullpage.html?res=9E03E4DA123FF935A35757COA9669C8B63&scp=1&sq=island+of+modernity+in+a+sea+of+civil+war+harden&st=nytAfrica's diamond wars. *New York Times*, 6 April 2000 (last accessed 27 February 2008)

71. Harris, C. 2004. How did colonialism dispossess? Comments from an edge of empire. *Annals of the Association of American Geographers*, 94(1): 165–182.

72. Hart, G. 2006. Denaturalizing dispossession: Critical ethnography in the age of resurgent imperialism. *Antipode*, 38(5): 977–1004.

73. Hart, M. 2001. *Diamond: A journey to the heart of an obsession*, New York: Walker.

74. Hartwick, E. 1998. Geographies of consumption: A commodity-chain approach. *Environment and Planning D: Society and Space*, 16: 423–37.

75. Hartwick, E. 2000. Towards a geographical politics of consumption. *Environment and Planning A*, 32(7): 1177–92.

76. Harvey, D. 1982. *Limits to capital*, Chicago: University of Chicago Press.

77. Harvey, D. 2003. *The new imperialism*, Oxford, U.K.: Oxford University Press.

78. Hayter, R. 2003. "The war in the woods": Globalization, post-fordist restructuring and the contested remapping of British Columbia's forest economy. *Annals of the Association of American Geographers*, 93: 706–29.

79. Hayter, R., Barnes, T. J. and Bradshaw, M. J. 2003. Relocating resource peripheries to the core of economic geography's theorizing: Rationale and agenda. *Area*, 35(1): 15–23.

80. Hayward, F. M. 1972. The development of a radical political organization in the bush: A case study in Sierra Leone.*Canadian Journal of African Studies/Revue Canadienne des Études Africaines*, 6(1): 1–28.

81. Heaps, T. 2006. The heart of darkness. *Corporateknights*, 16: 19–22.

82. Heemskerk, M. 2001. Do international commodity prices drive natural resource booms? An empirical analysis of small-scale gold mining in Suriname. *Ecological Economics*, 39: 295–308.

83. Henry, D. "Diamonds are not a boy's best friend: What happened when Foday met the rebels". In *The human tradition in modern Africa*, Edited by: Cordell, D. Lanham, MD: Rowman & Littlefield. forthcoming

84. Hentschel, T., Hruschka, F. and Priester, M. 2002. *Global report on artisanal and small-scale mining*, London: International Institute for Environment and Development.

85. Homer-Dixon, T. 1999. *Environment, scarcity and violence*, Princeton, NJ: Princeton University Press.

86. Human Rights Watch. 2005. *The curse of gold*, New York: Human Rights Watch.

87. Humphreys, M. 2005. Natural resources, conflict, and conflict resolution: Uncovering the mechanisms. *Journal of Conflict Resolution*, 49(4): 508–37.

88. Humphreys, M. and Weinstein, J. M. 2004. *What the fighters say: A survey of ex-combatants in Sierra Leone, June–August 2003*, Freetown, Sierra Leone: PRIDE.

89. Humphreys, M. and Weinstein, J. M. 2006. Handling and manhandling civilians in civil war. *American Political Science Review*, 100: 429–47.

90. Isham, J., Woolcock, M., Pritchett, L. and Busby, G. 2005. The varieties of resource experience: Natural resource export structures and the political economy of economic growth. *World Bank Economic Review*, 19(2): 141–74.

91. Jean, F. and Rufin, J.-C., eds. 1996. *Economies des Guerres Civiles (Economies of Civil Wars)*, Paris: Hachette.

92. Kaldor, M. 1999. *New and old wars: Organized violence in a global era*, Cambridge, U.K.: Polity Press.

93. Kandeh, J. 1999. Ransoming the state: Elite origins of subaltern terror in Sierra Leone. *Review of African Political Economy*, 26(81): 349–66. \

94. Keen, D. 1998. *The economic functions of violence in civil wars*, Oxford, U.K.: Oxford University Press.

95. Keen, D. 2005. *Conflict and collusion in Sierra Leone*, Oxford, U.K.: James Currey.

96. Khoury, S. J. 1990. *The valuation and investment merits of diamonds*, Greenwood, CT: Quorum Books.

97. Killick, T. 1973. The benefits of foreign direct investment and its alternatives: An empirical exploration. *Journal of Development Studies*, 9(3): 301–16.

98. Kivilu, S. 2001. "A la recherche de paradis terrestre: les Bana Luunda entre le diamant et le dollar (Searching for paradise on earth: The Bana Luunda between diamond and dollar)". In *Chasse au diamant au Congo/Zaïre (Hunting for diamonds in Congo/Zaire)*, Edited by: Monnier, L., Jewsiewicki, B. and de Villers, G. 127–69. Paris: Editions L'Harmattan.

99. Klare, M. 1980. Resource wars. *Harper's*, : 20–23.

100. Klare, M. 2001. *Resource wars: The new landscape of global conflict*, New York: Metropolitan Books.

101. Korf, B. and Fünfgeld, H. 2006. War and the commons: Assessing the changing politics of violence, access and entitlements in Sri Lanka. *Geoforum*, 37(3): 391–403.

102. Lacina, B. and Gleditsch, N. P. 2004. *Monitoring trends in global combat: A new dataset of battle deaths*, Oslo, Norway: Centre for the Study of Civil War, Peace Research Institute in Oslo.

103. Lanning, G. and Mueller, M. 1979. *Africa undermined: Mining companies and the underdevelopment of Africa*, Harmondsworth, U.K.: Penguin.

104. Le Billon, P. 2001. The political ecology of war: Natural resources and armed conflicts. *Political Geography*, 20(5): 561–84.

105. Le Billon, P. 2004. The geopolitical economy of "resource wars.". *Geopolitics*, 9(1): 1–28.

106. Le Billon, P. 2005a. *Fuelling war: Natural resources and armed conflicts*, London: Routledge.

107. 2005b. *The geopolitics of "resource wars": Resource dependence, governance and violence*, London: Frank Cass.

108. Le Billon, P. 2006. Fatal transactions: Conflict diamonds and the (anti)terrorist consumer. *Antipode*, 38(4): 778–801.

109. Le Billon, P. 2007. Ending 'resource wars': Revenue sharing, economic sanction or military intervention?. *International Peacekeeping*, 14(5): 613–32.

110. Le Billon, P. and Levin, E. 2008. *Merging security and development: Conflict diamonds and peace building in Sierra Leone*, Vancouver: Liu Institute for Global Issues. Working paper

111. Le Billon, P. and Nicholls, E. 2007. Ending 'resource wars': Revenue sharing, economic sanction, or military intervention?.*International Peacekeeping*, 14(5): 613–32.

112. Leclercq, H. 2001. "Le rôle économique du diamant dans le conflit congolais (The economic role of diamonds in the Congolese conflict).". In *Chasse au diamant au Congo/Zaïre (Hunting for diamonds in Congo/Zaire)*, Edited by: Monnier, L., Jewsiewicki, B. and de Villers, G. 47–78. Paris: Editions L'Harmattan.

113. Leite, C. and Weidmann, J. 1999. *Does mother nature corrupt: Natural resources, corruption, and economic growth*, Washington, DC: International Monetary Fund. Working Paper WP/99/85

114. Leslie, D. and Reimer, S. 1999. Spatializing commodity chains. *Progress in Human Geography*, 23(3): 401–20.

115. Lesser, I. O. 1989. *Resources and strategy: Vital materials in international conflict, 1600–present day*, Basingstoke, U.K.: Macmillan.

116. Levin, E. 2006. "Reflections on the political economy of artisanal diamond mining in Kono District, Sierra Leone". In *Small-scale mining, rural subsistence and poverty in West Africa*, Edited by: Hilson, G. Rugby, , U.K.: Practical Action.

117. Levinson, A. A. 1998. "Diamond sources and their discovery". In *The nature of diamonds*, Edited by: Harlow, G. E. 72–104. Cambridge, , U.K.: Cambridge University Press.

118. Love, R. 1994. Drought, Dutch disease and controlled transition in Botswana agriculture. *Journal of Southern African Studies*, 20(1): 71–83.

119. Lujala, P., Gleditsch, N. P. and Gilmore, E. 2005. A diamond curse? Civil war and a lootable resource. *Journal of Conflict Resolution*, 49: 538–62.

120. Luong, P. J. and Weinthal, E. 2006. Rethinking the resource curse: Ownerships structure, institutional capacity, and domestic constraints. *Annual Review of Political Sciences*, 9: 246–63.

121. MacGinty, R. 2004. Looting in the context of violent conflict: A conceptualisation and typology. *Third World Quarterly*, 25(5): 857–70.

122. Maclure, R. and Denov, M. 2006. "I didn't want to die so I joined them": Structuration and the process of becoming boy soldiers in Sierra Leone. *Terrorism and Political Violence*, 18(1): 119–35.

123. Maconachie, R. and Binns, T. 2007. Beyond the resource curse? Diamond mining, development and post-conflict in Sierra Leone. *Resource Policy*, 32(3): 104–15.

124. Marques, R. and Falcao de Campos, R. 2005. *Lundas, the stones of death: Angola's deadly diamonds*, New York: Open Society Institute.

125. McIlwaine, C. 1999. Geography and development: Violence and crime as development issues. *Progress in Human Geography*, 23(3): 453–63.

126. Mehlum, H., Moene, K. and Torvik, R. 2006. Institutions and the resource curse. *Economic Journal*, 116(508): 1–20.

127. Misser, F. and Vallée, O. 1997. *Les gemmocraties: L'économie politique du diamant africain (The gemmocraties: The political economy of the african diamond)*, Paris: Desclée de Brouwer.

128. Moore, M. 2001. Political underdevelopment. What causes "bad governance"?. *Public Management Review*, 3(3): 385–418.

129. Neumann, R. P. 2004. Moral and discursive geographies on the war for biodiversity in Africa. *Political Geography*, 23(7): 813–37.

130. Neumayer, E. 2004. Does the "resource curse" hold for growth in genuine income as well?. *World Development*, 22(10): 1627–40.

131. Newbury, C. 1989. *The diamond ring: Business, politics, and precious stones in South Africa, 1867–1947*, Oxford, U.K.: Clarendon Press.

132. Ng, F. and Yeats, A. 2002. *What can Africa expect of its traditional exports?*, Washington, DC: World Bank. Africa Region Working Papers Series No. 26

133. Nordstrom, C. 2004. *Shadows of war: Violence, power, and international profiteering in the twenty-first century*, Berkeley: University of California Press.

134. Northwest Territories. 2005. Yellowknife, , Canada: Northwest Territories Industry, Tourism and Investment. Diamond Facts 2004/5. NWT Diamond Industry Report

135. O'Lear, S. 2005. Resource concerns for territorial conflicts. *Geojournal*, 64(4): 297–306.

136. Olsson, O. 2006. Diamonds are a rebel's best friend. *The World Economy*, 29(8): 1133–50.

137. Olsson, O. 2007. Conflict diamonds. *Journal of Development Economics*, 82(2): 267–86.

138. Omasombo Tshonda, J. 2001. "Des diamants de Kisangani: de nouveaux seigneurs se taillent des fiefs sur le modèle de l'Etat Zaïrois de Mobutu (The diamonds of Kisangani: new lords carve fiefdoms on the model of Mobutu's Zairian state)". In*Chasse au diamant au Congo/Zaïre (Hunting for diamonds in Congo/Zaire)*, Edited by: Monnier, L., Jewsiewicki, B. and de Villers, G. 79–126. Paris: Editions L'Harmattan.

139. Oomes, N. and Vocke, M. 2003. *Diamond smuggling and taxations in Sub-Saharan Africa*, Washington, DC: International Monetary Fund. Working Paper WP/03/167

140. Papyrakis, E. and Gerlagh, Reyer. 2004. The resource curse hypothesis and its transmission channels. *Journal of Comparative Economics*, 32(2): 181–93.

141. Partnership Africa Canada. 2004. *Diamond industry annual review: Sierra Leone 2004*, Ottawa, Canada: Partnership Africa Canada.

142. Partnership Africa Canada. 2005. *Diamond industry annual review: Sierra Leone 2005*, Ottawa, Canada: Partnership Africa Canada.

143. Partnership Africa Canada and Global Witness. 2004. *Rich man, poor man. Development, diamonds and poverty: The potential for change in the artisanal diamond fields of Africa*, Ottawa, Canada and London: Partnership Africa Canada and Global Witness.

144. Pearce, J. 2004. *War, peace and diamonds in Angola: Popular perceptions in the diamond industry in the Lundas*, Pretoria, South Africa: Institute for Security Studies.

145. Peluso, N. L. 2005. Seeing property in land use: Local territorialization in West Kalimantan. *Geografisk Tidsskrift*, 105(1): 1–15.

146. Peluso, N. L. and Watts, M., eds. 2001. *Violent environments*, Ithaca, NY: Cornell University Press.

147. Perala, M. Resource flow concentration and social fractionalization: A recipe for a curse?. *Journal of Economic Studies*, 35(5) forthcoming

148. Perreault, T. 2006. From the Guerra del Agua to the Guerra del Gas: Resource governance, popular protest and social justice in Bolivia. *Antipode*, 38(1): 150–72.

149. Pinto Vallejos, J. 1985. Slave control and slave resistance in colonial Minas Gerais, 1700–1750. *Journal of Latin American Studies*, 17(1): 1–34.

150. Platt, R. 2006. *Commodity prices*, International Diamond Consultants Ltd. (received by e-mail, 31 May 2006)

151. Power, M. 2001. Patrimonialism and petro-diamond capitalism: Peace, geopolitics and the economics of war in Angola.*Review of African Political Economy*, 90: 489–502.

152. Raleigh, C. and Hegre, H. Introducing ACLED: An Armed Conflict Location and Event Dataset (selection of ACLED maps).Paper presented to the conference on Disaggregating the Study of Civil War and Transnational Violence, University of California Institute of Global Conflict and Cooperation. San Diego, CA. http://www.prio.no/cscw/acled/ACLED_Maps.pdf(last accessed 14 February 2008)

153. Regan, P. M. and Norton, D. 2005. Greed, grievance, and mobilization in civil wars. *Journal of Conflict Resolution*, 49(3): 319–36.

154. Reno, W. 1993. Economic reform and the strange case of "liberalization" in Sierra Leone. *Governance*, 6(1): 23–42.

155. Reno, W. 1995. *Corruption and state politics in Sierra Leone*, Cambridge, U.K.: Cambridge University Press.

156. Reno, W. 2003. Political networks in a failing state. The roots and future of violent conflict in Sierra Leone. *Internationale Politik und Gesellschaft*, 2: 44–66.

157. Revolutionary United Front (RUF). 1995. http://www.fas.org/irp/world/para/docs/footpaths.htm*Footpaths to democracy: Towards a new Sierra Leone.* Revolutionary United Front/Sierra Leone (last accessed 14 February 2008)

158. Richards, P. 1996. *Fighting for the rain forest: War, youth and resources in Sierra Leone*, Oxford, U.K.: James Currey.

159. Richards, P. 2001. "Are "forest wars" in Africa resource conflicts? The case of Sierra Leone". In *Violent environments*, Edited by: Peluso, N. L. and Watts, M. 65–82. Ithaca, NY: Cornell University Press.

160. Richards, P. 2003. *The political economy of internal conflict in Sierra Leone*, The Hague: Netherlands Institute of International Relations.

161. Richards, P. 2005. To fight or to farm? Agrarian dimensions of the Mano River conflicts (Liberia and Sierra Leone). *African Affairs*, 104: 571–90.

162. Riddell, J. B. 1985. Internal and external forces acting upon disparities in Sierra Leone. *Journal of Modern African Studies*, 23(3): 389–406.

163. Robbins, P. 2004. *Political ecology: A critical introduction*, Oxford, U.K.: Blackwell.

164. Roberts, R. S. and Emel, J. 1992. Uneven development and the tragedy of the commons: Competing images for nature-society analysis. *Economic Geography*, 68(3): 249–71.

165. Ross, M. 1999. The political economy of the resource curse. *World Politics*, 51(2): 297–322.

166. Ross, M. 2001a. Does oil hinder democracy?. *World Politics*, 53(3): 325–62.

167. Ross, M. 2001b. *Extractive sectors and the poor*, 1–28. Boston: Oxfam America.

168. Ross, M. 2003. "Oil, drugs, and diamonds: How do natural resources vary in their impact on civil war?". In *Beyond greed and grievance: The political economy of armed conflict*, Edited by: Ballentine, K. and Sherman, J. Boulder, CO: Lynne Rienner.

169. Ross, M. 2004a. How do natural resources influence civil war? Evidence from 13 cases. *International Organization*, 58(1): 35–67.

170. Ross, M. 2004b. What do we know about natural resources and armed conflicts. *Journal of Peace Research*, 41(3): 337–56.

171. Ross, M. 2006. A closer look at oil, diamonds, and civil war. *Annual Review of Political Science*, 9: 265–300.

172. Sachs, J. and Warner, A. 2001. The curse of natural resources. *European Economic Review*, 45(4–6): 827–38.

173. Sala-i-Martin, X. and Subramanian, A. 2003. *Addressing the natural resource curse: An illustration from Nigeria*, Cambridge, MA: National Bureau of Economic Research. NBER Working Paper Series

174. Sarraf, M. and Moortaza, J. 2001. *Beating the resource curse: The case of Botswana*, Washington, DC: World Bank Environmental Department.

175. Saylor, R. G. 1967. *The economic system of Sierra Leone*, Durham, NC: Duke University Press.

176. Scott, J. 1990. *Domination and the arts of resistance: Hidden transcripts*, New Haven, CT: Yale University Press.

177. Shapiro, M. 1997. *Violent cartographies: Mapping cultures of war*, Minneapolis: University of Minnesota Press.

178. Shearer, D. 1998. *Private armies and military intervention*, Oxford, U.K.: Oxford University Press.

179. Sheppard, S. 1998. Foot soldiers and the new world order: The rise of the corporate military. *New Left Review*, 228: 128–38.

180. Sidaway, J. D. 1998. "What is in a gulf? From the "arc of crisis" to the Gulf War". In *Rethinking geopolitics*, Edited by: Tuathail, G. O. and Dalby, S. 224–39. London: Routledge.

181. Sidaway, J. D. 2003. Sovereign excesses? Portraying postcolonial sovereigntyscapes. *Political Geography*, 22(2): 157–78.

182. Silberfein, M. 2004. The geopolitics of conflict and diamonds in Sierra Leone. *Geopolitics*, 9(1): 213–41.

183. Simmons, C. S. 2004. The political economy of land conflict in the Eastern Brazilian Amazon. *Annals of the Association of American Geographers*, 94(1): 183–206.

184. Sinding, K. 2005. The dynamics of artisanal and small-scale mining reform. *Natural Resources Forum*, 29: 243–52.

185. Smillie, I. 2005. "What lessons from the Kimberley Process Certification Scheme". In *Profiting from peace: Managing the resource dimensions of civil war*, Edited by: Ballentine, K. and Nitzschke, H. 47–67. Boulder, CO: Rienner.

186. Smillie, I., Gberie, L. and Hazleton, R. 2000. *The heart of the matter: Sierra Leone, diamonds and human security*, Ottawa, Canada: Partnership Africa Canada.

187. Smith, N. 1984. *Uneven development: Nature, capital and the production of space*, Oxford, U.K.: Blackwell.

188. Snyder, R. and Bhavnani, R. 2005. Diamond, blood, and taxes: A revenue-centered framework for explaining political order. *Journal of Conflict Resolution*, 49: 563–97.

189. Stanton, L. 2000. "Ten reasons why you should never accept a diamond ring from anyone". In *Ultimate field guide to the U.S. economy*, Edited by: Heintz, J. and Folbre, N. New York: New Press. Center for Popular Economics

190. Swyngedouw, E. 1999. Modernity and hybridity: Nature, regeneracionismo, and the production of the Spanish waterscape 1890–1930. *Annals of the Association of American Geographers*, 89(3): 443–65.

191. Tamm, I. J. 2004. Dangerous appetites: Human rights activism and conflict commodities. *Human Rights Quarterly*, 26: 687–704.

192. Taylor, I. 2003. As good as it gets? Botswana's "democratic development". *Journal of Contemporary African Studies*, 21(2): 215–31.

193. TRC. 2004. http://trcsierraleone.org/drwebsite/publish/v3a-c4.shtml*Final report*. Freetown, Sierra Leone: Truth and Reconciliation Commission of Sierra Leone (last accessed 2 May 2007)

194. Turner, M. D. 2004. Political ecology and the moral dimensions of "resource conflicts": The case of farmer-herder conflicts in the Sahel. *Political Geography*, 23(7): 863–89.

195. Turrell, R. V. 1981. The 1875 Black Flag revolt on Kimberley diamonds fields. *Journal of Southern African Studies*, 7(2): 194–235.

196. Turrell, R. V. 1987. *Capital and labour on the Kimberley diamond fields 1871–1890*, Cambridge, U.K.: Cambridge University Press.

197. United Nations. 2000a. *Report of the Panel of Experts Appointed Pursuant to Security Council Resolution 1306 (2000), Paragraph 19, in Relation to Sierra Leone*, New York: United Nations.

198. United Nations. 2000b. *Report of the Panel of Experts on Violations of Security Council Sanctions against UNITA*, New York: United Nations Secretariat.

199. United Nations. 2001. *Report of the Panel of Experts on the Illegal Exploitation of Natural Resources and Other Forms of Wealth of the Democratic Republic of Congo*, New York: United Nations Secretariat.

200. United Nations. 2002. *Report of the Monitoring Mechanism on Sanctions against UNITA*, New York: United Nations Secretariat.

201. United Nations. 2004. http://www.un.org/Depts/Cartographic/map/dpko/ctivoire.pdf Côte d'Ivoire. Map No. 4203 Rev.4. Department of Public Information (last accessed 2 May 2007)

202. United Nations Development Program. 2006. *Human development report*, New York: United Nations Development Program.

203. United Nations General Assembly. 2000. New York: United Nations General Assembly. United Nations General Assembly Resolution 55/56, 1 December 2000

204. United States Geological Survey. (USGS). 2005. http://minerals.usgs.gov/pubs/myb.html*Minerals yearbook*. United States Geological Survey(last accessed 2 May 2007)

205. Vandergeest, P. and Peluso, N. L. 1995. Territorialization and state power in Thailand. *Theory and Society*, 24(3): 385–426.

206. van der Laan, H. L. 1965. *The Sierra Leone diamonds: An economic study covering the years 1952–1961*, Oxford, U.K.: Oxford University Press.

207. Ward, M. D. and Gleditsch, K. S. 2002. Location, location, location: An MCMC approach to modeling the spatial context of war and peace. *Political Analysis*, 10: 244–60.

208. Watts, M. 2004. Antinomies of community: Some thoughts on geography, resources and empire. *Transactions of the Institute of British Geographers*, 29(2): 195–216.

209. Watts, M. J. and Bohle, H. G. 1993. The space of vulnerability: The causal structure of hunger and famine. *Progress in Human Geography*, 17(1): 43–67.

210. Weinstein, J. M. 2007. *Inside rebellion: The politics of insurgent violence*, New York: Cambridge University Press.

211. Wescoat, J. L. J. 1992. Resource management: Oil resources and the Gulf conflict. *Progress in Human Geography*, 16(2): 243–56.

212. Williams, J., Sutherland, D., Cartwright, K. and Byrnes, M. 2002. *Sierra Leone diamond policy study*, Colwyn Bay, U.K.: AMCO-Robertson Mineral Services Ltd.

213. World Bank. 2005. *World development indicators*, Washington, DC: World Bank.

214. Zack-Williams, A. 1995. *Tributors, supporters and merchant capital: Mining and underdevelopment in Sierra Leone*, Brookfield, MA: Avebury.

Chapter 5

Investigation of Low-Resistivity from Hydrogenated Lightly B-Doped Diamond by Ion Implantation

Cui Xia Yan[1,2], Ying Dai[1], Meng Guo[1], Lin Yu[1], Dong Hong Liu[1], Bai Biao Huang[1], Rui Qin Zhang[3], Wen Jun Zhang[3] and Igor Bello[3]

[1] *School of Physics, State Key Laboratory of Crystal Materials, Shandong University, Jinan 250100, People's Republic of China*

[2] *Department of Physics, Jining University, Qufu 273155, People's Republic of China*

[3] *Center of Super-Diamond and Advanced Films (COSDAF) and Department of Physics and Materials Science, City University of Hong Kong, Hong Kong, China*

ABSTRACT

We have implanted boron (B) ions (dosage: 5×10^{14} cm^{-2}) into diamond and then hydrogenated the sample by implantating hydrogen ions at room temperature. A p-type diamond material with a low resistivity of 7.37 mΩ cm has been obtained in our experiment, which suggests that the hydrogenation of B-doped diamond results in a low-resistivity p-type material. Interestingly, inverse annealing, in which carrier concentration decreased with increasing annealing temperature, was observed at annealing temperatures above 600 °C. In addition, the formation mechanism of a low-resistivity material has been studied by density functional theory calculation using a plane wave method.

Keywords: diamond, hydrogenation, low resistivity, plane wave method

INTRODUCTION

Diamond has been extensively investigated both theoretically and experimentally because of its unique structural properties and important technological applications. In particular, owing to its chemical immunity and negative electron affinity, it is rapidly becoming a candidate material for field emission devices [1, 2]. To realize diamond devices, low-resistivity p-type and n-type materials must be fabricated. In addition, electron field emission (FE) from diamond materials is interesting, which is the key element of cold-cathode field emission displays (FEDs) [3]. A study has found that the reduction in the resistivity of p-type diamond can increase emission efficiency [4]. Consequently, it was necessary to obtain a low-resistivity p-type diamond material.

The boron impurity in diamond has been studied for decades, because boron doping can lead to p-type diamond materials [5–13]. Low-resistivity p-type diamond with metallic-type conductivity has been obtained when the boron concentration exceeds the critical value of 2.0×10^{20} cm^{-3} [5, 7, 14]. Although it is known that B-doped diamond for forming p-type semiconductor needs an activation energy of about 0.37 eV, the formation mechanism of low-resistivity diamond materials remains unknown, which is an obstacle for the manufacture of low-resistivity materials [8, 15]. For example, the formation of the low-resistivity contact of field effect transistors (FETs) is still under study [16, 17]. A few studies of heavily B-doped diamond have been performed on the low resistivity and superconductivity of doped diamond [18–29]. Takano *et al* and Mukuda *et al* [30, 31] have investigated the superconductivity and the origin of superconductivity in heavily B-doped diamond by microwave plasma-assisted chemical vapor deposition and ^{11}B-NMR, respectively. Oguchi [32] has studied the electronic structure of heavily B-doped diamond with different doping sites by first-principles calculation. The results show that the superconductivity of heavily B-doped diamond is mediated by an electron-phonon interaction. The interstitial B dopant is unstable compared with the substitutional B dopant. Recently, Tsubouchi *et al* [29] have obtained a low-resistivity p-type layer (resistivity: 2.3 mΩ cm) by implanting a high dosage (8.4×10^{16} cm^{-2}) of B ions into diamond. However, to heavily implant B ions into diamond usually requires special methods such as a high-temperature or high-pressure technique, which raises difficulties practically [26, 28, 29, 33, 34]. However, to lightly implant B ions into diamond is much easier. To our knowledge, little work related to low-resistivity materials has been performed on lightly B-doped diamond.

In this work, we attempted to hydrogenate lightly B-doped diamond (estimated maximum B concentration: 8.62×10^{19} cm^{-3}) and then anneal it in the range of 300–630 °C. As a result, we have obtained a low-resistivity (7.37 mΩ cm) material at an annealing temperature of 560 °C, which has the same order of resistivity as heavily B-doped diamond [29].

EXPERIMENT

A commercial 5×5×0.2 mm^3 diamond film with a small amount of nitrogen impurity synthesized by high-power DF arc plasma jet CVD was used in the study. The grain diameter is about 50 μm. Prior to ion implantation, it was confirmed that the diamond film has high sufficient resistivity at ~10^{10} Ω cm. B ions with 400 keV at a dosage of 5×10^{14} cm^{-2} were implanted into the diamond film at a beam current density of 0. 16 μA cm^{-2} at room temperature. To prevent radiation damage induced by ion implantation, the film was annealed for 1.5 h in vacuum of 10^{-3} Pa at 1450 °C. Then it was washed immediately following the annealing in a boiling acid mixture (H_2SO_4:HNO_3=9:1) for 1.5 h to remove amorphous carbon. The procedure has been proved to be effective (it will be depicted in detail subsequently). The H ions at a dosage of 1.8×10^{15} cm^{-2} with 140 keV were implanted into B-doped diamond to hydrogenate the sample. The depth profiles of B and H are shown in figure 1 based on TRIM simulation [35]. The largest B concentration is 8.62×10^{19} cm^{-3} at 0.48 μm depth and the largest H concentration is 3.13×10^{20} cm^{-3} at 0.46 μm depth. The density of defects produced by the B and H doses in the sample is lower than the critical defect density $N_c=10^{22}$ vac cm^{-3} based on TRIM simulation [35]. To prevent damage caused by H ion implantation and to promote the interaction of the B and H ions, one diamond sample was annealed at different temperatures (300–630 °C) sequentially in a quartz tube with an Ar flux of 30 ml min^{-1}. The annealing is described in detail in table 1. After the annealing, the diamond was cleaned in a boiling H_2SO_4:HNO_3=9:1 solution to remove amorphous carbon. To bare the implanted layer and make contact between the electrode and the implanted layer, the hydrogenated B-doped diamond was etched by reactive ion etching technology. Then indium point contacts were formed at four corners of the sample surface to measure the electronic properties of the sample using a four-point probe method in the Van der Pauw configuration at room temperature. The electrode displayed good ohmic I–V characteristics. The sample was characterized by atomic force microscopy (AFM) and Raman spectroscopy. The Nd:YAG laser (532 nm) was used in Raman measurement.

Table 1. Parameters of annealing process of film in range of 300–630 °C in quartz tube with Ar flux of 30 ml min^{-1}.

Temperature (°C)	300	350	400	450	500	510	520	530	540
Time (min)	30	30	60	60	60	60	60	60	60
Temperature (°C)	550	560	570	580	590	600	610	620	630
Time (min)	60	60	60	60	60	60	60	60	60

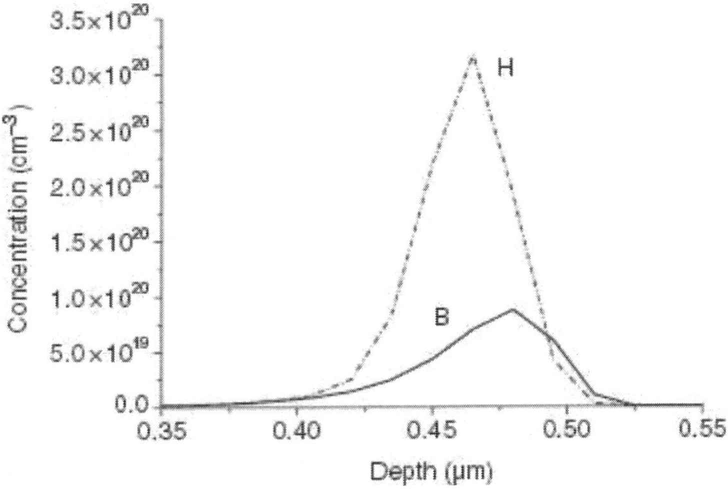

Figure 1 Depth profiles of B and H in diamond based on TRIM simulation. The solid line is the depth profile of B and the dot line is that of H.

RESULTS AND DISCUSSION

AFM

Figure 2(a) shows the surface of the original sample (as-grown diamond film), which is quite smooth with a few small thorns and a surface roughness Ra of 4.39 Å, suggesting the good quality of the sample. Figure 2(b) shows the surface of the B-doped sample, which is not smooth with a few small protuberances. The surface roughness Ra is 165.9 Å. The thorns may be intrinsic defects formed during the growth and the protuberances are the defects that resulted from the implantation. Although the sample had a few defects, we believe that the quality of B-doped diamond is good for further research.

Raman spectroscopy

To confirm the validity of removing amorphous carbon on the sample surface, we show the Raman spectra of diamond in various states in figure 3. The Raman spectrum of the original sample (as-grown diamond film) is shown in line *a*, in which two peaks appear: one at 1332 cm^{-1}, which is the typical eigenpeak of diamond, and the other at 1458 cm^{-1} The original sample has a sufficiently high resistivity at $\sim 10^{10}$ Ω cm (it is mentioned in the experiment section); thus, there is no graphitic amorphous carbon in the original sample. The broad peak at 1458 cm^{-1} in line *a* should be caused by intrinsic defects formed during the growth of the diamond film (as seen in figure 2(a)). According to $C_{nd}=1/[1 + 75(I_d / I_{nd})]$ [36, 37], the defect structure content

C_{nd} is about 0.89%, where I_d=1.62 and I_{nd}=1.1 are the intensities of the peaks at 1332 and 1458 cm^{-1}, respectively. Such a low percentage of defects will have little influence on the following experiment results and discussion.

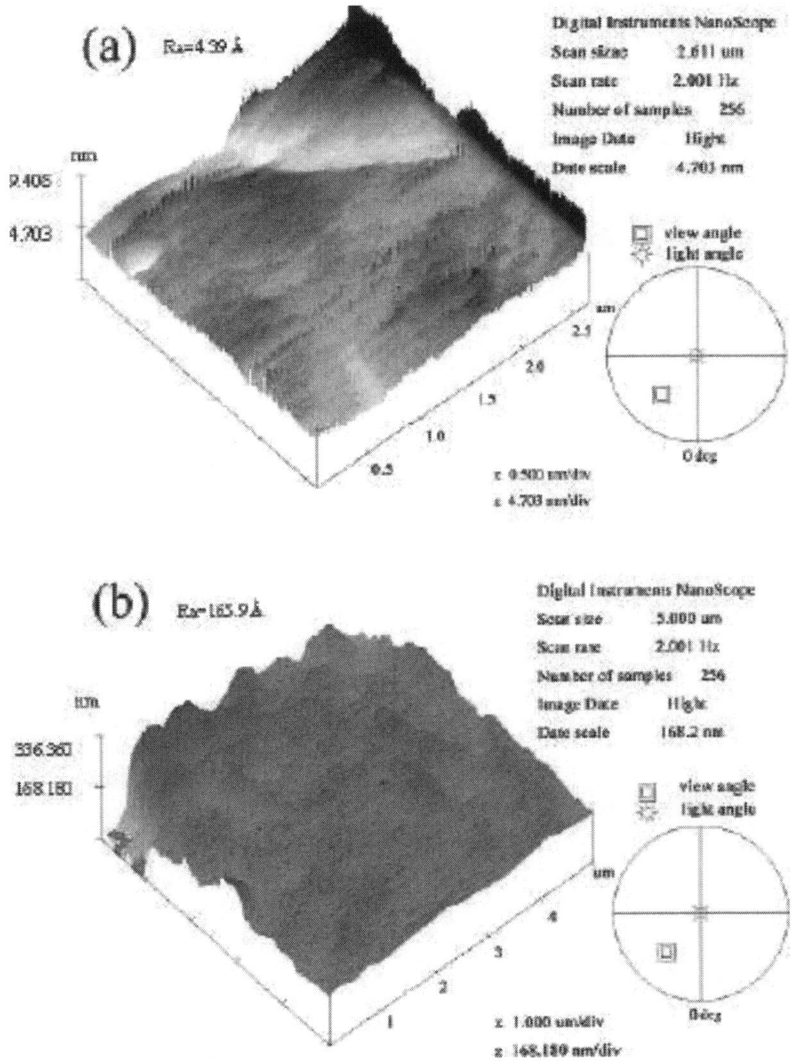

Figure 2 The surface morphologies of diamond (a) and B-doped diamond (b) are shown, as determined by the AFM. The surface of the original sample (a) is quite smooth with small thorns; that of boron-doped diamond (b) is not smooth with a few small protuberances.

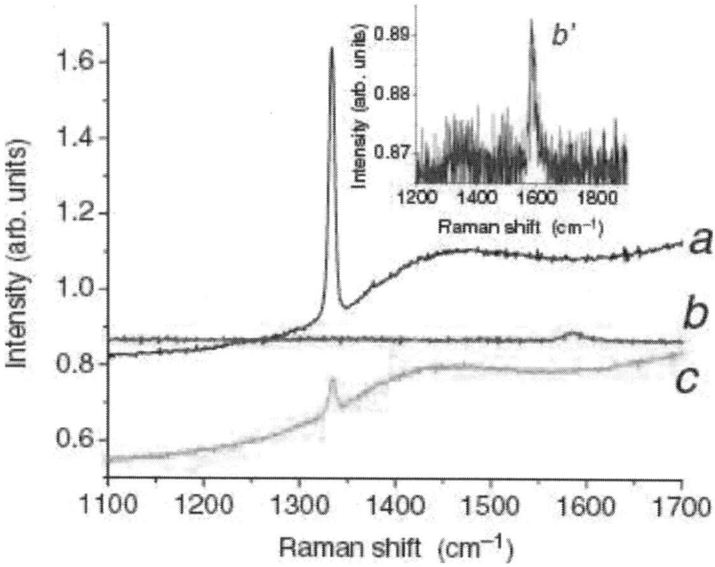

Figure 3 Raman spectra of diamond in various states. Line *a* is the Raman spectrum of the original sample. Line *b* is that of B-doped diamond (annealed at 1450 °C). Line *c* is that of hydrogenated B-doped diamond annealed at 560 °C and then washed in a boiling acid mixture and etched.

Line *b* is the Raman spectrum of the B-doped diamond sample annealed at 1450 °C. There is only a peak at 1587 cm^{-1} (see the 100-fold-enlarged peak *b'* in the inset) in the spectrum. Owing to long-time (1.5 h) annealing at a high temperature (1450 °C), C–C bonds on the surface are reconstructed with more stable sp^2 bonds. Therefore, the peak at 1332 cm^{-1} disappears and the peak at 1587 cm^{-1} appears. Because the characteristic peak of graphite appears at about 1580 cm^{-1}, the surface should be composed of amorphous carbon similar to graphite. The hydrogenated sample was annealed in the range of 300–630 °C and shows special properties at 560 °C (it will be shown and discussed in detail subsequently), so the Raman spectrum of the sample annealed at 560 °C and treated by washing in a boiling acid mixture and etching is shown in line *c*. The peak at 1587 cm^{-1} disappears and two peaks appear at 1332 and 1458 cm^{-1} in the spectrum. The results indicate that there is no amorphous carbon on the surface of hydrogenated diamond, namely, the amorphous carbon was completely removed after the sample was washed in a boiling acid mixture (H_2SO_4:HNO_3=9:1). The broad peak at 1458 cm^{-1} in line *c* is similar to that in line *a*, which indicates that the peak may be caused by intrinsic defects and not by the high graphitic amorphous content.

Hall effect measurement

The Hall effect measurement showed a positive sign of the sample, which indicates that the doped diamond film has a p-type conduction character. Figure 4 shows the annealing temperature (300–630 °C) dependence of the resistivity, which was determined by Hall effect measurement in the Van der Pauw configuration at room temperature. The resistivity is low in the range of 300–600 °C and increases sharply over 600 °C. To further investigate the annealing temperature dependence of the resistivity, the inset in figure 4 shows the resistivity of the sample in the range of 300–600 °C. The resistivity is not sensitive to temperature in the range of 300–500 °C; however, it changes evidently in the range of 550–600 °C and a minimum of 7.37 mΩ cm is reached at 560 °C.

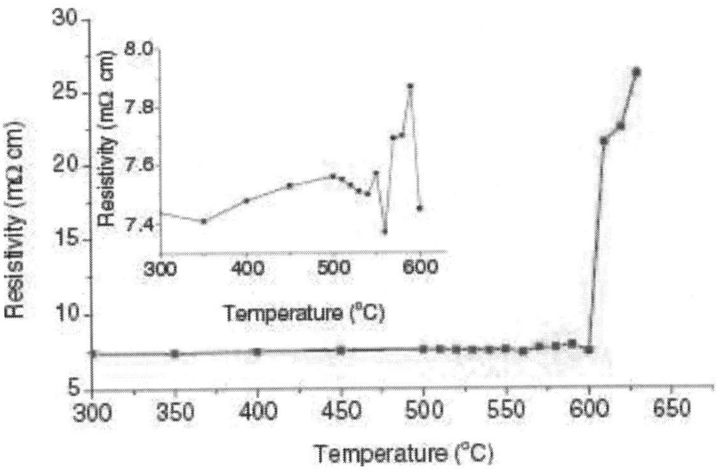

Figure 4 Annealing temperature dependence of resistivity in the range of 300–630 °C. The inset shows the annealing temperature dependence of resistivity in the range of 300–600 °C.

Figure 5 shows the annealing temperature (300–630 °C) dependence of the carrier concentration of hydrogenated diamond. It can be seen that the carrier concentration is quite low and hardly changes in the range of 300–500 °C; however, it changes evidently in the range of 500–600 °C and a maximum of 1.0×10^{18} cm^{-3} is reached at 560 °C. The carrier concentration decreases sharply above 600 °C. From the above analysis, we may conclude that the lowest resistivity 7.37 mΩ cm and the highest carrier concentration 1.0×10^{18} cm^{-3} are reached at 560 °C, which suggest that the use of hydrogenated B-doped diamond enables the formation of a low-resistivity p-type material [29, 38]. Technologically, the production of such low-resistivity p-type material by implantation and hydrogenation techniques would be useful for diamond device processing. In general, the carrier concentration increases with increasing annealing temperature. However, an abnormal phenomenon, that is, the carrier concentration decreases with increasing annealing temperature, may occur in a certain temperature range. This is the so-called 'inverse annealing' phenomenon [39]. For hydrogenated B-doped diamond, the `inverse annealing' phenomenon occurs when the annealing temperature is over 600 °C (as shown in figure 5). It can result from the exchange reaction between the substitutional B impurities and the interstitial C ones.

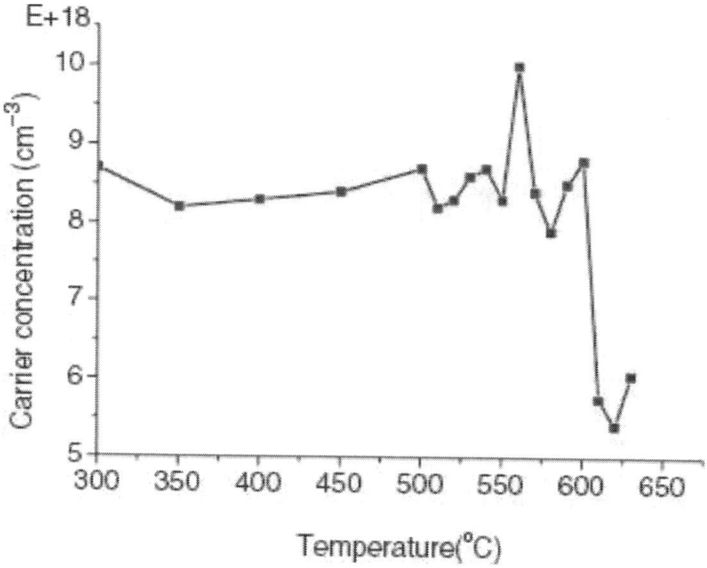

Figure 5 Annealing temperature dependence of carrier concentration in the range of 300–630 °C.

Formation mechanism of low-resistivity material

To determine the formation origin of low-resistivity materials during the hydrogenation of lightly B-doped diamond, we have calculated the electronic structures of diamond doped with various (B + H) configurations by density functional theory (DFT) calculation. For B and H-doped diamonds, a 64-atom host supercell is used. The calculation was performed using CASTEP codes [40], with conjugated gradient approximation (GGA) and ultrasoft pseudopotentials in reciprocal space. Exchange-correlation potential parametrized by the Perdew–Burke–Ernzerhof scheme (PBE) was used [41]. The energy cutoff for the plane wave expansion was 310 eV. The calculation was performed using the Monkhorst-Pack scheme for sampling in the Brillouin zone with a mesh of 4×4×4 special *k* points [42].

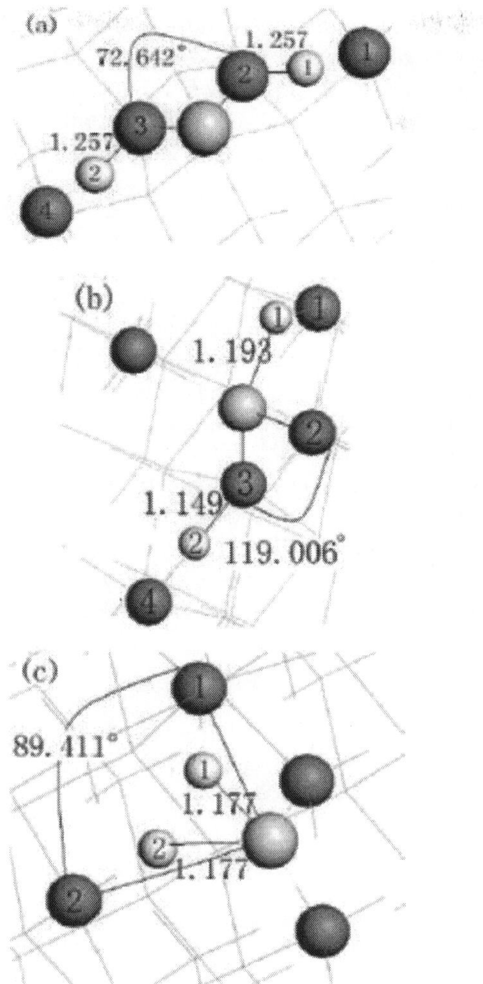

Figure 6 Optimized structures of diamond doped in (B+H) configurations: (a) B–$2H_{cc}$, (b) B–H_{bc}–H_{cc}, and (c) B–$2H_{bc}$. The black, grey and white balls are C, B and H atoms, respectively.

Here, we show three optimal (B + H) configurations in figure 6, which are stable according to formation energy (it will be shown in detail subsequently). In the three configurations, B atoms replace C and H atoms in various sites. The first one, B–2 H_{cc}, is shown in figure 6(a), in which two H atoms, H_1 and H_2, occupy the C_1–C_2 and C_3–C_4 buckled bonds, respectively. The bond lengths of C_2–H_1 and C_3–H_2 are about 1.257 Å, and the angle between the C_2–B bond and C_3–B bond is about 72.642°. Figure 6(b) shows the second one, B–H_{bc}–H_{cc}, in which the two H atoms, H_1 and H_2, occupy the C_1–B and C_3–C_4 buckled bonds, respectively. The bond lengths B–H_1 and C_3–H_2 are about 1.193 and 1.149 Å, respectively. The angle between the C_2–B and C_3–B bonds is about 119.006°. The third configuration, B–$2H_{bc}$, is shown in figure 6(c), in which two H atoms, H_1 and H_2, occupy the C_1–B and C_2–B buckled bonds. The bond lengths of B–H_1 and B–H_2 are about 1.177 Å, and the angle C_1–B–C_2 is about 89.411°.

Figure 7 shows the density of states (DOS) of diamond with the three (B+H) configurations. The DOS of pure diamond is shown in figure 7(a) as reference. The DOS of diamond with the B–2 H_{cc} configuration (as seen in figure 7(b)) shows that the Fermi level E_F locates in the valence band. It reveals clearly that the B–2 H_{cc} configuration induces impurity band mixing with the valence band and the Fermi level lies in the valence band, which results in the doped diamond having a p-type metallic conductivity character. However, figures 7(c) and (d) show that the B–H_{bc}–H_{cc} or B–$2H_{bc}$ configuration introduces deep gap states and that the Fermi level lies in the band gap or near the bottom of the conductivity band, respectively, which indicates that the doped diamond has an insulator character or n-type conductivity character. These results demonstrate that if the B–$2H_{cc}$ configuration is formed in doped diamond, a low-resistivity p-type conductivity will occur. Consequently, we can deduce that it is the formation of the B–$2H_{cc}$ configuration in our sample during the hydrogenation of B-doped diamond, that leads to low-resistivity phenomena.

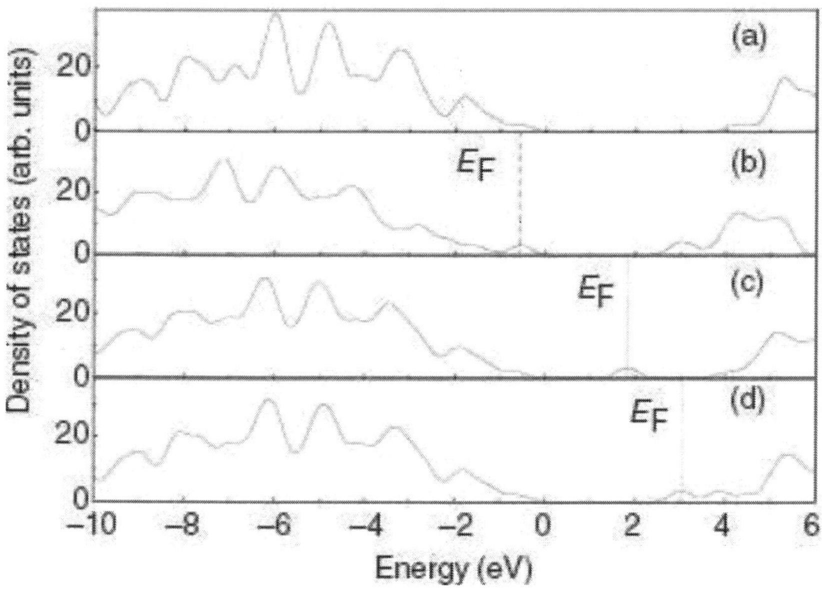

Figure 7 DOSs of (a) pure diamond, (b) diamond with B–$2H_{cc}$ configuration, (c) diamond with B–H_{bc}–H_{cc} configuration, and (d) diamond with B–$2H_{cc}$ configuration. The dot line represents the Fermi level E_F.

To investigate the relative stability of (B + H) configurations, the formation energy (E^f) of BH_2 configurations was estimated using [43]

$$E^f = E(C_{63}BH_2) - 63\mu_C - \mu_B - 2\mu_H,$$

where μ_C is the chemical potential of the atoms C obtained from pure diamond. μ_B and μ_H are the chemical potential of B and H impurities, which are obtained from B_2H_6 and H_2, respectively. $E(C_{63}BH_2)$ is the total energy of the supercell with the BH_2 configuration, which can be accurately obtained by first-principles techniques. The results show that the formation energies for the B–2 H_{cc} (figure 6(a)), B–H_{bc}–H_{cc} (figure 6(b)), and B–2H_{cc} (figure 6(c)) configurations are about -13.65, -14.16, and -12.82 eV, respectively. This means that the formation energies for various configurations have slight differences, which means each of them may be possibly formed under suitable experimental conditions. The hydrogenation of lightly B-doped diamond is a new and better method of obtaining p-type low-resistivity materials; therefore, many problems still need to be further studied, e.g., controlling the formation of configuration B–2 H_{cc} by researching suitable experimental conditions.

CONCLUSIONS

In summary, we obtained a p-type diamond material with the low resistivity of 7.37 mΩ cm during the hydrogenation of B-doped diamond. By Raman spectroscopy, we confirmed that there are no amorphous carbon on the surface of hydrogenated diamond after the sample is washed in a boiling acid mixture. Furthermore, we deduce that the low-resistivity material is hydrogenated B-doped diamond and not graphite. Our theoretical results suggest that the hydrogenation of B-doped diamond results in a metallic characteristic with low-resistivity p-type conductivity if some B–2 H_{cc} configurations form. Technologically, the production of such a low-resistivity p-type material would be useful for diamond device processing. The phenomenon of inverse annealing is also observed at annealing temperatures above 600 °C, which suggests that annealing above 600 °C is unnecessary.

ACKNOWLEDGMENTS

This work is financially supported by the National Basic Research Program of China (973 program, Grant No. 2007CB613302), National Natural Science Foundation of China under Grant No. 10774091, Natural Science Foundation of Shandong Province under Grant No. Y2007A18, and the Specialized Research Fund for the Doctoral Program of Higher Education 20060422023. We also thank the Institute Semiconductor of the Chinese Academy of Sciences and Dr J

Q Pan for help with the Hall measurement and Dr S Q Sun for the Raman measurement. The work described in this paper is also supported by a grant from the CAS – Croucher Funding Scheme for Joint Laboratories (project No. 9500004) and by the Centre for Applied Computing and Interactive Media (ACIM).

REFERENCES

1. HimpselF JKnappJ Avan VechtenJ AEastmanD E 1979 Phys. Rev. B 20
2. van der WeideJZhangZBaumannP KWensellM GBernholcJNemanichR J 1994 Phys. Rev. B 50 5803
3. AngusJ CCollinsA T 1994 Nature 370 601
4. ItahashiMUmeharaYKoideYMurakamiM 2001 Diamond Relat. Mater. 10 2118
5. VishnevskiiA SGontarA GTorishniiV IShul'zhenkoA A 1981 Sov. Phys. Semicond. 15 659
6. NishimuraKDasKGlassJ T 1991 J. Appl. Phys. 69 3142
7. WerherM et al 1994 Appl. Phys. Lett. 64 595
8. GlesenerJ W 1994 Appl. Phys. Lett. 64 217
9. LeeS TZhangR JLamY W 1996 Diamond Relat. Mater. 5 1288
10. GossJ PBriddonP RJonesRTeukamZBallutaudDJomardFChevallier JBernardMDeneuvilleA 2003 Phys. Rev. B 68 235209
11. LombardiEMainwoodAOsuchK 2004 Phys. Rev. B 70 205201 h
12. GossJ PBriddonP R 2006 Phys. Rev. B 73 085204
13. BarnardA SRussoS PSnookI K 2003 Philos. Mag. 83 1163
14. BustarretEGheeraertEWatanabeK 2003 Phys. Status Solidi a 199 9
15. YokoyaTNakamuraTMatsushitaTMuroTTakanoYNagaoMTakenouchiTKawaradaHOguchiT 2005 Nature 438 647
16. NegoroYKimotoTMatsunamiHSchmidFPenslG 2004 J. Appl. Phys. 96 4916
17. ZeidlerJ RHewettC AWilsonR G 1993 Phys. Rev. B 47 2065
18. PrinsJ F 1989 J. Phys. D: Appl. Phys. 22 1562
19. BraunsteinGKalishR 1983 J. Appl. Phys. 54 2106
20. VenkatesanVMaltaD MDasKBeluA M 1993 J. Appl. Phys. 74 1179

21. KalishRUzan-SaguyCSamolloffALocherRKoidlP 1994 Appl. Phys. Lett. 64 2532

22. PrawerSKalishR 1995 Phys. Rev. B 51 15711

23. KalishRUzan-SaguyCRanBFerberHGuettlerHZachaiR 1999 Diamond Relat. Mater. 8 877

24. HsiehH HChangY KPongW FTsaiM HChienF ZTsengP KLinI NChenH F 1999 Appl. Phys. Lett. 75 222

25. MaminR FInushimaT 2001 Phys. Rev. B 63 033201

26. BrazhkinV VEkimovE ALyapinA GPopovaS VRakhmaninaA VStishovS MLebedevV M 2006 Phys. Rev. B 74 140502(R)

27. SidorovV AEkimovE AStishovS MBauerE DThompsonJ D 2005 Phys. Rev. B 71 060502

28. EkimovE ASidorovV ABauerE DMel'nikN NCurroN JThompsonJ DStishovS M 2004 Nature 428 542

29. TsubouchiNOguraMHorinoYOkushiH 2006 Appl. Phys. Lett. 89 012101

30. TakanoYNagaoMSakaguchiITachikiMHatanoTKobayashiKUmezawaHKawaradaH 2004 Appl. Phys. Lett. 85 2851

31. MukudaHTsuchidaTHaradaAKitaokaYTakenouchiTTakanoYNagaoMSakaguchiIKawaradaH 2006 Sci. Technol. Adv. Mater. 7 S37

32. OguchiT 2006 Sci. Technol. Adv. Mater. 7 S67

33. VoronovO ARakhmaninaA V 1993 Inorg. Mater 29 707

34. YakovlevE NVoronovO ARakhmaninaA V 1984 Sverhtverdye Mater. 4 8 (In Russian)

35. ZieglerJ FBiersackJ PLittmarkU 1985 Stopping and Range of Ions in Matter New York Pergamon

36. ZhangM LXiaY BWangL JGuB BSuQ FLouY Y 2005 J. Cryst. Growth 274 21

37. SilveiraMBecucciMCastellucciEPolla MattiotFBarbarossaVTomacielloRGalluzziF 1993 Diamond Relat. Mater. 2 1257

38. BorstT HWeisO 1995 Diamond Relat. Mater 4 948

39. Institute of Low Energy Nuclear Physics Beijing Normal University, The Theory and Technology of Implantation, Beijing,1982 p 124

40. SegallM DLindanP J DProbertM JPickardC JHasnipP JClarkS JPayneM C 2002 J. Phys.: Condens. Matter 14 2717

41. PerdewJ PBurkeKErnzerhM 1996 Phys. Rev. Lett. 77 3865

42. MonkhorstH JPackJ D 1976 Phys. Rev. B 13 5188

43. ZhangS BNorthrupJ E 1991 Phys. Rev. Lett. 67 2339

Chapter 6

Ab Initio Structure Determination of N-Diamond

Da Li[1,2], Fubo Tian[1], Binhua Chu[1], Defang Duan[1], Xiaojing Sha[1], Yunzhou Lv[1], Huadi Zhang[1], Nan Lu[2], Bingbing Liu[1] & Tian Cui[1]

[1] *State Key Lab of Superhard Materials, College of Physics, Jilin University, Changchun 130012, P. R. China.*

[2] *State Key Lab of Supramolecular Structure and Materials, College of Chemistry, Jilin University, Changchun 130012, P. R. China.*

ABSTRACT

A systematic computational study on the crystal structure of n-diamond has been performed using first-principle methods. A novel carbon allotrope with hexagonal symmetry *R*32 space group has been predicted. We name it as *HR*-carbon. *HR*-carbon composed of lonsdaleite layers and unique C_3 isosceles triangle rings, is stable over graphite phase above 14.2 GPa. The simulated x-ray diffraction pattern, Raman, and energy-loss near-edge spectrum can match the experimental results very well, indicating that *HR*-carbon is a likely candidate structure for n-diamond. *HR*-carbon has an incompressible atomic arrangement because of unique C_3 isosceles triangle rings. The hardness and bulk modulus of *HR*-carbon are calculated to be 80 GPa and 427 GPa, respectively, which are comparable to those of diamond. C_3 isosceles triangle rings are very important for the stability and hardness of *HR*-carbon.

INTRODUCTION

The graphite composed of sp^2-hybridized bonds is one of the most stable carbon phases in nature. Under high pressure, the rehybridization from sp^2-

bonding to sp^3-bonding appears in graphite. It can transform to sp^3-bonding diamond under high pressure (15 GPa) and high temperatures (1600–2500 K) without using the catalysts1′2. Under room temperature, the graphite converts to distorted sp^3-bonding cold-compressed post graphite phase at above 14 GPa3. The transformation mechanisms of diamond/cold-compressed graphite phase transition have been studied systematacially4′5. The possible crystal structures of cold-compressed post graphite phases have been explored by both theoretical and experimental methods, such as M-carbon, W-carbon, C-carbon etc6′7′8′9′10′11′12′13′14′15′16′17′18′19′20′21′22′23′24′25. By using shock compression and rapid quenching methods, the graphite transforms to a new diamond-like phase (n-diamond)26 which has also been confirmed by several other experimental techniques27′28′29′30′31′32′33′34′35. Non-hydrostatic pressure shows a much more significant effect on the formation of n-diamond. The n-diamond is a metastable phase of graphite as diamond and can also be synthesized from other carbon forms such as carbon black, nanotubes, C_{60} films etc35′36′37. So far, although n-diamond has been found for more than thirty years, the crystal structure of n-diamond is still mysterious. X-ray powder diffraction (XRD) has already been used to characterize n-diamond. Previous observations indicate that the XRD of n-diamond is very similar to that of cubic diamond27. Interestingly, some forbidden reflections of cubic diamond have also been observed at the same time in experiments. The n-diamond was initially thought to be a face-central cubic modified diamond (Fcc carbon)31′32′38′39′40. However, theoretical calculations indicate that the Fcc carbon is less stable than the other candidate phases41′42′43. Moreover, the energy-loss near-edge spectrum (ELNES) of Fcc carbon do not agree with the measured ELNES spectrum44′45. The carbon-hydrogen zincblende structure, hydrogen-doped cubic diamond, intermediate phases between graphite and diamond with cubic $F43m$ or rhombohedral $R3$ space group and tetragonal glitter carbon have also been proposed as the candidates for n-diamond46′47′48′49. However, only glitter carbon is dynamically stable phase among the known candidate structures47. Moreover, to our knowledge, none of them can account for the experimental XRD very well. In addition, experimental samples are usually a mixture of different reactants, making the comparison between experimental and theoretical XRD even harder. Up to now, the crystal structure of n-diamond is the subject of continuing debate. Hear we predicted a likely candidate structure for n-diamond by using the genetic algorithm for crystal structure prediction. It has $R32$ space group and is more energetically favorable than the other candidates. The simulated XRD patterns, Raman and ELNES spectra can match the experimental data. We call this novel hexagonal carbon allotrope as *HR*-carbon which has an incompressible atomic arrangement due to uniqueC_3 isosceles triangle rings. It is stable in the range of 0 GPa up to 50 GPa at least. C_3 isosceles triangle rings of *HR*-carbon are critical for the stability and hardness of *HR*-carbon.

RESULTS AND DISCUSSION

A novel candidate structure for n-diamond, named it as *HR*-carbon, has been predicted by using USPEX method50'51'52 with 15 carbon atoms in the simulation cell. *HR*-carbon has hexagonal lattice *R*32 symmetry as depicted in Fig. 1(a). It is made of exclusively three-dimensional sp^3-hybridized covalent bonds. The equilibrium lattice parameters are $a = 4.305$ Å, $c = 16.291$ Å at ambient condition. Within this structure, five inequivalent atoms occupy the crystallographic 18*f* and 6*c* sites in the unit cell, which are (0.667, 0.009, 0.247), (0.667, 0.333, 0.114), (0.0, 0.0, 0.127), (0.333, 0.667, 0.117) and (0.871, 0.538, 0.333) positions. There are 45 atoms in the unit cell. In rhombohedral representation, it has a 15 atoms rhombohedral unit cell. The crystal structure of *HR*-carbon is distinct from that of diamond. The *HR*-carbon can be regarded as a modulated graphite phase composed of lonsdaleite layers and C_3isosceles triangle rings layers with stacking sequence of *ABCABCABC*...along the crystallographic *c* axis of hexagonal lattice. The close-packed*A* and *B* layers are lonsdaleite layers. The *C* layers composed of isolatedC_3 isosceles triangle rings are sandwiched between two *AB* layers with unique twisted sp^3 covalent bonds with bond angle 60°, 107.15°, and 140.04° which are much different from those of standard sp^3 covalent bonds of diamond (bond angle 109.47°). This finding suggests a novel combination form of carbon atoms to construct post graphite phases. The structural type of *HR*-carbon is also consistent with previous theoretical suggestion that n-diamond should have cubic or rhombohedral space group46. Enthalpy calculations suggest that *HR*-carbon is much stable than the previously proposed candidates of n-diamond Fig. 1(b)47. And *HR*-carbon becomes stable relative to graphite at above 14.2 GPa which is almost equal to the phase transition pressure of cold-compressed graphite phase (14 GPa)3. Its bulk modulus and shear modulus are 428 and 471 GPa, respectively. Furthermore, no imaginary frequencies are observed throughout the whole Brillouin zone in phonon dispersion Fig. 2(b), confirming dynamically structural stability of *HR*-carbon. The highest phonon frequency of *HR*-carbon (41.5 THz) is very close to that of diamond (40 THz)53, which reflects the diamond-like structural bonding character of *HR*-carbon. The calculated electronic band structure and density of state (DOS) at ambient pressure reveal that the *HR*-carbon is semiconductor with direct band gap 4.6 eV Fig. 2(a). Strong hybridization between *s* and *p*orbitals of DOS indicates the presence of strong covalent bonds in *HR*-carbon. The *HR*-carbon can be expected to have good mechanical properties among compressed graphite phases.

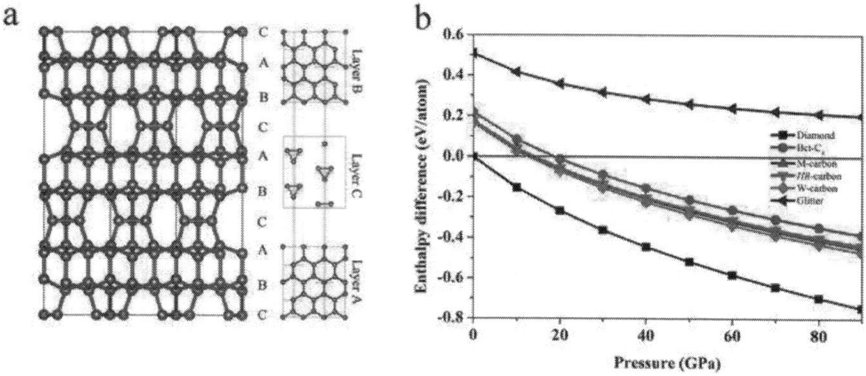

Figure 1. The structure of *HR*-carbon and enthalpy difference curves.(**a**) The crystal structure of *HR*-carbon and side views of *A, B* and *C* Layer from the [001] direction. (**b**) The enthalpy per atom of *HR*-carbon and other carbon polymorphs as functions of pressure with respect to graphite.

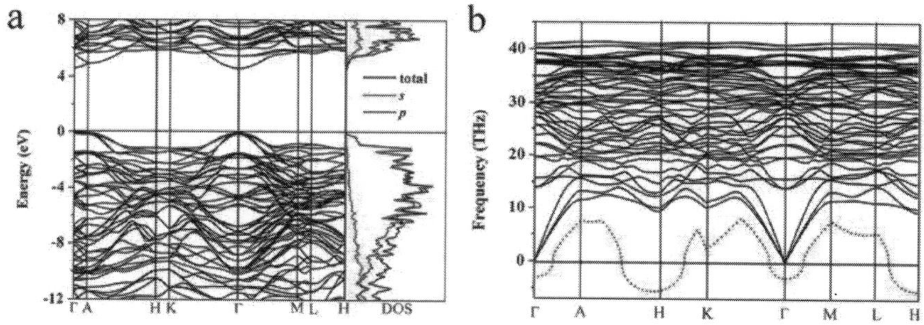

Figure 2. The electronic band structure, partial density of state and phonon spectra of *HR*-carbon.(**a**) Calculated electronic band structure and partial density of state of *HR*-carbon. (**b**) The phonon spectra of *HR*-carbon at ambient pressure. The red dash line is the imaginary phonon mode of modulated *HR-carbon*.

To confirm the consistency of *HR*-carbon and n-diamond, we simulate the XRD pattern of *HR*-carbon and compare with the experimental results Fig. 327′49. Our simulated XRD patterns for *HR*-carbon, reactants, and products are consistent with experimental XRD patterns. Three individual peaks of diamond can be easily indexed in the experimental XRD pattern Fig. 3(a). And the peaks of *HR*-carbon at ~46° and 51° can explain the experimentally observed forbidden diamond peaks. The relatively weaker peaks of *HR*-carbon in the range of 54°–95° have merged into the background. While the peaks at ~42°, 76° and 93° contribute to the peaks broadening of diamond at ~45°, 75° and

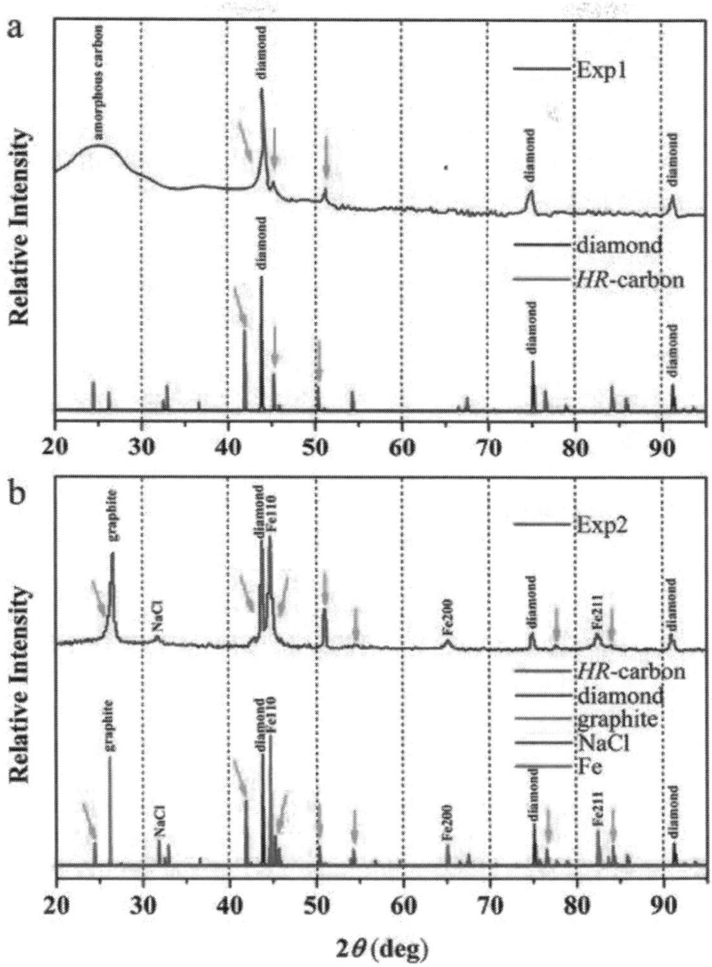

Figure 3. The comparison of theoretical and experimental XRD. (**a**) The simulated XRD patterns of *HR*-carbon, diamond, and experimental XRD pattern (from Ref. 27) (**b**) The simulated XRD patterns of *HR*-carbon, diamond, graphite, NaCl, α-Fe and experimental XRD pattern (from Ref. 49). The used x-ray wavelength is 1.5405 Å as employed in the experiments. Green arrows stand for those XRD peaks of *HR*-carbon observed by two experiments.

90°. The peaks of *HR*-carbon in the range of 20°–40° have been covered by the peaks of amorphous carbon (a-C). Furthermore, we also compare our simulated XRD pattern with that of n-diamond synthesized by Fe-catalysed carbon black methods50. The peaks of diamond, graphite, NaCl, and Fe have been indexed in the experimental XRD pattern Fig. 3(b). Four characteristic peaks of experimental XRD at 50°, 54°, 77°, and 84° can be explained very well by the peaks of *HR*-carbon in the same range. The peaks of *HR*-carbon at ~24° and 26°, ~33°, and ~45° contribute the peak broadening of graphite (~26°), NaCl (~32°) and Fe (~45°), respectively. The peaks of *HR*-carbon at ~42° can explain the experimentally observed shoulder peak at ~43°. The other peaks of *HR*-carbon have merged to the background. It is noted that the other candidate structures can also simulate the experimental XRD patterns. However, they are all thermodynamically unstable with respect to *HR*-carbon. We also compare the XRD pattern of *HR*-carbon with those of previously proposed post-graphite phases as shown in Fig. 4. Nineteen inequivalent structures (M-carbon7, W-carbon11, Bct-$C_4$8, X-carbon18, O-carbon21 (H-carbon20, R-carbon19), C-carbon16 (S-carbon20), F-carbon12(J-carbon22), T12-carbon13, Z-carbon15 (Cco-C89, oC16-II10), P-carbon19(Z4-A3B114), oC32-carbon23, M585-carbon25, R3-carbon24, mP16-carbon6, mS32-carbon6, oP20-carbon6, oP24-I-carbon6, oP24-II-carbon6 and oP28-carbon6) have been considered. It can be found that only the XRD of *HR*-carbon at ~45°, ~50°, ~54°, ~76° and ~84° can explain the experimentally observed characteristic peaks of n-diamond (45°, 51°, 55°, 78° and 84°)27′49. So *HR*-carbon is the most possible candidate structure of n-diamond.

To confirm reliability of our structure, the ELNES and Raman of *HR*-carbon was calculated and compared with the experimental results32′38. The ELNES spectra are very useful to distinguish the various hybridized covalent bonds in carbon materials. Strikingly, the XRD results indicate a coexistence of n-diamond, amorphous carbon, diamond and some other carbon allotropes, consistent with Konyashin's conclusion that the experimental sample is a mixture of n-diamond and other carbon allotropes44. Obvious π* feature of experimental ELNES indicates the presence of sp^2-hybridized covalent bonds. However, no π* features are found in the ELNES of diamond and *HR*-carbon, indicating the sample should contain graphite or a-C because those are composed of sp^2-hybridized covalent bonds. Moreover, the a-C was already detected by XRD because of the presence of broad peak in range of 20–30° Fig. 3(a). In order to compare to experimental data, we calculate the fitting average values (FAV) of *HR*-carbon and possible carbon allotropes (diamond, graphite and amorphous carbon) (Fig. 5). It can be found that the FAV of *HR*-carbon and diamond cannot match the experimental ELNES. The FAV of *HR*-carbon and graphite/a-C are very similar to each other. The two main differences between those two spectra are found in range of 280–290 eV and 315–330 eV. Obvious π* feature can be found in the range of 280–290 eV in those two situation. However, the intensity of π* peak of FAV of *HR*-carbon and a-C is stronger than that of π* peak of FAV of *HR*-carbon and graphite and can match the experimental data very well. But only the peaks of FAV of *HR*-carbon and graphite in the range of 315–330 eV can match the experimental third broad peak. It is noteworthy that the presence of graphite was not reported in Li's experimental XRD pattern27 because the main peaks of graphite and

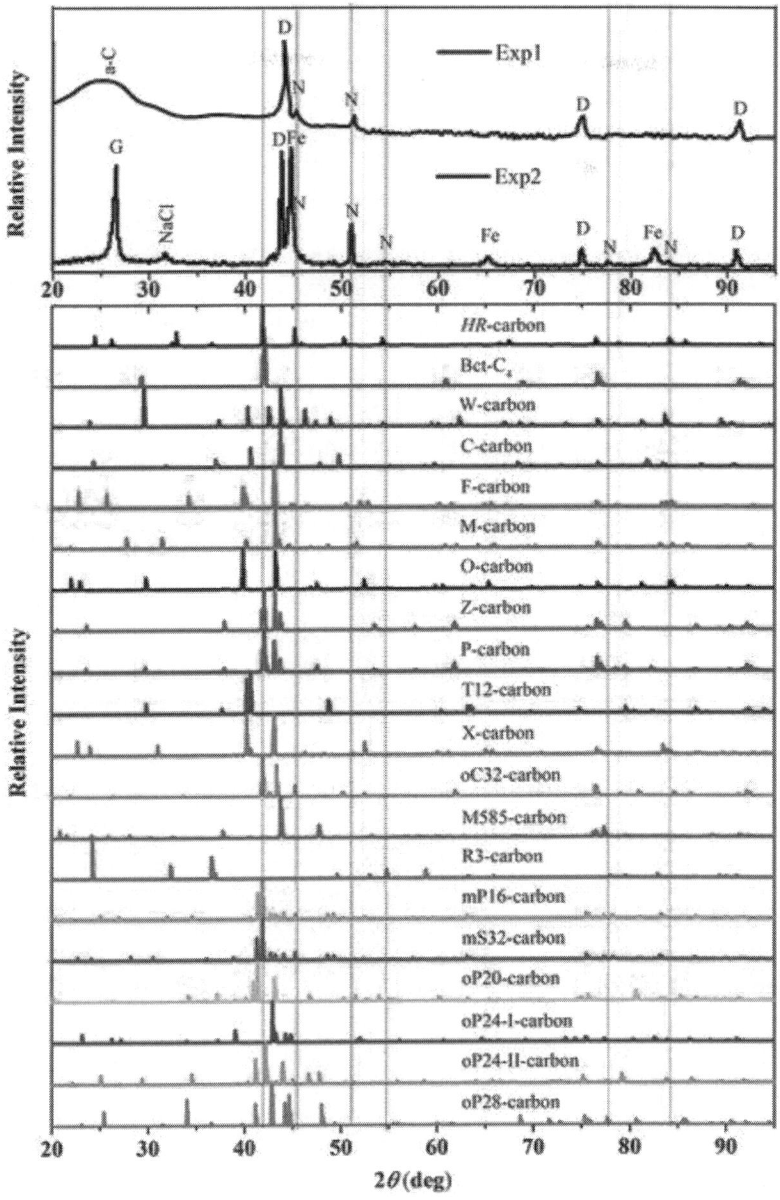

Figure 4. The XRD patterns of experiments, *HR*-carbon and previously proposed post-graphite phases.The XRD of experiment 1 comes from Ref. 27. The XRD of experiment 2 comes from Ref. 49. a-C: amorphous carbon; G: graphite; N: n-diamond; D: diamond; Fe: iron.

amorphous carbon merge together in the XRD pattern and some weak peaks have been covered by the background. Moreover, only the valley of ELNES of *HR*-carbon in range of 305–310 eV can explain the experimental valley between the second and third board peaks at 310 eV very well. This valley is an important key to confirm the existence of *HR*-carbon. So the ELNES of *HR*-carbon together with those of amorphous carbon, diamond, and graphite can explain the experimental ELNES very well. The comparison of theoretical and experimental Raman spectra is shown in Fig. 6. Previous experiments confirmed that the experimental sample is a mixture of n-diamond and other carbon allotropes. So it is difficult to distinguish the Raman bands of n-diamond from the experimental Raman spectrum. Because the diamond-like carbon allotropes have characteristic peaks in the range of ~1300–1400 cm^{-1} which are located in the range of Raman bands of diamond54. However, the *HR*-carbon has three independent characteristic peaks at ~1300 cm^{-1} which can match the experimentally observed three peaks at about 1350 cm^{-1}. These also confirm that *HR*-carbon is the most possible candidate structure of n-diamond.

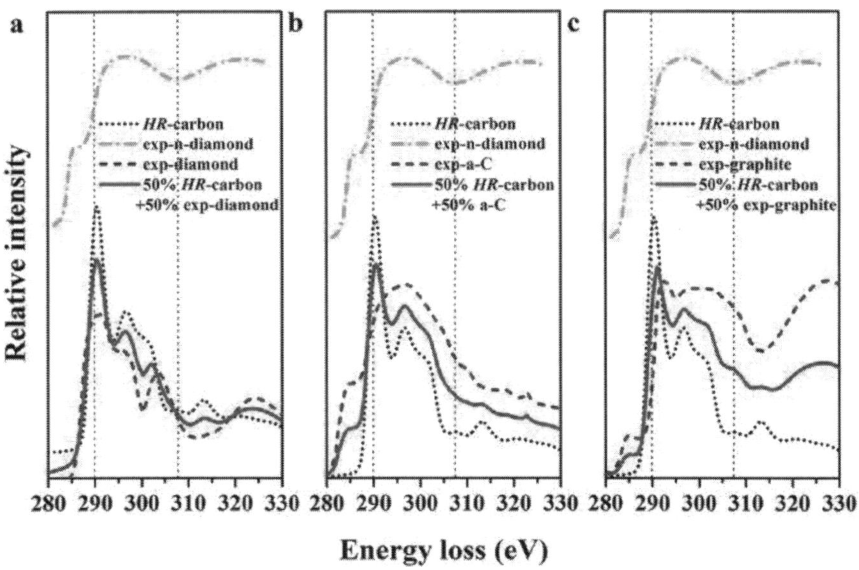

Figure 5. The comparison of theoretical and experimental ELNES. Theoretical ELNES of *HR*-carbon, experimental ELNES of diamond, graphite, n-diamond and a-C and (**a**) the fitting ELNES of the mixture of 50% *HR*-carbon + 50% diamond (**b**) the fitting ELNES of the mixture of 50% *HR*-carbon + 50% amorphous carbon. (**c**) the fitting ELNES of the mixture of 50% *HR*-carbon + 50% graphite. Experimental data come from Ref. 32.

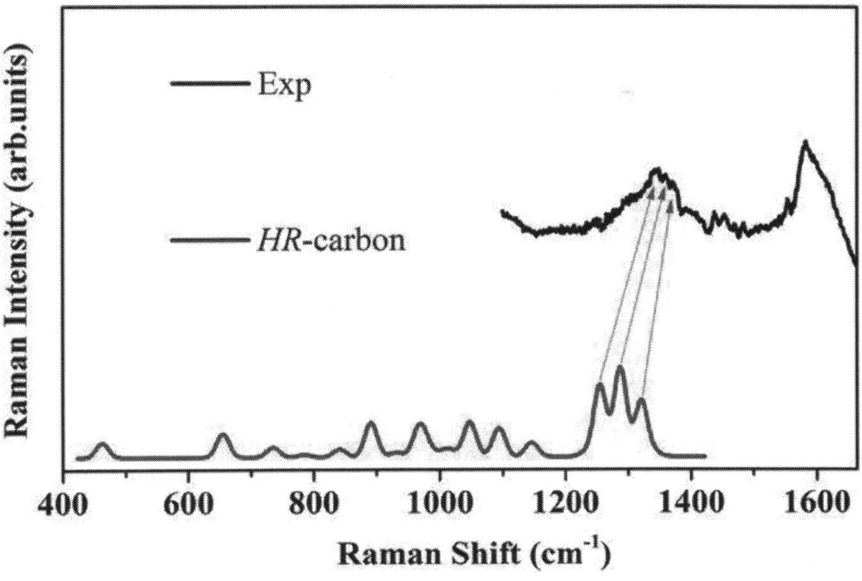

Figure 6. The comparison of our simulated Raman of *HR*-carbon and experimental data. The experimental data come from Ref. 38.

While the graphite transforms to sp^3-hybridized carbon allotropes such as diamond and cold-compressed graphite phase, the hardness of materials enhances much more. So it is worth expecting that *HR*-carbon has high hardness among compressed graphite phases. The calculated bulk modulus (B_0) of *HR*-carbonis 427 GPa which is very close to that of diamond (466 GPa). The Vickers hardness H_v of *HR*-carbon (80 GPa), estimated by the Chen's hardness model55, is much larger than that of c-BN (~62 GPa). However, it is smaller than that of diamond (~95 GPa). *HR*-carbon has very similar atomic arrangement to that of diamond except the C_3 isosceles triangle rings. So the hardness gap of*HR*-carbon and diamond could be attributed to the unique C_3 isosceles triangle rings. To confirm the effect of C_3 isosceles triangle rings for structural stability, we remove the C_3 isosceles triangle rings from the crystal structure of *HR*-carbon. This new structure was named as modulated *HR*-carbon (see supplementary Figure S1 and Table S1). We check the mechanical and dynamical properties of modulated *HR*-carbon. The bulk modulus of modulated *HR-carbon* is only 344 GPa which is much smaller than that of *HR*-carbon. So due to unique C_3isosceles triangle rings in the *C* layer of *HR*-carbon, the *HR*-carbon has stronger incompressibility. We also calculated the phonon dispersion of modulated *HR-carbon* to compare with that of *HR*-carbon Fig. 2(b). The red dashed line of the phonon spectrum is the imaginary phonon mode of modulated *HR*-carbon. The imaginary frequencies at *H* and *Γ* points indicate that the C_3 isosceles triangle rings are very important for the dynamic stability of *HR*-carbon (see Supplementary Figure S1).

In an effort to assess the effect of C_3 isosceles triangle rings for the hardness of *HR*-carbon, we calculated the Mulliken overlap population (MOP) and bond length which can give a quantitative description for the covalent bonds of *HR*-carbon as summarized in Table I. The average bond length and MOP of *HR*-carbon is 1.53 Å and 0.81, respectively, which is comparable with those of diamond (1.531 Å, 0.75). It is noteworthy that the bond length of C_3 isosceles triangle rings is 1.53 Å which is equal to the average bond length. However, the MOP of C_3isosceles triangle rings is only 0.61 which is much smaller than the average MOP of *HR*-carbon. So the bond strength of C_3 isosceles triangle rings is weaker than that of the other bonds in *HR*-carbon. The bond strength of C_3 isosceles triangle rings is not the mainly reasons for the high hardness of *HR*-carbon. However, it can be found that the*bond*3 has the smallest bond length (1.508 Å) and largest MOP (0.88), indicating the bond strength of *bond*3 is the strongest one in *HR*-carbon. It makes the C_3 isosceles triangle ring stable between Layer *A*and *B* and limits the mobility of Layer *A* and *B* along *a* and *b* directions, resulting in high hardness of *HR*-carbon.

Table 1: The bond length (*d*) and Mulliken overlap population (*MOP*) of bonds in *HR*-carbon.

Bond	MOP	d (Å)
*Bond*1 (layer A or B)	0.83	1.533
*Bond*2 (between layer A and B)	0.79	1.557
*Bond*3 (between layer A/B and C)	0.88	1.508
*Bond*4 (layer C)	0.61	1.530

CONCLUSION

In conclusion, we have found a novel diamond-like carbon allotrope,*HR*-carbon, which is a likely candidate structure for n-diamond. The simulated XRD, Raman and ELNES of *HR*-carbon can reproduce the experimental results. *HR*-carbon is a semiconductor with an incompressible atomic arrangement. It is stable over graphite above 14.2 GPa. This allotrope possesses high hardness (80 GPa) and high bulk modulus (427 GPa), which are comparable to those of diamond.

UniqueC_3 isosceles triangle rings are very important for the stability and hardness of *HR*-carbon.

METHODS

Using the USPEX method, the variable-cell structure prediction of n-diamond was performed at 10, 20, 30, 50 and 100 GPa, respectively50'51'52. The simulation cell containing 2 ~ 20 carbon atoms were selected. The structural relaxation were performed within the density functional theory, carried out within the Vienna *ab initio*simulation package (VASP)56'57. The projector augmented wave method was used58. The $2s^2 2p^2$ electrons are treated as valence electrons. The local density approximation (LDA) was employed59. The tested plane-wave cutoff energy was taken as 1100 eV. A Gamma-point-centered *k*-mesh of 7 × 7 × 2 *k*-point sampling was used for the calculations. The geometries were optimized when the remanent Hellmann-Feynman forces on the ions are less than 0.01 eV/Å. The ELNES, Raman and Mulliken population calculations were performed by CASTEP code60. The ELNES calculation was performed within the supercell-core-excited approach. A supercell containing 64 atoms was used to avoid the interaction of adjacent core holes. The phonon spectra were calculated using the direct supercell method, performed by PHONON software.

ACKNOWLEDGEMENTS

We are thankful for the financial support from the National Basic Research Program of China (No. 2011CB808200), Program for Changjiang Scholars and Innovative Research Team in University (No. IRT1132), National Natural Science Foundation of China (Nos. 51032001, 11074090, 11404134, 10979001, 51025206, 11104102, and 11204100), National Found for Fostering Talents of basic Science (No. J1103202). Parts of calculations were performed in the High Performance Computing Center (HPCC) of Jilin University.

AUTHOR CONTRIBUTIONS

T.C. initiated the project. D.L. performed the first principle calculations and prepared all figures. D.L. and T.C. analyzed the data and wrote the manuscript text. D.L., F.B.T., B.H.C., D.F.D., X.J.S., Y.Z.L., H.D.Z., N.L., B.B.L. and T.C. reviewed the manuscript.

COMPETING INTERESTS

The authors declare no competing financial interests.

REFERENCES

1. Sumiya, H. & Irifune, T. Hardness and deformation microstructures of nano-polycrystalline diamonds synthesized from various carbons under high pressure and high temperature. J. Mater. Res. 22, 2345–2351 (2007).
2. Irifune, T., Kurio, A., Sakamoto, S., Inoue, T. & Sumiya, H. Materials: Ultrahard polycrystalline diamond from graphite. Nature 421, 599–600 (2003).
3. Mao, W. L. et al. Bonding changes in compressed superhard graphite. Science 302, 425–427 (2003).
4. Boulfelfel, S. E., Oganov, A. R. & Leoni, S. Understanding the nature of "superhard graphite". Sci. Rep. 2, 00471–00479 (2012),
5. Boulfelfel, S. E., Selli, D. & Leoni, S. Novel Carbons: Habits and Oddities. Z. Anorg. Allg. Chem. 640, 681–688 (2014).
6. Baburin, I. A., Proserpio, D. M., Saleev, V. A. & Shipilova, A. V. From zeolite nets to sp3 carbon allotropes: a topology-based multiscale theoretical study. Phys. Chem. Chem. Phys. 17, 1332–1338 (2015).
7. Li, Q. et al. Superhard Monoclinic Polymorph of Carbon. Phys. Rev. Lett. 102, 175506–175509 (2009).
8. Umemoto, K., Wentzcovitch, R. M., Saito, S. & Miyake, T. Body-Centered Tetragonal C4: A Viable sp3 Carbon Allotrope. Phys. Rev. Lett. 104, 125504–125507 (2010).
9. Zhao, Z. et al. Novel Superhard Carbon: C-Centered Orthorhombic C8. Phys. Rev. Lett. 107, 215502–215506 (2011).
10. Selli, D., Baburin, I. A., Martoňák, R. & Leoni, S. Superhard sp3 carbon allotropes with odd and even ring topologies. Phys. Rev. B 84, 161411(R)–161415(R) (2011).
11. Wang, J.-T., Chen, C. & Kawazoe, Y. Low-Temperature Phase Transformation from Graphite to sp3 Orthorhombic Carbon. Phys. Rev. Lett. 106, 075501–075504 (2011).
12. Tian, F., Dong, X., Zhao, Z. S., He, J. L. & Wang, H. T. Superhard F-carbon predicted by ab initio particle-swarm optimization methodology. J Phys.: Condens. Matter 24, 165504–165509 (2012).

13. Zhao, Z. et al. Tetragonal Allotrope of Group 14 Elements. J. Am. Chem. Soc. 134, 12362–12365 (2012).

14. He, C. et al. Four superhard carbon allotropes: a first-principles study. Phys. Chem. Chem. Phys. 14, 8410–8414 (2012).

15. Amsler, M. et al. Crystal Structure of Cold Compressed Graphite. Phys. Rev. Lett. 108, 065501–065504 (2012).

16. Li, D. et al. Lowest enthalpy polymorph of cold-compressed graphite phase. Phys. Chem. Chem. Phys. 14, 4347–4350 (2012).

17. Wang, Y., Panzik, J. E., Kiefer, B. & Lee, K. K. Crystal structure of graphite under room-temperature compression and decompression. Sci. Rep. 2, 00520–00526 (2012),

18. Zhu, Q., Zeng, Q. & Oganov, A. R. Systematic search for low-enthalpy sp3 carbon allotropes using evolutionary metadynamics. Phys. Rev. B 85, 201407(R)–201412(R) (2012).

19. Niu, H. et al. Families of Superhard Crystalline Carbon Allotropes Constructed via Cold Compression of Graphite and Nanotubes. Phys. Rev. Lett. 108, 135501–135505 (2012).

20. He, C. et al. New superhard carbon phases between graphite and diamond. Solid State Commun 152, 1560–1563 (2012).

21. Wang, J.-T., Chen, C. & Kawazoe, Y. Orthorhombic carbon allotrope of compressed graphite: Ab initio calculations. Phys. Rev. B 85, 033410–033413 (2012).

22. Wang, J.-T., Chen, C. & Kawazoe, Y. Phase conversion from graphite toward a simple monoclinic sp3 -carbon allotrope. J. Chem. Phys. 137, 024502–024506 (2012).

23. Zhang, M. et al. Orthorhombic C32: a novel superhard sp3 carbon allotrope. Phys. Chem. Chem. Phys. 15, 14120–14125 (2013).

24. Jiang, X., Århammar, C., Liu, P., Zhao, J. & Ahuja, R. The R3-carbon allotrope: a pathway towards glassy carbon under high pressure. Sci. Rep. 3, 01877–01885 (2013),

25. He, C. & Zhong, J. M585, a low energy superhard monoclinic carbon phase. Solid State Commun 181, 24–27 (2014).

26. Hirai, H. & Kondo, K.-I. Modified Phases of Diamond Formed Under Shock Compression and Rapid Quenching. Science 253, 772–774 (1991).

27. Li, Y. A Reduction-Pyrolysis-Catalysis Synthesis of Diamond. Science 281, 246–247 (1998).

28. Amaratunga, G., Putnis, A., Clay, K. & Milne, W. Crystalline diamond growth in thin films deposited from a CH4/Ar rf plasma. Appl. Phys. Lett. 55, 634–635 (1989).

29. Frenklach, M. et al. Homogeneous nucleation of diamond powder in the gas phase. J. Appl. Phys. 66, 395–399 (1989).

30. Endo, S., Idani, N., Oshima, R., Takano, K. & Wakatsuki, M. X-ray diffraction and transmission-electron microscopy of natural polycrystalline graphite recovered from high pressure. Phys. Rev. B 49, 22–27 (1994).

31. Jarkov, S. M., Titarenko, Y. N. & Churilov, G. N. Electron microscopy studies of FCC carbon particles. Carbon 36, 595–597 (1998).

32. Konyashin, I. et al. A new carbon modification: 'n-diamond' or face-centred cubic carbon? Diamond Relat. Mater. 10, 99–102 (2001).

33. Orwa, J. O. et al. Diamond nanocrystals formed by direct implantation of fused silica with carbon. J. Appl. Phys. 90, 3007–3018 (2001).

34. Peng, J. L., Bursill, L. A., Jiang, B., Orwa, J. O. & Prawer, S. Growth of c-diamond, n-diamond and i-carbon nanophases in carbon-ion-implanted fused quartz. Philos. Mag. B 81, 2071–2087 (2001).

35. Wen, B. et al. Preparation of diamond nanocrystals from catalysed carbon black in a high magnetic field. J Phys.: Condens. Matter 15, 8049–8054 (2003).

36. Wen, B., Li, T., Dong, C. & Jin, J. Transformation Mechanism from Carbon Nanotubes to n-diamond. J. Mater. Res. 20, 1485–1489 (2011).

37. Yoo, C. S., Nellis, W. J., Sattler, M. L. & Musket, R. G. Diamondlike metastable carbon phases from shock-compressed C60 films. Appl. Phys. Lett. 61, 273–275 (1992).

38. Konyashin, I. et al. A new hard allotropic form of carbon: Dream or reality? Int. J. of Refract. Met. Hard Mater. 24, 17–23 (2006).

39. Dadsetani, M. & Pourghazi, A. The theoretical consistency of the experimental EELS of the fcc phase of carbon. Diamond Relat. Mater. 15, 1695–1699 (2006).

40. L. S. Palatnik, M. B. Guseva, V. G. Babaev, N. F. Savchenko & Fal'ko, I. I. γ carbon Sov. Phys. JETP 60, 520–522 (1984).

41. Pickard, C. J., Milman, V. & Winkler, B. Is there theoretical evidence for a metallic carbon polymorph with space group symmetry Fm3- m at ambient conditions? Diamond Relat. Mater. 10, 2225–2227 (2001).

42. Mailhiot, C. & McMahan, A. K. Atmospheric-pressure stability of energetic phases of carbon. Phys. Rev. B 44, 11578–11591 (1991).

43. Yin, M. & Cohen, M. Will Diamond Transform under Megabar Pressures? Phys. Rev. Lett. 50, 2006–2009 (1983).

44. Dadsetani, M., Titantah, J. T. & Lamoen, D. Ab initio calculation of the energy-loss near-edge structure of some carbon allotropes: Comparison with n-diamond. Diamond Relat. Mater. 19, 73–77 (2010).

45. Murrieta, G., Tapia, A. & de Coss, R. Structural stability of carbon in the face-centered-cubic (Fmm) phase. Carbon 42, 771–774 (2004).

46. Hirai, H., Kondo, K. i. & Sugiura, H. Possible structural models of n-diamond: A modified form of diamond. Appl. Phys. Lett. 61, 414–416 (1992).

47. Baldissin, G. & Bull, D. J. n-Diamond: Dynamical stability of proposed structures. Diamond Relat. Mater. 34, 60–64 (2013).

48. Bucknum, M. & Castro, E. On the n-diamond and i-carbon nanocrystalline forms. J Math. Chem. 50, 1034–1038 (2012).

49. Wen, B., Zhao, J., Li, T. & Dong, C. n-diamond: an intermediate state between rhombohedral graphite and diamond? New J. Phys. 8, 62 (2006),

50. Oganov, A. R., Glass, C. W. & Ono, S. High-pressure phases of CaCO3: Crystal structure prediction and experiment. Earth. Planet. Sci. Lett. 241, 95–103 (2006).

51. Oganov, A. R. & Glass, C. W. Crystal structure prediction using ab initio evolutionary techniques: Principles and applications. J. Chem. Phys. 124, 244704–244718 (2006).

52. Glass, C. W., Oganov, A. R. & Hansen, N. USPEX—Evolutionary crystal structure prediction. Comput. Phys. Commun. 175, 713–720 (2006).

53. Schwoerer-Böhning, M., Macrander, A. T. & Arms, D. A. Phonon Dispersion of Diamond Measured by Inelastic X-Ray Scattering. Phys. Rev. Lett. 80, 5572–5575 (1998).

54. Bai, Y. et al. First-principles investigation in the Raman and infrared spectra of sp3 carbon allotropes. Carbon 78, 70–78 (2014).

55. Chen, X.-Q., Niu, H., Franchini, C., Li, D. & Li, Y. Hardness of T-carbon: Density functional theory calculations. Phys. Rev. B 84, 121405(R)–121409(R) (2011).

56. Kresse, G. & Joubert, D. From ultrasoft pseudopotentials to the projector augmented-wave method. Phys. Rev. B 59, 1758–1775 (1999).

57. Kresse, G. & Furthmüller, J. Efficient iterative schemes for ab initio total-energy calculations using a plane-wave basis set. Phys. Rev. B 54, 11169–11186 (1996).

58. Blöchl, P. E. Projector augmented-wave method. Phys. Rev. B 50, 17953–17979 (1994).

59. Perdew, J. P. & Zunger, A. Self-interaction correction to density-functional approximations for many-electron systems. Phys. Rev. B 23, 5048–5079 (1981).

60. Segall, M. D. et al. First-principles simulation: ideas, illustrations and the CASTEP code. J Phys.: Condens. Matter 14, 2717–2744 (2002).

Chapter 7

Nanosecond Formation of Diamond and Lonsdaleite by Shock Compression of Graphite

D. Kraus[1] , A. Ravasio[2], M. Gauthier[2], D.O. Gericke[3], J. Vorberger[4,5,] S. Frydrych[6], J. Helfrich[6], L.B. Fletcher[2], G. Schaumann[6], B. Nagler[2], B. Barbrel[1] , B. Bachmann[7], E.J. Gamboa[2], S. Go¨de[2], E. Granados[2], G. Gregori[8], H.J. Lee[2], P. Neumayer[9], W. Schumaker[2], T. Do¨ppner[7], R.W. Falcone[1] , S.H. Glenzer[2] & M. Roth[6]

[1]*Department of Physics, University of California, Berkeley, California 94720, USA.*
[2]*SLAC National Accelerator Laboratory, Menlo Park, California 94025, USA.*
[3]*Centre for Fusion, Space and Astrophysics, Department of Physics, University of Warwick, Coventry CV4 7AL, UK.*
[4]*Max-Planck-Institut fu¨r Physik Komplexer Systeme, No¨thnitzer Strasse 38, 01187 Dresden, Germany.*
[5]*Institute of Radiation Physics, Helmholtz-Zentrum Dresden-Rossendorf, Bautzner Landstrasse 400, 01328 Dresden, Germany.*
[6]*Institut fu¨r Kernphysik, Technische Universita¨t Darmstadt, Schlossgartenstrasse 9, 64289 Darmstadt, Germany.*
[7] *Lawrence Livermore National Laboratory, Livermore, California 94550, USA.*
[8]*Department of Physics, University of Oxford, Parks Road, Oxford OX1 3PU, UK.*
[9]*GSI Helmholtzzentrum fu¨r Schwerionenforschung GmbH, Planckstrasse 1, 64291 Darmstadt, Germany. Correspondence and requests for materials should be addressed to D.K.*

ABSTRACT

The shock-induced transition from graphite to diamond has been of great scientific and technological interest since the discovery of microscopic diamonds in remnants of explosively driven graphite. Furthermore, shock synthesis of diamond and lonsdaleite, a speculative hexagonal carbon polymorph with unique hardness, is expected to happen during violent meteor impacts. Here, we show unprecedented in situ X-ray diffraction measurements of diamond formation on nanosecond timescales by shock compression of pyrolytic as well as polycrystalline graphite to pressures from 19 GPa up to 228 GPa. While we observe the transition to diamond starting at 50 GPa for both pyrolytic and polycrystalline graphite, we also record the direct formation of lonsdaleite above 170 GPa for pyrolytic samples only. Our experiment provides new insights into the processes of the shock-induced transition from graphite to diamond and uniquely resolves the dynamics that explain the main natural occurrence of the lonsdaleite crystal structure being close to meteor impact sites.

INTRODUCTION

The formation of diamond after applying pressure and heat on graphite is highly relevant to the artificial synthesis of diamond[1, 2, 3, 4], and also for a general understanding of high pressure phase transitions in environments comparable to meteor impacts or planetary interiors[5, 6]. In 1967, a hexagonal form of diamond[7], later called lonsdaleite[8], in its pure form supposedly harder than diamond[9], was identified for the first time inside fragments of the Canyon Diablo meteorite, the asteroid which created the Barringer Crater with an impact velocity of ~10 km s^{-1} generating pressures exceeding 200 GPa (ref. 5). Since then, occurrences of lonsdaleite and also nanometer-sized diamonds have been suggested to serve as marker for meteor impacts all over the world, for example, the Tunguska explosion in the early twentieth century[10], the Ries crater[11], the Younger Dryas event[12] or the impact which has been correlated to the extinction of the dinosaurs[13]. However, due to difficulties in creating a pure lonsdaleite phase under static compression conditions, serious doubts on the overall existence of lonsdaleite were published recently[2, 14]. Yet, static compression is not well suited to mimic the fast processes during impact events.

Dynamic shock compression experiments, driven by explosives[15, 16], gas guns[4, 17] or pulsed high-energy lasers[18], are much closer to such conditions despite the smaller dimensions and thus shorter timescales (nanoseconds to microseconds) compared with meteor impacts (seconds). For this reason, laboratory shock experiments are limited to detect phase transitions happening on corresponding timescales and cannot explore some complex structural changes, which are thought to require nucleation times up to seconds[19]. The shock-induced transition from graphite to diamond is thought to be a fast martensitic transition[4], but up to now, shock compression experiments on graphite have lacked a direct *in situ* measurement of the microscopic crystal structure. In fact, all conclusions regarding structural transitions were based on studies from the material that was recovered after applying the shock drive or dynamic measurement of macroscopic quantities, such as density and pressure. As a consequence, there is a speculation in understanding the dynamic processes for shock synthesis of diamond and lonsdaleite[20, 21, 22, 23,24, 25, 26, 27] and conclusions regarding meteor impacts[2, 5]. Particularly, the different concepts involve nucleation[24], plane-sliding and buckling[22], wave-buckling and slipping[27] or the intermediate formation of so-called layered diamond[20], which is a mixture of hexagonal and cubic diamond[23]. These models give very different results under which conditions, or if at all, diamond or lonsdaleite are formed from graphite. Moreover, while a fast martensitic process on nanosecond timescale has been accepted for compressing highly oriented graphite samples along the *c* axis[4,17], much slower diffusion-based mechanisms are thought to be more effective for highly defective graphite samples[28].

In the following, we address these questions in an unprecedented way by applying ultrafast X-ray diffraction for *in situ* structure measurements of the shock-induced transition from graphite to diamond. Our results show nanosecond formation of the diamond structure after shock compressing pyrolytic as well as porous polycrystalline graphite samples. At the highest pressures, we find evidence for a relatively pure lonsdaleite structure when compressing pyrolytic graphite, whereas the porous polycrystalline samples are transferred to a complex liquid in this regime.

RESULTS

Experimental method

The experiments reported here were performed at the Matter at Extreme Conditions endstation of the Linac Coherent Light Source of Stanford National Accelerator Laboratory[29]. Here, the newly available combination of high-energy lasers ($\gtrsim$10 J per pulse) with an X-ray free electron laser allows for femtosecond X-ray diffraction to directly access the kinetics of the shock-induced transition. A schematic of the experiment is shown in Fig. 1.

The recorded diffraction angle ϑ is directly connected to the lattice spacing d via the Bragg relation

$$\sin(\theta) = \frac{n\lambda}{2d}, \qquad (1)$$

where n is the diffraction order, and λ the wavelength of the incident light source. For the symmetric cubic diamond crystal structure, the lattice spacing given by the (111) diffraction peak directly determines the density $\rho \propto d^{-3}$. Knowing the density of the shocked state ρ_1 as well as the shock transit time,

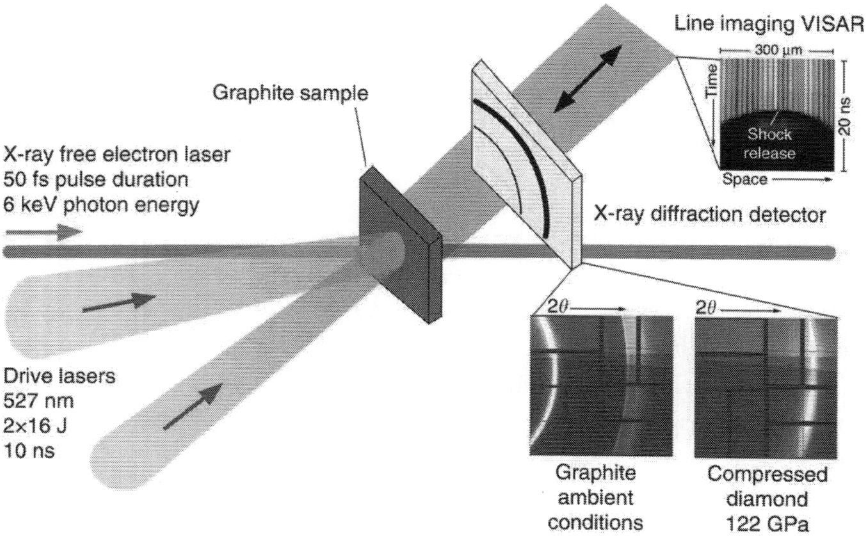

Figure 1. Schematic of the experimental setup at the Matter at Extreme Conditions endstation of the Linac Coherent Light Source. Two high-energy laser beams drive shock waves into graphite samples generating pressures from 20 to 230 GPa. The shock transit times of few nanoseconds are recorded by a VISAR system, which detects the shock-induced reflectivity drop of a 100-nm thick aluminum coating when the shock exits on the target rear side. The microscopic state is probed by a single X-ray pulse with 6 keV photon energy and 50 fs pulse duration. X-ray diffraction is recorded by a large area X-ray detector.

and thus the shock velocity v_S, allows for inferring the pressure p using the Rankine–Hugoniot relations, which simply apply mass, momentum and energy conservation[30]

$$p = v_S^2 \rho_0 \frac{\rho_1 - \rho_0}{\rho_1}. \tag{2}$$

Here, ρ_0 is the initial density of the graphite sample.

Pyrolytic graphite

Figure 2 shows X-ray diffraction results for oriented pyrolytic graphite (ρ_0=2.21 g cm^{-3}) driven parallel to the graphite c axis with different drive pressures. For ~20 GPa, we observe the (002) Bragg reflection shifting to larger diffraction angles, thus providing a signature of compressed graphite. For pressures above ~55 GPa, the graphite (002) signature has completely vanished, whereas a broad peak at ~60 degrees appears. This new feature is in agreement with the (111) Bragg reflection of diamond. The large width of the diffraction peak at this pressure most probably documents remaining disorder (that is, very small diamond crystallites) and thus implies that the transition from graphite to diamond is not fully completed within the shock transit time of ~9 ns. Higher drive pressures result in faster transition times, reducing the width of the (111) diamond diffraction peak due to less disorder, that is, creation of larger diamond crystallites. For drive pressures above 100 GPa, we observe the formation of diamond on timescales less than 1 ns (inset in Fig. 2d). For these pyrolytic samples at shock pressures higher than 170 GPa, we see two diffraction ring fragments emerging that are clearly caused by different crystal planes showing some degree of texture (see Fig. 3), conserving the preferred crystallite orientation of the pyrolytic samples (see Supplementary Figs 1 and 2). The inferred spacing is compatible with diffraction from the (010) and (002) planes of lonsdaleite. For this structure, we obtain an average <u>*c/a*</u> lattice constant ratio of ~1.69, which is slightly above the ideal value of $\sqrt{8/3}$ for a wurtzite lattice with equidistant atoms. This reasonably follows the trend given by reported values for ambient conditions[8] (*c/a*=1.64) and simulations for 20 GPa (ref. 23; *c/a*=1.66). Density functional theory coupled with molecular dynamics (DFT-MD) simulations of the lonsdaleite structure at 4.4 g $cm^{-3}$and 6,000 K, resulting in a pressure of 205 GPa, show a very good agreement with our diffraction measurements.

The observed diffraction pattern is only compatible with lonsdaleite and not with any other structure that has been predicted for this pressure regime. Faulted and twinned cubic diamond, as shown by Nemeth *et al.*[2], would result in a cubic diamond main peak with small side peaks due to the defects. Furthermore, a mix of cubic and hexagonal diamond, the so-called layered diamond struture, would not result in the observed clear doublet of peaks[20]. Other more complicated polymorphs show many more diffraction maxima[23] and can certainly be excluded by our measurements. Hexagonal lonsdaleite being favoured at higher drive pressures is conceivable since the direct transition path to cubic diamond requires more lattice distortions[22, 24, 26]. Recent MD calculations[27] demonstrate that plane sliding of the graphite basal planes to rhombohedral graphite, which in turn would lead to the formation of diamond, is prevented by the emergence of few interplanar sp^3 bonds, when the spacing between the graphite basal planes falls below a certain distance. For this case, these simulations predict the formation of lonsdaleite[27]. Our experiments indicate that this limit is reached by dynamic shock compression of pyrolytic graphite above 170 GPa.

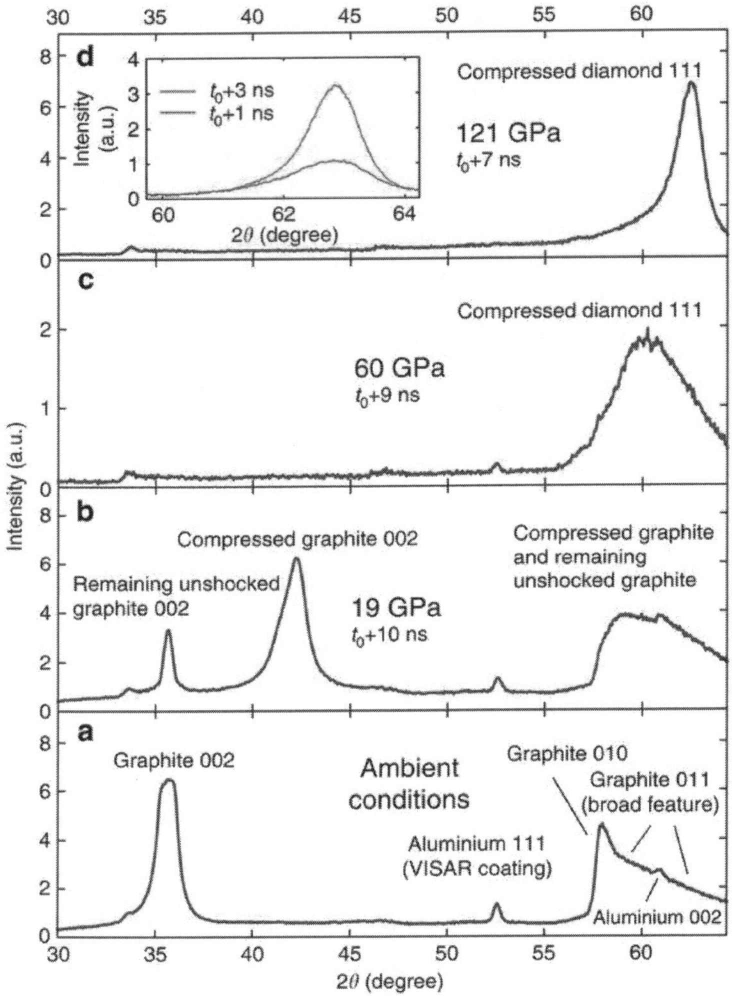

Figure 2. Transition from pyrolytic graphite to diamond.X-ray diffraction data of (**a**) cold and (**b–d**) compressed pyrolytic graphite samples, driven parallel to the graphite *c* axis; t_0 is denoting the start of the drive laser pulse. For 19 GPa, diffraction shows graphite, which is mainly compressed along the *c* axis, together with some cold material from the rear side of the sample not reached by the shock. For 60 GPa at exactly the moment when VISAR records the shock having traversed the whole sample, every signature from graphite vanishes. In this case, only a broad cubic diamond (111) diffraction peak remains. For higher pressures (for example, 122 GPa), the width of the peak decreases due to the reduced transition time and we observe the formation of a sharp Bragg reflection within 1 ns.

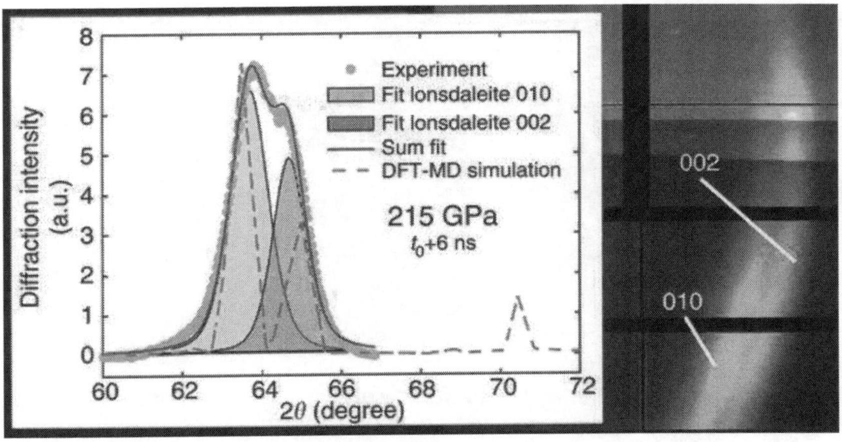

Figure 3. Diffraction from lonsdaleite. When compressing pyrolytic graphite to pressures above 170 GPa, we observe two strong diffraction peaks consistent with lonsdaleite (010) and (002) planes. The grey-dashed line shows a DFT-MD simulation of the lonsdaleite phase at 4.4 g cm^{-3} and 6,000 K, resulting in a pressure of 205 GPa, which is in qualitative agreement with the observed diffraction pattern. As the range of detectable diffraction angles ends at 2ϑ=68 degrees in our experiments, the (011) peak at 2ϑ=70.5 degrees is not recorded. The occurrence of partial diffraction rings shows a preferred orientation of the crystallites that is compatible with the preferred orientation of the initial pyrolytic graphite sample.

Figure 4 combines the recorded microscopic crystal structure with the macroscopic quantities density and pressure. Applying the simple relation between Bragg angle and lattice spacing, and thus density, in combination with equation (2), we can create an experimental pressure–density diagram, giving a line of states achievable by shock compression, the shock Hugoniot curve (all experimental values including error estimates can be found in Supplementary Tables 1 and 2). Classical Hugoniot measurements usually determine only macroscopic quantities like density and pressure via velocity measurements. As such, the shock Hugoniot is one of few experimentally accessible quantities capable of benchmarking high-pressure equation of state theories, from which, for example, the temperature of the created states can be deduced. Only based on these models, it is then possible to draw conclusions on microscopic properties. In contrast, our measurements directly record the microscopic structure of the sample in addition to determining density and pressure. Thus, they give a more complete picture joining both microscopic and macroscopic observations to an absolute equation of state measurement.

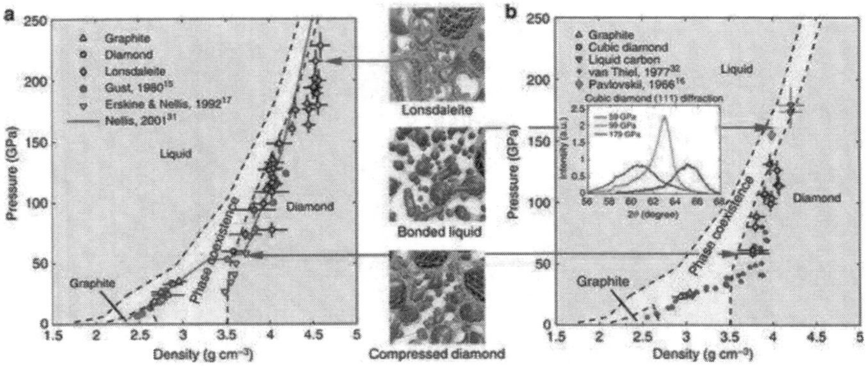

Figure 4: Summary of the experimental results. (**a**) Recorded pressure–density diagram for compressed pyrolytic graphite (ρ_0=2.21 g cm^{-3}) compared with literature data without structure information and a first principles phase diagram[33]. At lower pressures, there is very good agreement, whereas at higher pressures, due to the formation of lonsdaleite, we observe higher densities than predicted by a proposed shock Hugoniot[31], which instead suggests a transition to the liquid. (**b**) Recorded pressure–density diagram for porous polycrystalline graphite (ρ_0=1.84 g cm^{-3}) compared with experiments without structure information[16, 32] and a first principles phase diagram. Comparable to the pyrolytic samples, the diamond formation is not fully completed within few nanoseconds at lower pressures, resulting in a broad peak which sharpens up to ~100 GPa. At higher pressures, however, the increasing temperature leads to melting of the diamond structure, resulting in broader and fainter diffraction peaks in agreement with a bonded liquid[18]. No signature of lonsdaleite is observed when compressing porous graphite.

For compressed pyrolytic graphite, our results connect and overlap with established literature data at lower pressures[15, 16, 17, 31, 32] as well as first principles simulations of the carbon phase diagram[33] (see Fig. 4a). At higher pressures, our data extend to a regime which has not been well characterized before. For states which remain in the graphite phase, it needs to be considered that graphite mainly compresses along the *c* axis of the unit cell[34]. The (002) diffraction peak position directly determines the lattice spacing d_c along the *c* axis and using the relation $\rho \propto d_c^{1.14}$ from static compression experiments[35], we can infer the density of the compressed graphite states. The rearrangement in the microscopic structure, that is, the graphite–diamond transition, is clearly visible due to a density jump at the transition pressure of ~50 GPa and a lower compressibility of the diamond states. We do not observe a two-wave structure as, for example, Erskine and Nellis[17] did for this transition after shock compression of pyrolytic graphite samples with a higher grade of orientation. These experiments achieved the transition already at 27 GPa, while the two-wave structure was overdriven at 50 GPa. In agreement with our experiments, Erskine and Nellis did not record a two-wave structure for graphite samples with lower grade of orientation. At higher pressures, we observe a second smaller jump in density at ~170 GPa, pointing at the transition to lonsdaleite consistent with a slightly increased energy that is required for the structural transition compared with diamond[36]. In turn, we obtain higher densities than predicted by a proposed shock Hugoniot[31] that does not take the formation of lonsdaleite into account in this regime. However, this phase exhibits even higher stiffness, supporting predictions that the hardness of cubic diamond is exceeded by its hexagonal polymorph[9] and possibly leading to a re-connection with the proposed Hugoniot line in the liquid phase. Using a very recent multi-phase equation of state model for carbon[37], we expect the temperatures of the states, which remain in the graphite phase in the order of 4,000 K, 4,000–6,000 K for the diamond data and 6,000–7,000 K for the measurements, which show the lonsdaleite structure.

Porous polycrystalline graphite

For porous polycrystalline samples, we obtain slightly different results (see Fig. 4b). When compressing porous samples, the voids close instantly nearly without resistance, resulting in an entropy increase. Further compression then leads to states of higher entropy and thus temperature compared with shock compression of the higher density pyrolytic samples to the same pressure[30]. For porous polycrystalline graphite, we record nanosecond formation of diamond with a transition threshold at slightly lower pressure compared with the pyrolytic samples. The width of the diffraction peak is marginally larger than for pyrolytic samples. This indicates that only slightly smaller diamond crystallites are formed for the polycrystalline samples. Starting from ~100 GPa, the diffraction again broadens and weakens due to melting and, instead of reaching the hexagonal phase, a bonded liquid is formed, which is in good agreement with the first principles phase diagram. It has been shown that short-time bonding remains active in liquid carbon, probably up to temperatures of more than 20,000 K (ref. 38). As the liquid structure is diamond-like, the diffraction signatures do not immediately vanish or shift above the melting line. Again using a recent multi-phase equation of state model[37], we expect the temperatures of the states which remain in the graphite phase in the order of 4,000–5,000 K, 5,000–8,000 K for the diamond data and temperatures exceeding 8,000 K for the data points, which show the transition to the liquid.

DISCUSSION

The data of dynamic diamond formation presented here resolve controversies on the shock-induced graphite to diamond transition: a fast transition from graphite to diamond on nanosecond timescale does not require highly oriented graphite samples and shock loading perpendicular to the *c* axis. Furthermore, recording the first clear diffraction signal of the meta-stable hexagonal diamond polymorph, that is not obscured by dominant cubic diamond or graphite, demonstrates that this state exists. Finally, observing lonsdaleite only for pyrolytic graphite samples compressed along the *c* axis to very high pressures suggests that sliding of the graphite basal planes and thus diamond formation can be prevented under these conditions. Hence, the picture that lonsdaleite synthesis is highly difficult applying static pressure but can be formed in violent impact events is strongly supported.

METHODS

Laser-driven shock compression

The graphite samples (pyrolytic graphite, GoodFellow C000720, ρ_0=2.21 g cm^{-3}, mosaicity ~10 degrees and porous polycrystalline graphite, SGL Carbon R6650, ρ_0=1.84 g cm^{-3}, 10% volume porosity, 5 μm average grain size) were compressed using two pulsed high-energy lasers (527 nm, up to 16 J per beam, 10 ns pulse duration, 150–200 μm focal spot diameter, smoothed with random phase plates). The samples were laser-cut out of larger graphite pieces and then polished to achieve a surface roughness of 0.1 μm. The sample thickness was varied between 60 and 100 μm and had individually been characterized for each sample with an accuracy of ~0.5 μm. The shock transit time was recorded with a VISAR system, which monitored the drop in reflectivity of a 100 nm aluminium coating on the sample rear side at the moment of shock release. The VISAR system at MEC has two streak cameras and the transit time was determined as the average of the two cameras. The arrival time of the drive laser was determined by the same streak cameras as used for the VISAR system by scattering light of the drive laser beams from the exact position of sample interaction into the VISAR system. Typical transit times varied between 6 and 12 ns, depending on sample thickness and drive pressure. Recording transit times for different sample thickness and using the spatial resolution of the VISAR, sufficient shock steadiness as well as shock planarity could be guaranteed for the region that was irradiated by the X-ray beam. This was confirmed by measurements using graphite samples with a LiF window on the rear side, where a very steady interface velocity could be monitored. Using this interface velocity and the shock velocity in graphite, as well as assuming the LiF equation of state to be well known, density and pressure in the shocked graphite sample can be inferred in principle. These results are in agreement with the X-ray diffraction method, but by far not as accurate for our experiments (particularly in terms of density) and cannot provide additional insights.

Ultrafast X-ray diffraction

The graphite samples were probed using the LCLS XFEL beam in SASE mode (5.9–6.1 keV photon energy, 0.3% spectral bandwidth, 50 fs pulse duration, 20 μm spot size, ~3 mJ per pulse corresponding to 3×10^{12} photons per pulse). The input spectrum was recorded for each run with a high-resolution silicon crystal spectrometer. The timing of the X-ray pulse relative to the drive laser was chosen to probe the sample slightly before or exactly at the moment when the shock wave had traversed the whole sample. X-ray diffraction was measured by an $8 \times 8\ cm^2$ Cornell-Stanford Pixel Array detector (CSPAD) at 11 cm distance to the sample covering diffraction angles from 27 to 68 degrees on top of the XFEL beam axis. In this way, interfering effects due to the horizontal polarization of the X-ray beam could be avoided. Angular calibration for the recorded 2ϑ range of 20–68 degrees was obtained by diffraction of a LaB_6 powder sample.

ACKNOWLEDGEMENTS

Use of the Linac Coherent Light Source (LCLS), SLAC National Accelerator Laboratory, is supported by the U.S. Department of Energy, Office of Science, Office of Basic Energy Sciences under Contract No. DE-AC02-76SF00515. This work was performed at the Matter at Extreme Conditions (MEC) instrument of LCLS, supported by the US Department of Energy Office of Science, Fusion Energy Science under contract No. SF00515. D.K., B. Barbrel and R.W.F. acknowledge support by the US Department of Energy, Office of Science, Office of Fusion Energy Sciences and by the National Nuclear Security Administration under Award Numbers DE-FG52-10NA29649 and DENA0001859. SLAC HED is supported by DOE Office of Science, Fusion Energy Science under FWP 100182. S.F., J.H. and M.R. were supported by German Bundesministerium fu¨r Bildung und Forschung project Nos. 05P12RDFA1 and 06DA9043I. The work of B. Bachmann and T.D. was performed under the auspices of the US Department of Energy by Lawrence Livermore National Laboratory under Contract No. DE-AC52-07NA27344.

AUTHOR CONTRIBUTIONS

D.K., R.W.F., B.N., H.J.L., T.D., S.H.G., G.S., D.O.G., J.V., G.G., P.N. and M.R. were involved in the project planning. D.K., A.R., M.G., S.F., J.H., L.B.F., B.N., B. Barbrel, B. Bachmann, E.J.G., S.G., E.G., H.J.L., W.S. and T.D. carried out the experiment. G.S., J.H., S.F., M.R. and D.K. designed and built the samples. Experimental data were analysed and discussed by D.K., S.H.G., A.R., M.G., D.O.G., J.V. and T.D. The manuscript was written by D.K., S.H.G. and D.O.G

REFERENCES

1. DeCarli, P. S. & Jamieson, J. C. Formation of diamond by explosive shock. Science 133, 1821–1822 (1961).
2. Nemeth, P. et al. Lonsdaleite is faulted and twinned cubic diamond and does not exist as a discrete material. Nat. Commun. 5, 5447 (2014).
3. Bovenkerk, H. P., Bundy, F. P., Hall, H. T., Strong, H. M. & Wentorf, R. H. Preparation of diamond. Nature 184, 1094–1098 (1959).
4. Erskine, D. J. & Nellis, W. J. Shock-induced martensitic phase transformation of oriented graphite to diamond. Nature 349, 317–319 (1991).
5. Hanneman, R. E., Strong, H. M. & Bundy, F. P. Hexagonal diamonds in meteorites: implications. Science 166, 995–997 (1967).
6. Ross, M. The ice layer in Uranus and Neptune-diamonds in the sky? Nature 292, 435–436 (1981).
7. Bundy, F. P. & Casper, J. S. Hexagonal diamond-a new form of carbon. J. Chem. Phys. 46, 3437 (1967).
8. Frondel, C. & Marvin, U. B. Lonsdaleite, a hexagonal polymorph of diamond. Nature 217, 587–589 (1967).
9. Pan, Z., Sun, H., Lang, Y. & Chen, C. Harder than diamond: superior indentation strength of wurtzite BN and lonsdaleite. Phys. Rev. Lett. 102, 055503 (2009).
10. Kvasnytsya, V. et al. New evidence of meteoritic origin of the Tunguska cosmic body. Planet. Space Sci. 84, 131–140 (2013).
11. Hough, R. M. et al. Diamond and silicon carbide in impact melt rock from the Ries impact crater. Nature 378, 41–44 (1995).

12. Kennet, D. J. et al. Shock-synthesized hexagonal diamonds in Younger Dryas boundary sediments. Proc. Natl Acad. Sci. USA 106, 12623–12628 (2009).

13. Carlisle, D. B. & Braman, D. R. Nanometre-size diamonds in the Cretaceous/ Tertiary boundary clay of Alberta. Nature 352, 708–709 (1991).

14. Salzmann, C. G., Murray, B. J. & Shephard, J. J. Extent of stacking disorder in diamond. Diamond Relat. Mater. 59, 69–72 (2015).

15. Gust, W. H. Phase transition and shock-compression parameters to 120 GPa for three types of graphite and for amorphous carbon. Phys. Rev. B 22, 4744–4756 (1980).

16. Pavlovskii, B. J. & Drakin, V. P. Concerning the metallic phase of carbon. JETP Lett. 4, 116 (1966).

17. Erskine, D. J. & Nellis, W. J. Shock-induced martensitic phase transformation of highly oriented graphite to diamond. J. Appl. Phys. 71, 4882–4886 (1992).

18. Kraus, D. et al. Probing the complex ion structure in liquid carbon at 100 GPa. Phys. Rev. Lett. 111, 255501 (2013).

19. Ohtani, E. et al. Formation of high-pressure minerals in shocked L6 chondrite Yamato 791384: constraints on shock conditions and parent body size. Earth Planet. Sci. Lett. 227, 505–515 (2004).

20. Mundy, C. J. et al. Ultrafast transformation of graphite to diamond: an ab initio study of graphite under shock compression. J. Chem. Phys. 128, 184701 (2008).

21. Winey, J. M. & Gupta, J. M. Shock-compressed graphite to diamond transformation on nanosecond time scales. Phys. Rev. B 87, 174104 (2013).

22. Scandolo, S., Bernasconi, M., Chiarotti, G. L., Focher, P. & Tosatti, E. Pressureinduced transformation path of graphite to diamond. Phys. Rev. Lett. 74, 4015–4018 (1995).

23. Ribeiro, F. J., Tangney, P., Louie, S. G. & Cohen, M. L. Structural and electronic properties of carbon in hybrid diamond-graphite structures. Phys. Rev. B 72, 214109 (2005).

24. Khaliullin, R. Z., Eshet, H., Ku¨hne, T. D., Behler, J. & Parrinello, M. Nucleation mechanism for the direct graphite-to-diamond phase transition. Nat. Mater. 10, 693–697 (2011).

25. Kurdyumov, A. V., Britun, V. F., Yarosh, V. V., Danilenko, A. I. & Zelyavskii, V. B. The influence of the shock compression conditions on

the graphite transformations into lonsdaleite and diamond. J. Superhard Mater. 34, 19–27 (2012).

26. Pineau, N. Molecular dynamics simulations of shock compressed graphite. J. Phys. Chem. C 117, 12778–12786 (2013).

27. Xie, H., Yin, F., Yu, T., Wang, J.-T. & Liang, C. Mechanism for direct graphiteto-diamond phase transition. Sci. Rep. 4, 5930 (2014).

28. Hirai, H., Kukino, S. & Kondo, K. Predominant parameters in the shock induced transition from graphite to diamond. J. Appl. Phys. 78, 3052–3059 (1995).

29. Gauthier, M. et al. New experimental platform to study high density laser-compressed matter. Rev. Sci. Instrum. 85, 1E616 (2014).

30. Zeldovic, Y. B. & Raizer, Y. P. Physics of Shock Waves and High-Temperature Hydrodynamic Phenomena (Academic Press, New York and London, 1966).

31. Nellis, W. J., Mitchell, A. C. & McMahan, A. K. Carbon at pressures in the range 0.1-1 TPa (10 Mbar). J. Appl. Phys. 90, 696–698 (2001).

32. van Thiel, M. Compendium of shock wave data. Lawrence Livermore Laboratory Report UCRL-50108 (1977).

33. Wang, X., Scandolo, S. & Car, R. Carbon phase diagram from ab initio molecular dynamics. Phys. Rev. Lett. 95, 185701 (2005).

34. Nakayama, A. et al. Compression of polyhedral graphite up to 43 GPa and x-ray diffraction study on elasticity and stability of the graphite phase. Appl. Phys. Lett. 84, 5112–5114 (2004).

35. Zhao, Y. X. & Spain, I. L. X-ray diffraction data for graphite to 20 GPa. Phys. Rev. B 40, 993–997 (1989).

36. Wu, B. R. & Xu, J. Total energy calculations of the lattice properties of cubic and hexagonal diamond. Phys. Rev. B 57, 13355–13358 (1998).

37. Benedict, L. X. et al. Multiphase equation of state for carbon addressing high pressures and temperatures. Phys. Rev. B 89, 224109 (2014).

38. Kraus, D. et al. The complex ion structure of warm dense carbon measured by spectrally resolved x-ray scattering. Phys. Plasmas 22, 056307 (2015).

Chapter 8

Mechanical Characterization of Black Diamond (Low-k) Structures for 3D Integrated Circuit and Packaging Applications

Vasarla Nagendra Sekhar

*[1] Institute of Microelectronics, A*STAR (Agency for Science, Technology and Research), Science Park II, Singapore*

1. INTRODUCTION

As Integrated Circuit (IC) technology scales down into the nanotechnology regime, it allows millions of active components to be fabricated on a single chip in accordance with the historical trend of Moore's law. Integration of all these active elements on a single IC, multilevel on-chip interconnect system must be developed in BEOL (Back End of the Line) technology [1]. Accordingly, to meet the industry requirements of scaling for improved performance, semiconductor technologies are forced to move from well-established Al/SiO_2 interconnect technology to Cu/low-k (Copper/low dielectric constant) technology [2-3]. The main objective of this transition is to reduce cross talk noise between metal lines, propagation delays and power dissipation from RC delay. Copper and low-k inter metal dielectric layers are used as multilevel interconnects to enhance the speed of logic devices. Amongst the available low-k materials, Black diamond ™ (BD, low-k, SiOC:H) has been considered as potential inter metal dielectric material for integration in ULSI (Ultra Large Scale Integration) due to its better electrical and dielectric properties. But in dielectric material processing the key issue is the trade-off between dielectric property and mechanical strength. Hence it is very important to study the mechanical properties of BD films and Cu/BD stacks.

There are four interdependent properties which are responsible for the mechanical reliability of thin film structures and they are elastic modulus, hardness, fracture toughness and interfacial adhesion [4]. Both theoretical and experimental studies recommend that hardness and elastic modulus are the two key material characteristics affecting the CMP process [5]. Generally, mechanical properties of thin films differ from those of the bulk materials. This is mainly attributed to their microstructure, interfacial mismatch stresses and molecular restructuring of the films. Parameters used to characterize the mechanical strength of BEOL interconnect materials include hardness, elastic modulus and adhesion/cohesion strength. Interfacial adhesion is the most important property to ensure the thermo-mechanical integrity of Cu/low-k stacks. Interfacial adhesive failure may occur during fabrication processes such as CMP and high temperature curing steps. In addition, delamination or cracking can also be observed during electronic packaging processes like reflow process, flip-chip bonding and back grinding due to thermo-mechanical stresses. This chapter further discusses on mechanical behavior of BD thin films of different thicknesses, single and dual dielectric stacks with Ta & TaN barriers. Nanoindentation experiments with continuous stiffness measurement (CSM) attachment have been performed on different BD film stacks to assess the multi-layer effect on the mechanical behavior of BD films. Effect of wafer backgrinding on the active side of the low-k structures have been studied using the Nanoindentation method as wafer backgrinding is one of the key technologies paving way to high performance three-dimensional (3D) microelectronic packages.

2. LOW-K DIELECTRIC MATERIALS FOR BEOL (BACK END OF THE LINE) MULTILEVEL INTERCONNECT AND 3D INTEGRATION APPLICATIONS

The contribution of the R-C delay by conventional multilevel interconnects scheme increases as the IC fabrication technology moving into further miniaturization. The introduction of Cu/low-k interconnect technology into BEOL, has progressively enhanced this condition when compared to the conventional Al/SiO_2 technology by reducing R-C delay in between interconnect lines. [6-7]. In addition to migrating to Cu/low-k multilevel interconnects at chip level, microelectronic industry is more focused on 3D IC integration and 3D packaging of chips at system level. In recent times, three-dimensional (3D) IC integration and packaging is gaining more attention because of its innovativeness, high performance, high functionality, and ability to reduce size of the final product. The major applications of the 3-D packaging include digital and mixed-signal electronics, wireless, electro-optical, MEMS and other integration technologies. At this juncture, the key technologies supporting the 3-D packaging, are as: through silicon vias (TSVs), wafer thinning or back grinding, precision alignment of wafer to wafer or chip to wafer, and wafer to wafer or chip to wafer bonding [8]. Among these key technologies, wafer thinning plays a vital role in the 3D packaging integration, as it allows to accommodate or stack more dies in one package and ultimately results in the reduction of package size. Besides the reduction of package size, the stacking of thin chips provides many other advantages, such as, more functionality per package and improved heat dissipation. Electronic packaging industry has to put a lot of R&D efforts and spend millions of dollars on the wafer thinning technologies as there is no manufacturing technology available for directly producing the ultra-thin wafers [9]. That is why in the past years, many wafer thinning methods, such as mechanical grinding, chemical mechanical polishing (CMP), wet etching and atmospheric downstream plasma (ADP), dry chemical etching (DCE) have been evolved. So far many researchers have extensively studied wafer thinning/back grinding processes in terms of die strength by assessing the quality of the grinded surface [10].

In the recent years, many researchers have extensively assessed the quality of back grinding process with the help of die strength evaluation. Die strength of the thinned wafer can be evaluated by using the different mechanical testing techniques such as three point bend test, four point bend test, ball on ring test, ball breaker tests and ring-on-ring tests. The mechanical tests are greatly influenced by several processes and material parameters such as surface roughness or finish, degree of thinning, stress relief process, quality of the dicing edge [11-15]. However the literature available related to the effect of grinding processes on the active side of the die/chip is limited, and this necessitates a focused study on the effect of wafer back grinding on the active side of chip. For the first time we have studied the effect of back grinding processes on the active side of low-k stack by using Nanoindentation method. Usually active side of the chip is a few microns in thickness, and it cannot be studied using those methods which are being used for conventional die strength evaluation. Therefore, we have chosen sophisticated method like nanoindentation technique for mechanical characterization of low-k films.

2.1. Different Types Of Low-K Materials

In the recent past, many low-k materials have been developed, and they can be broadly classified into Si-based and non-Si based. Si-based materials can be further classified into Si-based and silsesquioxane (SSQ) based, which include hydrogen-SSQ (HSSQ) and methyl-SSQ (MSSQ). Non-Si based low-k materials can be further divided into two groups, polymer based and amorphous carbon. Several types of low-k materials with varied low-k values and different deposition methods are being used in the IC fabrication technology. Table 1 lists the contemporary deposition techniques together with the k value of various silicon based, non-silicon based and polymer dielectric material candidates for the 0.13µm and 0.1µm technology nodes.

In silica based dielectric materials, usually tetrahedral silica is the elementary unit. Each silicon atom is at the center of the tetrahedron of oxygen atoms as shown in Fig. 1 (a). Typically Si based dielectric materials are dense structures with higher chemical and thermal stability. In silica based materials, dielectric constant (k) value can be lowered by replacing the Si-O bonds with Si-F bond, producing fluorinated silica glasses (FSG) or doping with C by introducing CH_3 groups [7].

Black diamond (BD) is one of the popular low-k materials and it is a trade mark of CVD processed dielectric material, introduced by Applied Materials Inc. [17]. It is silica based dielectric material, obtained by doping of silica with -CH_3 groups as shown in figure Fig. 1 (b) and it has chemical formula SiOC:H. It is also called hybrid dielectric material as it contains both organic (-CH_3) and inorganic (Si-O) constituents. The dielectric constant (k) value of silica based dielectrics ranges from 2.6 to 3. Typically BD thin films are fabricated by using the Chemical Vapor Deposition (CVD) method near room temperature using organosilane precursor in the presence of oxygen as oxidant. The lower density of the BD films is achieved by introducing network terminating species (-CH_3) into the Si-O matrix [19]. The density and dielectric value of the BD films can be altered by the selection of terminating groups in silica network. Empirically, a larger terminating group gives lower density, because it acts as a network terminating group only and it is not part of the Si-O network. Therefore BD films retain many of the useful thermo-mechanical properties of silicon oxide. The summary of properties of BD films is given in Table 2. BD films can achieve bulk dielectric constant of around 2.5 to 2.7, and integrated ILD stack dielectric constant of <3 [20]. The glass transition temperature of the BD is well above 450°C. The dielectric constant of the BD films can be lowered mainly by introducing constitutive porosity into the microstructure [21]. By proper selection of the compatible barriers layers, thermal and mechanical properties of BD films provide evolutionary pathway to Cu/low-k interconnect technology.

2.2. Required Properties Of Low-K Materials For Integration

For successful integration of low-k materials into the BEOL interconnects, besides having low dielectric constant it should be chemically, mechanically and thermally stable in the system. Every IC fabrication node technology demands low-k materials with lower dielectric values and with optimum physical properties as summarized in Table 3. Choosing a new low-k material with optimal electrical, thermal and mechanical properties for current interconnects and IC fabrication technology is very crucial. As mentioned earlier, lower dielectric constants are obtained by modifying of the molecular structure of the materials, which ultimately affects the mechanical and thermal properties of the low-k materials. Good thermal stability and low coefficient of thermal expansion is needed to prevent both, damage to the film and, property changes during subsequent thermal processing. The bulk dielectric constant of the ILD stack, when low-k film is stacked with barrier layers and liners (SiC and SiN) should be less than 3.0.

Table 1. Dielectric constants of various contemporary low-k materials are the interest of 0.13μm and 0.1μm technology nodes [7, 16-18].

Dielectric Materials	Fabrication Technique	K value
SiO_2	CVD	3.9 – 4.5
Fluorosilicate glass (FSG)	CVD	3.2 – 4.0
Polyimides	Spin-on	3.1 – 3.4
Hydrogen silsesquioxane(HSQ)	Spin-on	2.9 – 3.2
Diamond-like Carbon (DLC)	CVD	2.7 – 3.4
Black Diamond™ (SiCOH)	CVD	2.7 – 3.3
Parylene-N	CVD	2.7
B-staged Polymers (CYCLOTENE™ and SiLK™)	Spin-on	2.6 – 2.7
Fluorinated Polyimides	Spin-on	2.5 – 2.9
Methyl silsesquioxane (MSQ)	Spin-on	2.6 – 2.8
Poly(arylene ether) (PAE)	Spin-on	2.6 – 2.8
Fluorinated DLC	CVD	2.4 – 2.8
Parylene-F	CVD	2.4 – 2.5
PTFE	Spin-on	1.9
Aerogels/Xerogels (porous silica)	Spin-on	1.1 – 2.2
Porous HSQ	Spin-on	1.7 – 2.2
Porous SiLK	Spin-on	1.5 – 2.0
Porous MSQ	Spin-on	1.8 – 2.2
Porous PAE	Spin-on	1.8 – 2.2
Air Gaps	-	1.0

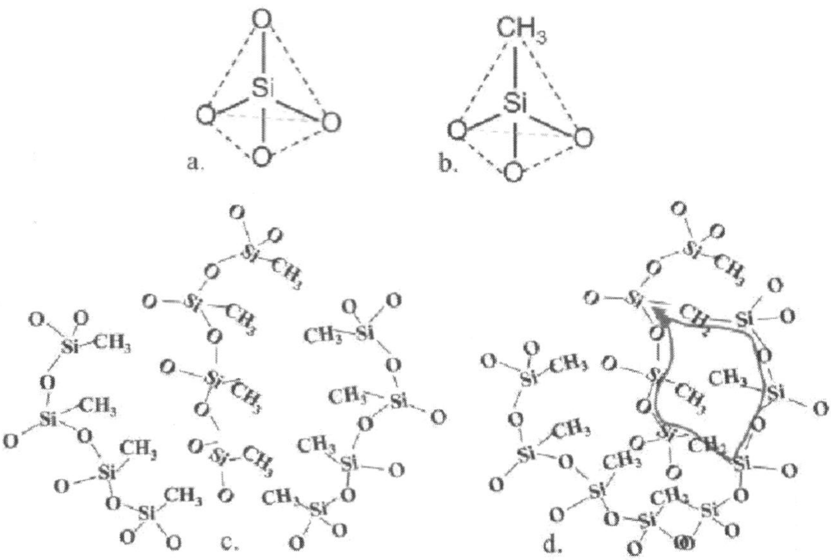

Figure 1. Elementary units of (a) SiO2 (b) carbon doped Silica, called as Black Diamond (SiOC:H) (c) C doped silica without cross linking (d) with cross linking [7]

Table 2. Summary of Black Diamond ™ film properties [17,21]

Property	Value of the Blanket film
Dielectric Constant-Bulk film(Hg Probe)	2.5-2.7 @ 1 MHz
Uniformity (%, 1σ)	<1.5
Stress (MPa)	40-60 Tensile
Stress Hysteresis (MPa)	<20 (RT-450ºC)
Cracking Threshold (µm; blanket film)	>1.5
ASTM scratch tape test on SiN, SiON, Ta, TaN	Passed
Leakage Current (Amps/cm≤)	10-9 @ 1MV/cm
Glass transition temperature	>450ºC
Hardness (GPa)	1.5 – 3.0
Modulus (GPa)	10 - 20

Table 3. Summary of required properties of low-k materials [16]

Electrical isotropic k < 3 @ 1MHz	Chemical No material change when exposed to acids, bases and strippers	Mechanical Thickness uniformity <10% within and <5% wafer to wafer for 8" wafer at 3σ	Thermal T_g >400 °C
Low Dissipation	Etch rate and selectivity better than oxide	Good adhesion to metal and other dielectrics	Coefficient of thermal expansion <50ppm/°C
Low leakage current	<1% moisture absorption at 100% relative humidity	Residual stress <(+)100MPa	Low thermal shrinkage
Low charge trapping	Low solubility in H_2O	High hardness	< 1% weight loss
High electric field strength	Low gas permeability	Low shrinkage	High thermal conductivity
High reliability	High purity	Crack resistance	
High dielectric breakdown voltage > 2-3 MV/cm	No metal corrosion	Tensile modulus >1GPa	
	Long shelf life	Elongation at break >5%	
	Low cost of ownership	Compatible with CMP	
	Commercially available		
	Environmentally safe		

2.3. Barriers And Adhesion Promoters For Cu/Low-K Structures

BEOL multilevel interconnect structures comprise varieties of materials, such as Cu, low-k materials, oxides and nitrides. Copper interconnects in BEOL technology have some challenges such as, poor adhesion to dielectric materials and diffusion of copper into silicon substrates. Copper easily reacts with silicon and forms copper silicide at low temperatures [22-23]. Traces of copper in silicon substrate will cause adverse effect on device operation. Most of the Interconnect metals (e.g. Cu, Ag, Au and W) except Al; do not bond well to underlying substrate and ILD. Therefore, the use of diffusion barrier layers and adhesion promoters between copper and underlying silicon substrate in ICs is mandatory. An ideal diffusion barrier material should also act as good diffusion barrier and adhesion promoters. Diffusion barriers under consideration are metal nitrides, carbides and borides, and metals such as Ti, Ta, and W. Diffusion barrier should be immiscible and non-reactive with copper. As stated by the studies, sputtered Ta and TaN films act as excellent diffusion barrier layers in Cu/low-k multilevel interconnects [24-25]. In the present investigation Ta, TaN, SiC and SiN are studied as barrier layers/cap layers in Cu/low-k stacks. SiC and SiN diffusion cap layers also acts as etch stops during the BEOL processes.

Silicon nitride is widely used as cap layer; first, because it acts as an excellent barrier to copper and second, because of its etch selectivity to oxides.

3. MECHANICAL CHARACTERIZATION OF LOW-K STRUCTURES BY USING NANOINDENTATION

Currently most of IC fabrication technology has been migrated to Cu/low-k interconnect technology, as there is an increase of RC delay associated with the conventional Al/SiO2 interconnects used in the miniaturization of IC technology. This technology transition brings several integration challenges for the Cu/low-k interconnects because of poor thermal and/or mechanical properties of low-k thin films and Cu/low-k stacks [1]. Hence, it is very important to study the mechanical properties of different Cu/low-k structures to evaluate the device reliability. In the present investigation, mechanical properties of various Cu/low-k structures are studied by using nanoindentation technique.

The desired 'k' value of the low-k films can be attained by modifying the molecular structure of the low-k film or by introduction of organic or inorganic groups into the base structure [7]. Mechanical reliability and dielectric constant of the low-k materials are mutually dependent and have inverse relation. In case of SiO_2 based low-k materials, dielectric constant strongly depends on the density of the material, which in turn also depends upon the amount of porosity introduced, as shown in Fig 2[27]. In chemical vapor deposition (CVD) based low-k materials, dielectric constant is decreased by the introduction of terminal methyl ($-CH_3$) groups, that will break the Si-O network and create nanopores. As the concentration of Si-O bonds decreases, percentage of pores and density non-uniformity increases, ultimately increasing the probability of mechanical failure of low-k films.

In IC fabrication technology, low-k material selection and its integration into BEOL is very crucial and it should withstand chemical mechanical polishing (CMP) without any failures. In microelectronic industry many researchers have been actively working toward finding threshold values of hardness and elastic modulus that can provide Cu/low-k system the ability to withstand CMP and wire bonding processes [28-30]. Researchers at Motorola [43] have concluded that passing the CMP process of low-k material is not a simple factor of modulus, hardness, adhesion or toughness, but a combination of all of these properties. As

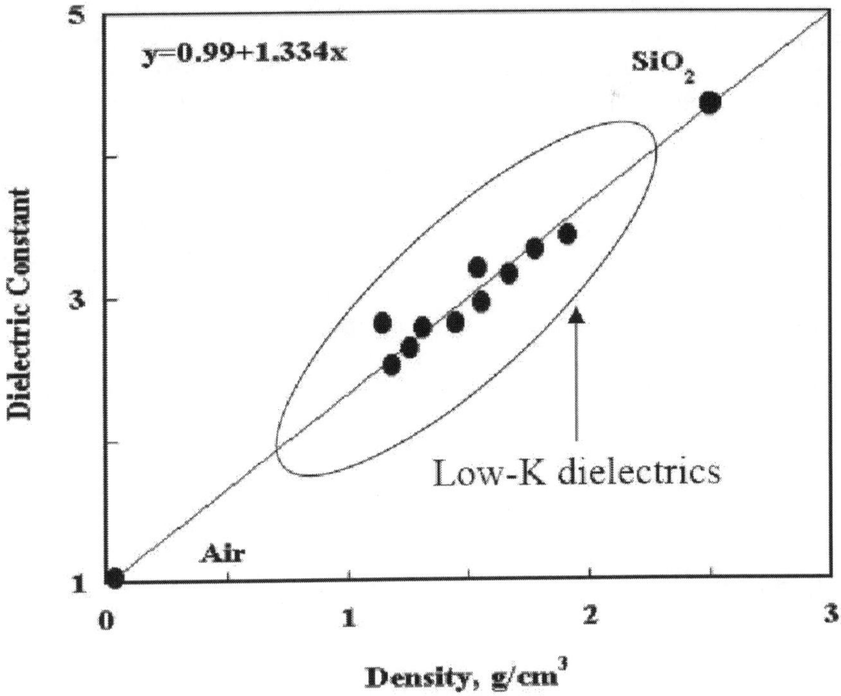

Figure 2. Dielectric constant dependence on low-k material density [27]

Table 4. Mechanical properties of various low-k materials studied by nanoindentation technique.

Dielectric thin films	Thickness (nm)	Elastic Modulus (GPa)	Hardness (GPa)	Researcher
Organo Silicate Glass (OSG)	2000	6.6-8.4	1.2-1.7	A.A. Volinsky et al [4]
USG (undoped silicate glass)	200-1000	79.06-80.66	5.65-7.52	Lu Shen et al [51]
SiLK ™	600	0.4	6.65	Lu Shen et al [55]
Porous SiLK ™	600	0.26	5.34	
MSQ –Hard	500-1000	12.5	0.936	S. Y. Chang et al [56]
MSQ-Soft	500-1000	2.7	0.19	
Low-k/barrier/Si (Anonymous)	1000	0.5	0.05	I. S. Adhihetty et al [57]
Porous low-k (carbon base)	250-540	4	0.15	Y. H. Wang et al [58]
Porous low-k (silica base)	250-540	0.35	0.45	

stated before, the mechanical properties of low-k films depends on chemical structure, amount of porosity and composition, Elastic modulus and hardness values of the different dielectric thin films varies from 2 to 14 GPa and 0.5 to 7 GPa respectively [41-45]. Volinsky et al [1] have found a linear relationship between hardness and elastic modulus for silicate low-k dielectric films in nanoindentation testing with continuous stiffness measurement (CSM) attachment. In nanoindentation testing, mechanical response of low-k films is different from the metallic films and usually exhibits little or no plasticity [27]. Hardness and elastic modulus values of various low-k films and Cu/low-k stacks were tested with nanoindentation technique by many researchers and some of those results have been summarized in table 4. Based on the extensive literature survey and the present work, it is observed that the mechanical properties of various films and Cu/low-k stacks depend on many factors, more importantly the amount of porosity (Constitutive and Subtractive), composition, molecular structure, thickness, type of stack and diffusion barrier.

3.1. Sample Preparation

Materials used in the present work include BD thin films, oxide, nitride and barrier layers. Black diamond, oxide, SiC and SiN films were prepared by using PECVD technique. Sputtering technique was employed to deposit the copper seed (150 nm), Ta and TaN layers. Copper film of 1μm thickness was deposited by electroplating process. Different stacks of these layers were deposited by different experimental techniques. All thin film samples were prepared on 8" Si (100) wafer in semiconductor fabrication plant of class 1000 clean room environment. A thin oxide layer of thickness about 5 nm is deposited on the surface of silicon substrates to improve the adhesion between the substrate and the low-k thin films.

3.2. Continuous Stiffness Measurements (CSM)

This technique was introduced in the year 1989 by Oliver and Pethica [36-37]. The CSM technique developed over the last decade by researchers offers numerous advantages. It has the unique advantage of providing mechanical properties as a function of penetration depth. Calibration and testing procedure take very less time as there is no need for multiple loading and unloading. At high frequencies it allows to avoid obscure effects of the samples like creep, viscoelasticity and thermal drift, which cause much problem in the conventional calibration method. It allows us to measure the effect of contact stiffness changes and damping changes at the point of initial contact [38]. In CSM nanoindentation technique the contact stiffness is measured during loading of the indentation test and there is no need for separate unloading cycles. This is an ideal method to determine contact stiffness and it can measure at very small penetration depths. Hence this method is unique to measure mechanical properties of thin films of few tens of nanometers. It has an additional advantage that if the specimen shows viscoelastic behavior, the phase difference between the force and displacement signals gives idea about the storage and loss modulus of the specimen [26].

In nanoindentation experiment, the CSM technique is carried out by applying a harmonic force at relatively high frequency (69.3 Hz), which is added to increasing load, P, on the loading coil of the indenter as shown in Fig. 3. The applied current that determines the nominal load of the indenter is very small, which results oscillations to the indenter with a frequency related contact area and stiffness of the sample [39].

This technique accurately measures displacements as small as 0.001 nm using frequency specific amplification. To determine the contact stiffness of the sample, the dynamic response of the nanoindenter has to be determined. A dynamic model which is used in CSM system is shown in Fig. 4. Major components of the dynamic model are the mass of the indenter, the spring constant of the leaf springs that support the indenter, the stiffness of the indenter frame and the damping constant due to the air in the gaps of the capacitor displacement sensing system.

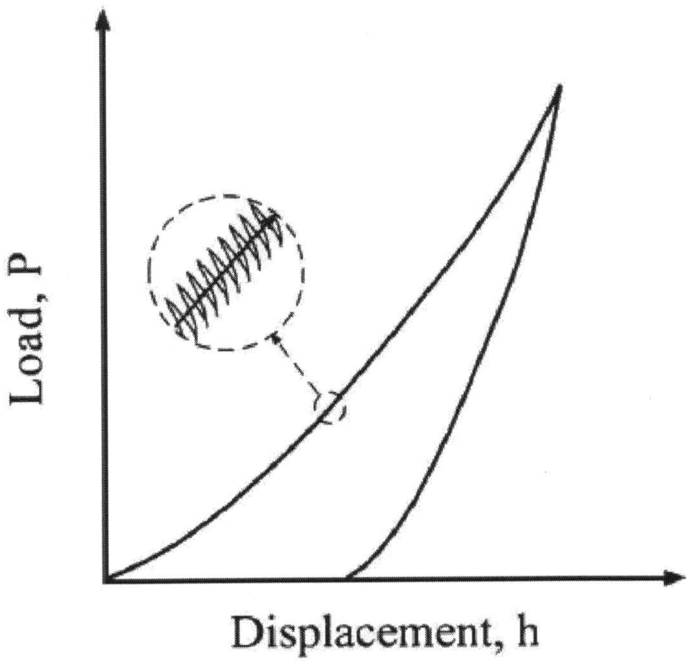

Figure 3. Schematic of nanoindentation CSM load-displacement curve [39;82].

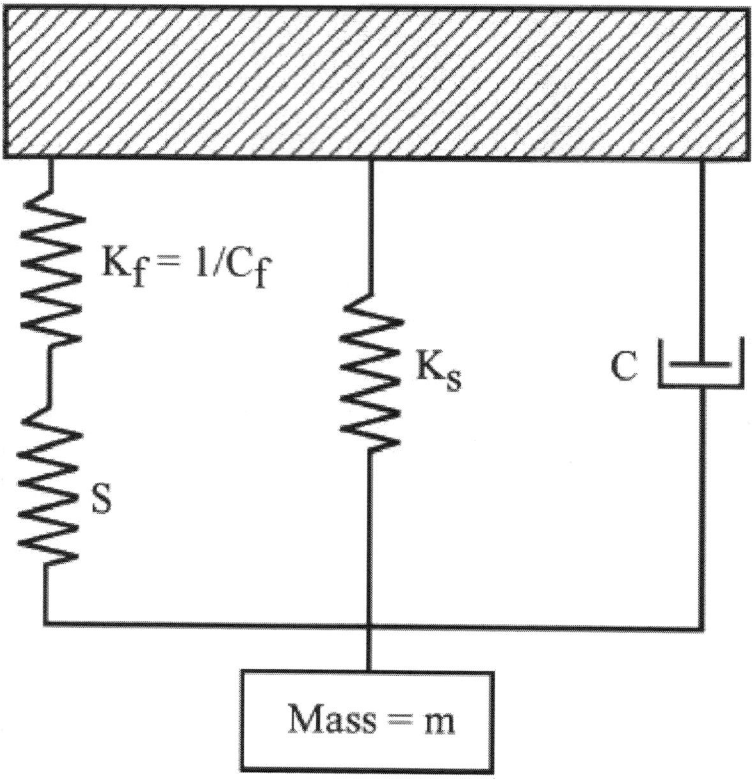

Figure 4. Schematic of components of dynamic model for the indentation CSM system [32]

By analyzing this model, the contact, S, can be calculated from the amplitude of the displacement signal from [39; 82],

$$\left|\frac{P_{OS}}{h(\omega)}\right| = \sqrt{\{(S^{-1} + C_f)^{-1} + K_S - m\omega^2\}^2 + \omega^2 D^2}$$

(1)

And the phase angle, ϕ between the driving force and the displacement response is

$$\tan(\phi) = \frac{\omega D}{(S^{-1} + C_f)^{-1} + K_s - m\omega^2}$$

(2)

Where

Cf= the compliance of the load frame (~1.13 m/MN)

Ks= the stiffness of the column support springs (~60 N/m)

D= the damping coefficient (~54 N s/m)

Pos= the magnitude of the force oscillation

h(ω)= the magnitude of the resulting displacement oscillation

ω= frequency of the oscillation

ϕ= the phase angle between the force and displacement signals

m= mass (~4.7 gm)

4. MECHANICAL CHARACTERIZATION OF BD THIN FILMS WITH VARYING THICKNESS

Thickness of the thin film structures used in the BEOL interconnects continue to decrease as the chip size decreases. Usually, thickness of the dielectric film has small effect on electrical properties [40], but it has significant influence on the mechanical properties of the film. Hence, it is very important to study the mechanical properties of these films of minute thicknesses. For this study, BD thin films of six different thicknesses, 100, 300, 500, 700, 1000 and 1200 nm have been prepared on 8" silicon substrate. Nanoindentation tests were performed on these films to study the effect of thickness on the nanomechanical behaviors. Fig.5 shows typical load-displacement curves of the BD thin films (100, 300, 500, 700, 1000 and 1200 nm) using nanoindentation CSM technique. By using this CSM technique, hardness and modulus can be determined as a function of indentation penetration depth with a single nanoindentation load-unload cycle. Berkovich indenter was employed in all experiments and a series of ten indentation tests were performed on each sample. The Poisson's ratio of the BD films of different thicknesses in nanoindentation experiment data is taken as 0.25, because the Poisson's ratio has a negligible effect on the indentation results [41]. During indentation, almost all BD films exhibit pop-in events as indicated by arrows in Fig. 5 due to failures in the films and it is evident from the optical micrographs (Fig. 6) and the corresponding pop-in events are marked on hardness and modulus vs. displacement graphs in Fig. 7. As the thickness increases, the degree of failure or the extent of damage increases in nanoindentation testing and it can be observed from the optical images shown inFig. 6, where cracks are indicated by arrows on the micrographs. Many researchers have observed this kind of fracture behavior (pop-in event) in various types of low-k materials, bulk glasses and silica foams [42-46] Table 5 gives the information about the BD films fracture/delamination during nanoindentation testing, in terms of threshold load, threshold indentation depth and % of thickness at which film cracking occurs. For 100 nm thick film, failure is observed when the indenter tip is in the substrate and it can be seen in Fig. 5. In case of BD films with thickness 300-1200 nm, failure is observed within the films at different loads and indentation depths as shown in Fig5 and Table 5. It is found that as the BD film thickness increases the threshold load of cracking and the threshold indentation depth increases. One common trend observed in the fracture among BD films (300-1200 nm) is that, the film failure (crack or/and delmaination) occurs at around 60-65% of the film thickness in indentation testing. This data is very useful in conventional nanoindentation processes to determine the physical properties of BD films, i.e. the threshold load for the BD films without cracking, threshold loads in CMP process and to measure the fracture toughness of the BD films.

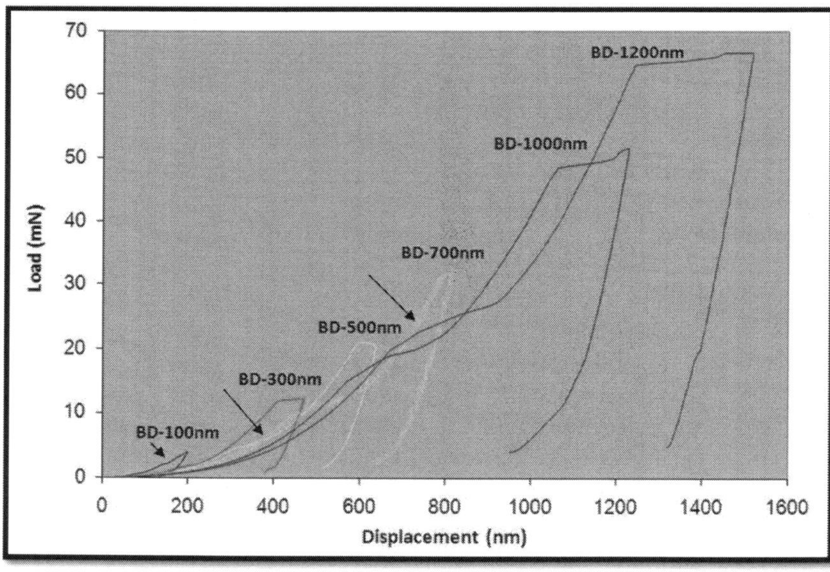

Figure 5. Load-displacement curves for Black Diamond™ (BD) films of six different thicknesses under nanoindentation, (a) 100, (b) 300, (c) 500, (d) 700, (e) 1000 and (f) 1200 nm. Arrows on the curves indicate the pop-in events.

The hardness and elastic modulus as a function of displacement (indentation depth into the film thickness) of all BD samples are shown in Fig. 7. Results at the initial indentation depths show large fluctuations and high values of hardness and elastic modulus as the tip touches the film surface which might be due to the equipment noise and the inaccuracy of the indenter tip function at very shallow depth of indentation. As the depth of the indentation increases, both the hardness and elastic modulus reaches minima region as a property plateau and then tip starts sensing the effect of silicon substrate which results in higher hardness and elastic modulus. This phenomenon is observed only in the case of soft films on hard substrate. Usually, the initial part of the data is usually discarded in nanoindentation analysis and this kind of problem has been observed by other investigators [47-48]. Here, the averaged hardness and elastic modulus of this minima plateau region is used to define the properties of the films. This phenomenon has been validated by comparing the properties at 1/10th of the film thickness and properties at the minima property region as shown in Table 6. There are no significant differences observed in between properties of the films at 1/10th of the thickness and properties from minimum plateau region. Hence in this scenario computing properties from minimum plateau region can be more practicable. Fig. 7 compares the mechanical properties of six different thicknesses of BD samples. Hardness and elastic modulus of the BD films (100 to1200 nm) are in the range of 1.66 to 2.02 GPa and 9.27 to16.48 GPa respectively.

Table 5. Summary of fracture behavior data for all BD samples in nanoindentation testing.

BD sample thickness (nm)	Threshold load of cracking (mN)	Threshold indentation depth of cracking (nm)	% of thickness at which cracking occurs (nm)
100	2.17	142	142
300	1.78	190.69	63.56
500	4.28	312.84	62.57
700	11.00	460.00	65.71
1000	18.83	644.00	64.40
1200	25.23	730.00	60.83

Table 6. Summary of mechanical properties of BD films of six different thicknesses

Thickness of BD films (nm)	Property at 10% of film thickness (GPa)		Property at Plateau Region	
	H	E	H	E
100	1.99±0.25	18.53±1.5	2.02±0.36	16.48±2.1
300	1.83±0.32	13.76±1.3	1.85±0.12	11.54±1.2
500	1.63±0.14	12.91±1.4	1.70±0.05	10.58±1.5
700	1.58±0.24	11.62±1.5	1.78±0.02	9.93±1.1
1000	1.76±0.35	13.52±1.3	1.73±0.07	10.41±1.4
1200	1.62±0.84	12.72±1.5	1.66±0.09	9.27±1.7

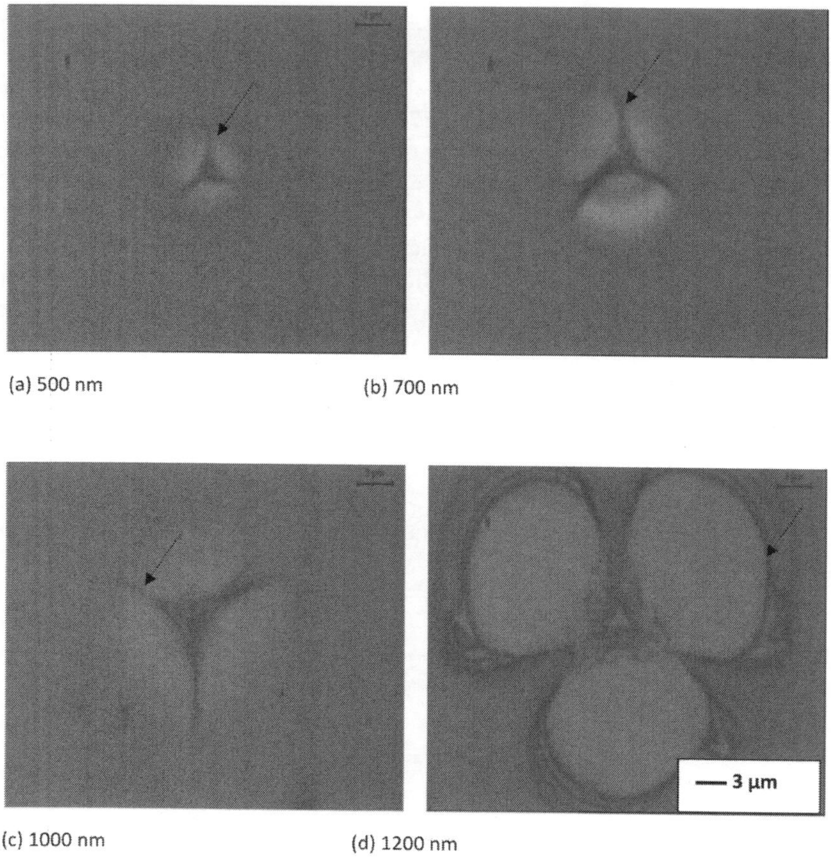

Figure 6. Optical images of residual nanoindentation impressions of (a) 500, (b) 700, (c) 1000 and 1200 nm BD thin films. Radial cracks in the films observed for all thicknesses as shown by arrows.

As the indentation depth increases initially slight decrease in hardness and elastic modulus values are observed and as the indentation depth further increases these property values reach minima, then tip starts sensing the effect of the substrate which results in higher hardness and elastic modulus values. The initial decrease of properties as the indentation displacement increases is observed during nanoindentation of very thin films [49]. The averaged values of properties in minima plateau region are considered to be the real properties of the films and, in the present study, these values are used to define the properties of each film as shown table 6. The hardness values of all BD films are in the range of 1.66 to 2.02 GPa and the elastic modulus values are in the range of 9.27 to 16.48 GPa. Sharp deviations are observed in the property vs. displacement (see Fig. 7) graphs of all BD films, which corresponds to the pop-in events shown in Fig. 5, resulting from film cracking and/or delamination at the BD film-silicon substrate interface. Significant differences in mechanical properties are observed when the BD film thickness is less than 500 nm (100-500 nm), mainly in the case of elastic modulus. When the BD film thickness is greater than the 500 nm (500-1200 nm) no significant variation in mechanical properties are observed and it can be assumed that these properties are representative of the bulk properties of the BD films.

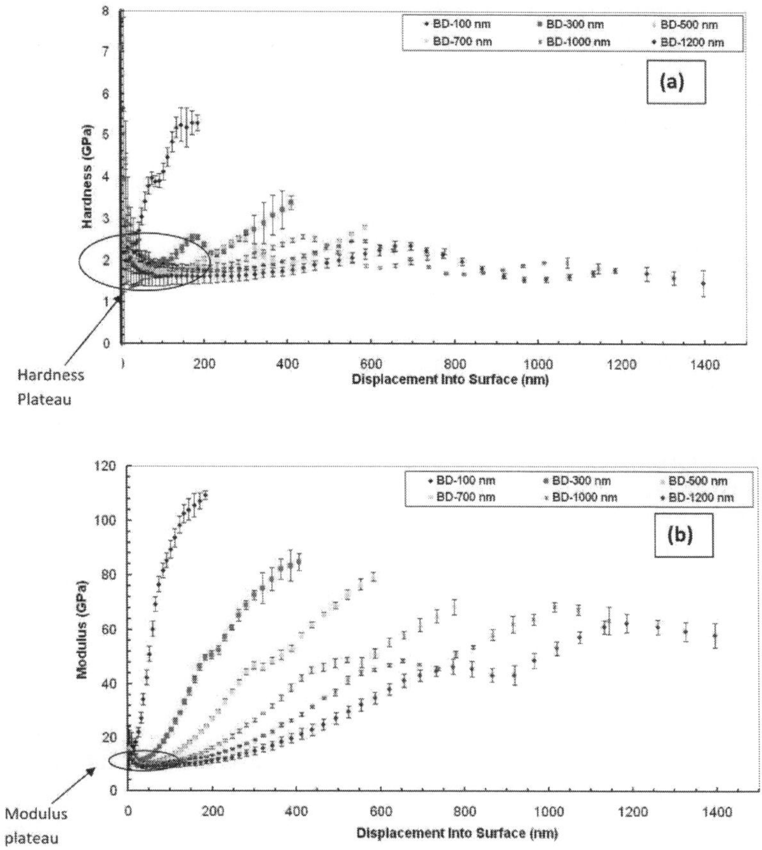

Figure 7. Hardness and (b) elastic modulus as functions of the displacement for BD films with six different thicknesses

The minima plateau region for the hardness is considerably large and there is nearly no change with respect to different thicknesses, when compared with elastic modulus plateau region. The minima elastic modulus plateau region decreases as the BD film thickness decreases because the effect of substrate is more on elastic modulus for thinner films. In Fig. 7, it is observed that the sharp increase in modulus from the minima plateau region, mainly due to the effect of substrate, is much more on the elastic modulus than on the hardness of the BD films. This is because the elastic modulus is associated with the elastic deformation during nanoindentation, and in contrast, the hardness response of the material is associated with plastic deformation. Extensive simulation studies show that, the effective elastic modulus of a film experiences greater substrate effect than the hardness value [44]. BD -100 nm film shows significantly higher elastic modulus (E=16.48 GPa) when compared to higher thickness BD films and it is expected due to molecular restructuring in very thin BD films (≤100 nm), since the elastic modulus is an intrinsic material property, which largely depends on interatomic or molecular bonds [49]. Hence the higher elastic modulus of BD-100 nm film is probably expected due to stronger molecular bonding between organic (-CH_3) and inorganic (Si-O) constituents.

5. EFFECT OF DIFFUSION BARRIERS ON BD STACKS INTEGRITY

For this study, four samples have been prepared with Ta and TaN barrier layers and dual stacks also have been prepared as BD/Ta/Si, BD/Ta/BD/Ta/Si, BD/TaN/Si and BD/TaN/BD/TaN/Si. Mechanical properties were assessed by nanoindentation technique. For single dielectric stack, the silicon substrate is first coated with Ta or TaN barrier layer of 25 nm thickness by using self-ionized metal plasma (SIP) technique at room temperature followed by the deposition of BD film of 1000 nm thickness by PECVD (plasma Enhanced Chemical Vapor Deposition) technique. For the deposition of dual dielectric stack this procedure was repeated one more time with BD films and barrier deposition. The application of Ta and TaN barrier layers to BD films improves stiffness in addition to other mechanical properties.

The thickness of single and dual dielectric stacks studied in the present work is 1025 and 2050 nm respectively. Fig. 8 shows the typical load-displacement curves of single and dual dielectric stacks. Both single and dual dielectric stacks show pop-in events. From Fig. 9 optical micrographs of residual nanoindentation impressions of single and double dielectric stacks reveal that massive failure of the films is more prominent in dual dielectric stacks. Dual dielectric stacks demonstrate massive failure which can be observed as more pop-in events in load-displacement curves especially at the interfaces, which are shown as dotted lines in Fig. 8. This can be confirmed by observing the residual nanoindentation marks as shown in Fig. 9. Hence, it is expected that crack formation and/or delamination may occur at the interfaces due to indenter penetration as observed by J. Vitiello [50]. In the current study, stacks having same barrier layer (Ta or TaN) are compared with respect to mechanical properties. Hardness and elastic modulus of single dielectric stacks are in the range of 1.43 to 1.91 GPa and 8.35 to 10.03 GPa respectively. No significant difference is observed in the case of double dielectric stacks. Mechanical properties of both single and dual stacks are given in Table 7.

5.1. Mechanical Properties Of The Bd Stacks With Ta Barrier Layer

Hardness and elastic modulus of both single and dual dielectric stack with Ta barrier layer is shown inFig. 10. High hardness values at the film surface for both stacks are observed mainly due to the inaccuracy of indenter tip functions and surface roughness [47]. Large fluctuations in hardness values throughout the film thickness measurements are observed due to the failure of stacks during the nanoindentation. Substrate effect is more prominent in single dielectric stack when compared with double dielectric stack, this is because the thickness of both the stacks are different and there is difference in the number of interfaces. The hardness and elastic modulus values of single dielectric stack is higher than the values for dual dielectric stack and these properties are consistent throughout the thickness of the stacks as shown in Fig. 10. The averaged minima property plateau is used to define the property of each stack and these values are given in Table 7. Single dielectric stack has hardness of 1.91 GPa and modulus of 10.03 GPa, whereas for dual stack, the values are 1.38 and 7.98 GPa respectively. From Fig 10, sudden deviations in properties are observed mainly due to the film failure during nanoindentation process which corresponds to pop-in events in load-displacement curves as discussed earlier. This mechanical property data of single and dual stacks are very useful in CMP performance, for example, single dielectric stack has slower removal rate when compared to dual stack for the same CMP pressure [51-52]. This is mainly due to the higher hardness value of the single dielectric stack. This mechanical characterization data will be very helpful in deciding the CMP loads as per the dielectric stacks.

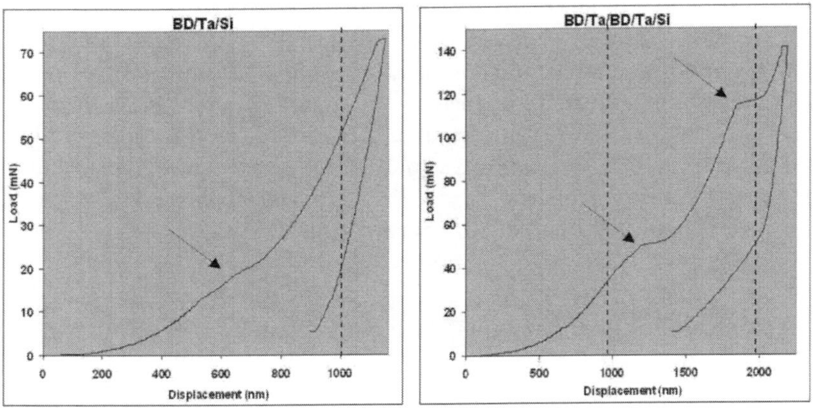

(a) (b)

Figure 8. Typical load-displacement curves of (a) BD/Ta/Si (b) BD/Ta/BD/Ta/Si, (c) BD/TaN/Si and (d) BD/TaN/BD/TaNs/Si stacks. Arrows indicate the pop-in events.

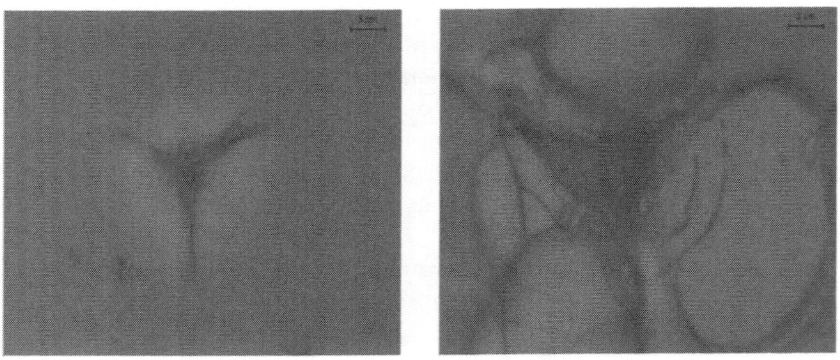

(a) BD/Ta/Si (Single dielectric stack) (b)BD/Ta/BD/Ta/Si (double dielectric stack)

Figure 9. Optical micrographs of residual nanoindentation impression on single and double dielectric stacks.

5.2. Mechanical Properties Of The Bd Stacks With Tan Barrier Layer

The total thickness of these stacks are maintained same as in previous section, i.e. 1025 nm for single dielectric stack and 2050 nm for dual dielectric stack. Hardness and elastic modulus as a function of displacement of these stacks measured using the nanoindentation CSM technique is presented in Fig. 11. Single dielectric stacks have slightly higher mechanical properties when compared with dual dielectric stacks as summarized in table 7. In both cases, the performance of single dielectric stacks is better than that for the dual dielectric stacks. By comparing the single dielectric stacks, BD/Ta/Si has higher hardness and elastic modulus than for BD/TaN/Si stack (Table 7). Thus barrier layer greatly affects the mechanical properties of the single stack. There is no significant difference observed in the case of BD/Ta/BD/Ta/Si and BD/TaN/BD/TaN/Si dual dielectric stacks. It is anticipated that the lower mechanical properties of dual stacks is mainly due to the presence of residual stresses. Usually, dual stacks have more residual stresses when compared with single dielectric stacks as dual-stacked samples are subjected to more processing steps. Compressive stresses in the stacks result in an increased hardness values but tensile stresses cause a decrease in hardness. Therefore, dual stacks are expected to have more tensile stresses when compared with single dielectric stacks.

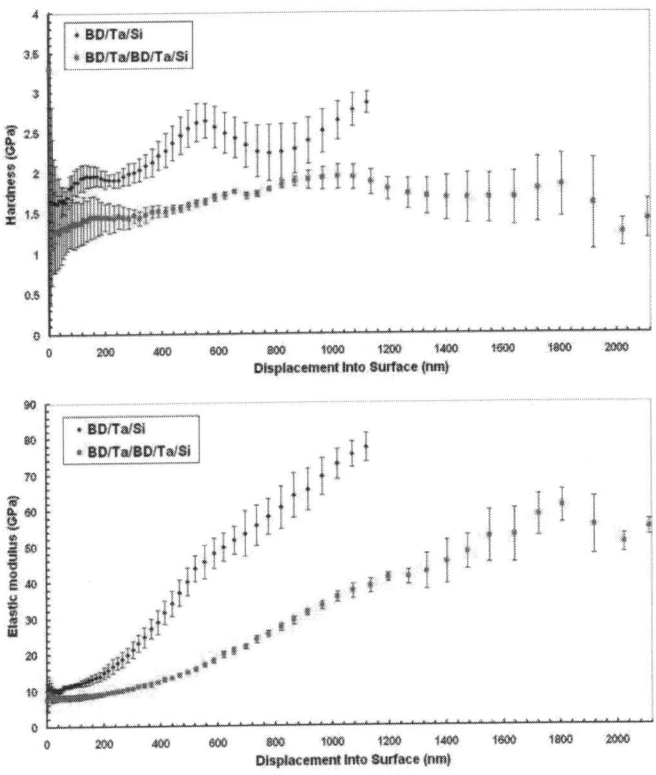

Figure 10. Hardness and elastic modulus as a function of displacement for BD/Ta/Si and BD/Ta/BD/Ta/Si stacks measured using the nanoindentation CSM technique.

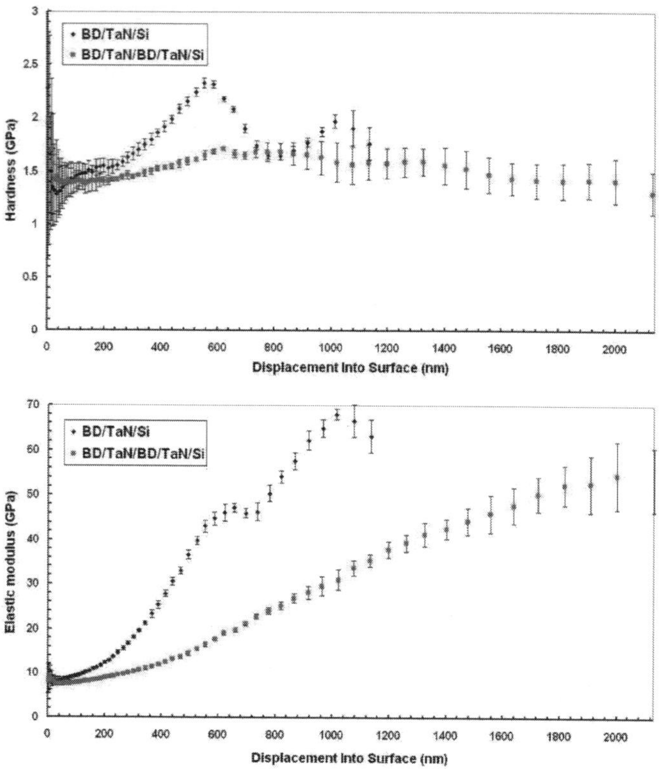

Figure 11. Hardness and elastic modulus as a function of displacement for BD/TaN/Si and BD/TaN/BD/TaN/Si stacks measured using the nanoindentation CSM technique

Table 7. Summary of mechanical properties of single and double dielectric stacks.

Multilayer Stack	Stack Thickness (nm)	Hardness (GPa)	Elastic Modulus (GPa)
BD/Ta/Si (Single dielectric stack)	1025	1.91±0.3	10.03±1.2
BD/Ta/BD/Ta/Si (double dielectric stack)	2050	1.38±0.1	7.98±1.5
BD/TaN/Si (Single dielectric stack)	1025	1.43±0.5	8.35±1.4
BD/TaN/BD/TaN/Si (dual dielectric stack)	2050	1.4±0.2	7.58±1.3

6. EFFECT OF THE WAFER THINNING/BACKGRINDING ON THE INTEGRITY LOW-K STACKS

For this study, multilayer low-k stack with fifteen different thin films comprising SiN, USG, Blok (SiC), BD (Black Diamond ™, low-k), has been fabricated as shown in Fig. 12 All thin film samples were prepared on 8" Si(100) wafer in a semiconductor processing clean room of class 1000 environment. Samples fabricated in this study are exclusively designed to study the BD (low-k) integrity in the BEOL interconnects and it resembles the Cu/low-k structure of the three metallizations. So the structure has three BD low-k layers at different levels according to the BEOL (Back End of the Line) interconnect design specifications. The total thickness of the multilayer low-k stack is around 3400 nm. Therefore present study gives an outlook about the response of the low-k test structures during back grinding process. To study the backgrinding/thinning effect on the Low-k stacks, all samples have been subjected to backgrinding process by using the commercial backgrinding system. Empirically backgrinding process involves, first coarse grinding using grit #300, then fine grinding using #2000 followed by dry polishing. A total five set of samples have been prepared for this study as, normal sample (without backgrinding), Back grinded samples of four different thicknesses (BG-500, 300, 150 & 75um).

Failure loads, hardness (H) and elastic modulus (E) normal (no back grinding) and back grinded low-k stacks (BG-500, 300, 150 & 75 μm) are computed by analyzing the nanoindentation load-displacement curves as shown in Fig. 13. From Fig. 13, it is obvious that all stacks have shown pop-in event and this event can be taken as the failure load/fracture strength of the stack. These pop-in events in the nanoindentation curves are resulting from the film cracking and delamination of the stack in the form of blisters [53]. The normal stack shows pop-in event (failure load/fracture strength of the stack) at lower loads and indentation depths, as compared to the back grinded stacks. Normal stack failed at 456.25±21.22 mN load and 2422.41±58.53 nm indentation depth, whereas back grinded stacks failed in the range of 482 to 661mN load and 2405-2979 nm indentation depth. The failure load and depth values of all types of samples are summarized in the Table. 8 and. From the nanoindentation curves and optical imaging (from Fig.14), analysis it is clearly evident that the nanoindentation response of the normal and back grinded stacks is different in terms of failure load depth and fracture behavior. Normal stack and BG-500μm show extensive delamination and chipping, whereas other back grinded stacks (BG 300, 150, 75 μm) show delamination blister and this behavior is in good agreement with the nanoindentation pop-in event. BG-500μm exhibits the mixed response as it shows chipping-off during nanoindentation and moderate pop-in failure load of 482.17 mN and this might be due to the moderate degree of back grinding. In case of the other back grinded stacks (BG 300, 150, 75 μm), even higher nanoindentation loads are not able to damage/chip-off the low-k stack and cause interfacial delamination only. No significant difference in fracture strength (pop-in event) is observed among BG-300, 150, and 75 μm back grinded stacks and all these grinded stacks show higher facture strength/loads than the normal stack and BG-500μm. Accordingly the increase in failure load depends on the degree of the back grinding, but not much difference is observed when the wafers are grinded to 300, 150 and 75 μm. After back grinding, the strength of the low-k stack is enhanced, mainly in terms of nanoindentation load and indentation depth and this increase is understood mainly due to the application of mechanical pressure and thermal stresses during back grinding action. The back grinding pressure or load may improve the adhesion, especially Vander walls forces at the multilayered interfaces and cause the densification of the individual films of the stack. It is being investigated by many researchers in the packaging field that the back grinding processes are deteriorating the die strength, but this is not the same phenomenon with the active side of the chip stack.

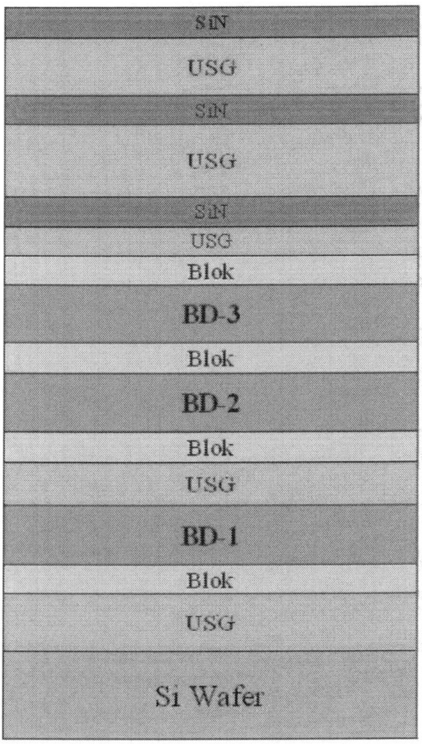

Figure 12. Low-k stack with fifteen multilayers for wafer backgrinding study

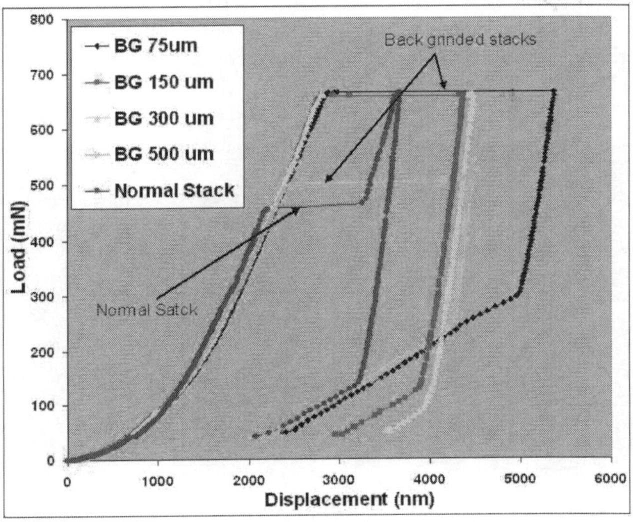

Figure 13. Typical load-displacement curves of normal and back grinded samples and its comparison.

Table 8. Summary of nanomechanical properties of normal and back grinded samples

BD samples	Fracture Behavior	
	Load (mN)	Indentation Depth (nm)
Normal stack	456.25±21.22	2422.41±58.53
BG-500μm	482.17±25.25	2405.86±70.47
BG-300μm	661.20±7.57	2979.79±21.58
BG-150μm	658.45±4.74	2809.01±30.60
BG-75μm	658.60±12.21	2942.52±71.20

Fig. 15 shows the hardness and elastic modulus as a function of the indentation depth for normal and back grinded stacks measured by using the nanoindentation CSM technique. Mechanical properties of all samples are not constant, but strongly influenced by the contact depth and this is mainly due to the presence of the different types of thin films with diverse physical properties and 15 interfaces. FromFig. 15, it is clear that initially all samples exhibit high hardness and modulus values due to the presence of the SiN layer on top of the stack. However, back grinded stacks show higher hardness values throughout the indentation depth and this difference is significant till ~1500 nm. Even though BG-500 μm samples shows moderate failure load values when compared to other back grinded stacks, it still shows higher mechanical properties than normal stack. There is no difference in hardness

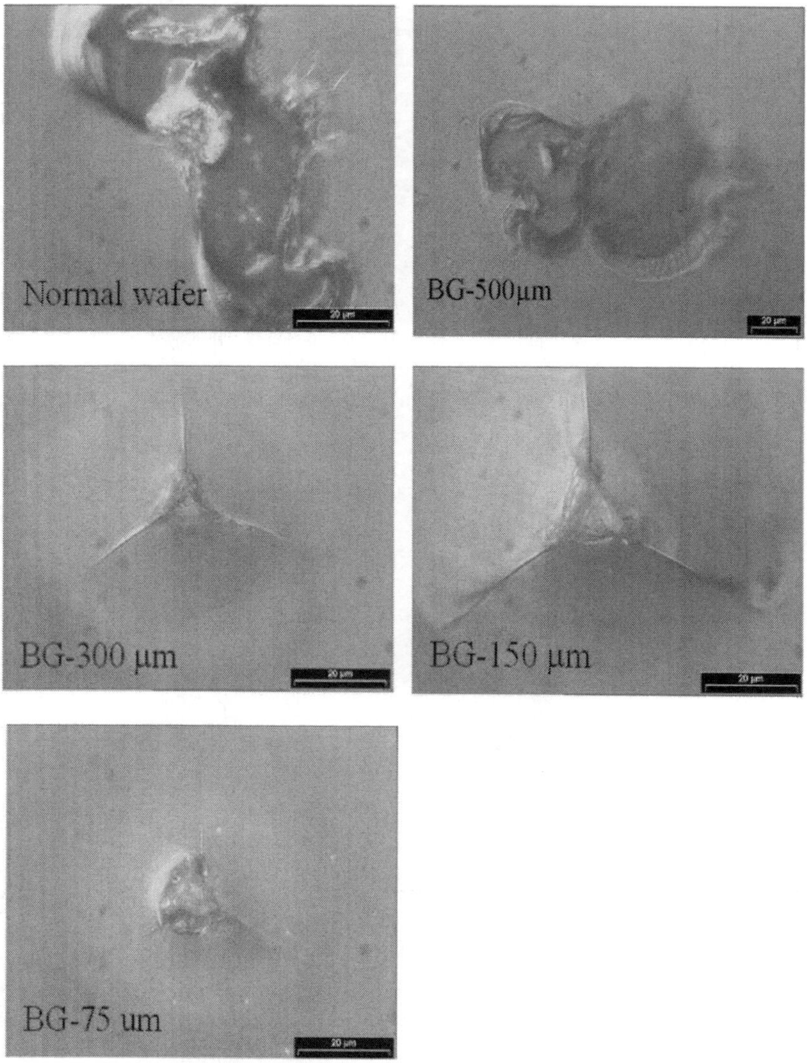

Figure 14. Optical images of residual nanoindentation impressions of normal and back grinded samples.

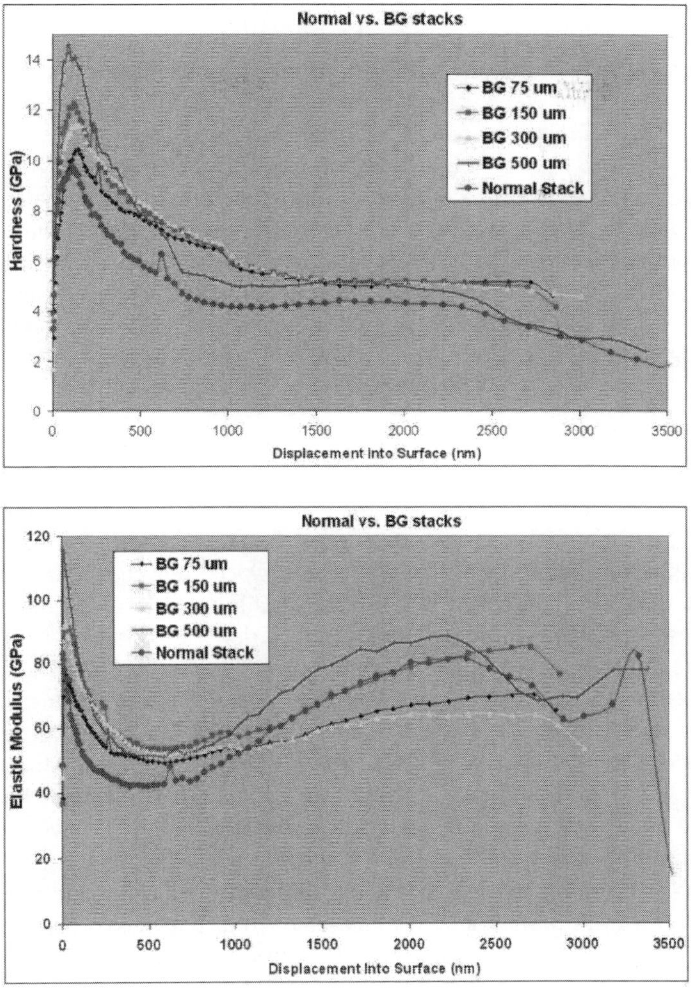

Figure 15. Hardness and elastic modulus as a function of displacement for normal and back grinded samples

values among the back grinded stacks. In case of the elastic modulus, the overall trend is mixed, in which initially back grinded stacks exhibit high values and from 1000 nm depth, BG-150 μm stack follows the trend of normal stack and BG-75 & 300 μm stacks show lower modulus values. The mixed trend of modulus values of all samples is mainly due to the fact that, the elastic modulus is highly sensitive and intrinsic property. Elastic modulus is greatly influenced by layers beneath the testing films and substrate. In the case of hardness, the difference between normal stack and back grinded stacks is very clear when compared with the modulus values. As a whole, back grinded stacks exhibit the higher hardness and elastic modulus values, and this trend is quite clear in the low-k region. This might be due to back grinding loads or pressures influencing the interfaces and causing the densification of the films, especially in the low-k region. Sekhar Et al., have extensively studied the nanoscratch behavior on the backgrinded stacks by analyzing the fracture behavior by critical loads [54]. In this study the fracture or failure strength, hardness and elastic modulus of the normal and grinded stacks are analyzed and compared at overall level as the nanoindentation analysis is very complicated and not well established for the multilayered stacks.

7. SUMMARY

This chapter presents the systematic mechanical characterization of the BD thin films for BEOL interconnects and 3D IC/packaging applications. For this purpose several thin film systems have been chosen which comprises, Black diamond (SiOC:H), USG (SiO_2),SiC, SiN and Ta & TaN thin films which have been prepared by using PECVD technique and sputtering technique respectively. Nanoindentation tests with continuous stiffness measurement (CSM) attachment have been performed on all samples to assess hardness (H) and elastic modulus (E) properties. The CSM attachment is preferable because it provides continuous measurement of the hardness and elastic modulus as a function of indentation depth. In this study several Low-k systems have been prepared and characterized as, mechanical characterization of low-k films with different thicknesses, influence of Ta & TaN barrier layers, and effect of wafer backgrinding/thinning on the complete low-k stack which is similar to actual BEOL with three metallization. In case of the thickness dependence of mechanical properties of BD films, hardness and elastic modulus values obtained of all BD films (100-1200 nm) are in the range of 2.02to 1.66 GPa and 16.48 to 9.27 GPa respectively. Considerable thickness dependence of the properties is observed when thickness is less than 500 nm. To study the effect of barrier layers on low-k stack, single and dual dielectric stacks have been fabricated with Ta and TaN barrier layers. In nanoindentation testing, the performance of single dielectric stacks is better than the dual dielectric stacks is mainly attributed to differences in the number of interfaces, total film thickness and residual stresses.

For the first time the effect of the back grinding process on active side of the low-k chip stack has been studied by using sophisticated nanoindentation technique. It is further investigated by analyzing the fracture or failure strength (pop-in event in nanoindentation), hardness, elastic modulus of the low-k stack for both normal and back grinded samples. After back grinding process, the overall integrity of low-k stack is enhanced and thus the back grinded stacks exhibit higher nanomechanical properties than the normal stacks. This can be attributed mainly to the straightening of the low-k interfaces and densification of the BD (low-k) layers during back grinding processes. Based on the results and their detail analysis it can be said that the thermo-mechanical stresses that applied and/or generated during wafer back grinding processes affect the interfaces and nanomechanical behavior of the low-k stack, in turn enhances the overall integrity of the back grinding stacks.

ACKNOWLEDGEMENT

Author would like to acknowledge the support from Dr. B. S. S. Chandra Rao for suggestions and proof reading. Many heartfelt thanks to Mrs. Sindhura for kind co-operation and encouragement.

REFERENCES

1. S. Wolf, 'Multilevel interconnects for ULSI', 'Silicon processing for the VLSI era': Vol 4-Deep submicron technology Chap.12, pp. 573-602, 2002, lattice press, Sunset beach CA.
2. International Technology Roadmap for Semiconductors,2002Update (SIA San Jose, CA, 2003), http://public.itrs.net
3. The international technology roadmap for semiconductors,2000update, (semiconductor industry association, San Jose, CA, 2000).
4. Alex. A. Volinsky, M. B. Palacio and W. W. Gerberich, 'Incompressible pore effect on the mechanical behavior of low-k dielectric films' Mat. Res. Symp. Proc., Vol. 750 (2003).
5. W. T. Tseng, C. W. liu, B. T. Dai and C. F. Yeh, 'Effects of mechanical characteristics on the chemical-mechanical properties of the thin films' Thin Solid Films, 290-291 (1996) 458-463.
6. J. D. Plummer, M. Deal and P. B. Griffin, 'Silicon VLSI Technology: Fundamentals, Practice and Modeling' Prentice Hall, NJ, USA.
7. D. Shamiryan, T. Abell, F. Lacopi, K. Maex, Jan-2004Low-k dielectric materials' Materials Today, 713439
8. www.semiconductor.net
9. Reiche, Manfred & Wagner, Gerald "Wafer thinning: techniques for Ultra-thin Wafers", Advanced Packaging, March, 2003.

10. Hoh Huey Jiun et al, "Effect of wafer thinning methods towards fracture strength and topography of silicon die" Microelectronics Reliability,462006200683684S

11. Shoulung Chen et al.The Evaluation of wafer thinning and Singulating Processes to Enhance Chip" IEEE Elelctronic Components and Technology Conference, 200515261530

12. J. Chen, I. De Wolf, Study of damage and stress induced by backgrind in Si wafers" Semicond Sci Technol 200318261268

13. Gaulhofer E, Oyer H., "Wafer thinning and strength enhancement to meet emerging packaging requirement" IEMT Europe Symposium 200 pp.1-4.

14. C. Mc Hatton, C. M. ". Gumbert, Eliminating backgrind defects with wet chemical etching" Solid-State Technol 1998November): 8590

15. Blech I and Dang D "Silicon Wafer Deformation After Backside Grinding", Solid State Tech., 37(8), 1994, pp.74-76.

16. Michael E. Clarke, Introducing low-k dielectrics into semiconductor processing: Application Notes, www.mykrolis.com

17. www.appliedmaterials.com

18. L. Xia, Handbook of Semiconductor Manufacturing Technology, pp 309-356.

19. K. Maex, M. R. Baklanov, D. Shamiryan, F. lacopi, S. H. Brongersma, Z. S. Yanovitskaya, Low dielectric constant materials for microelectronics, J. Appl. Phys. 93 (11): 8793-8841.

20. L. Xia, Handbook of Semiconductor Manufacturing Technology, pp 309-356.

21. Ben Pang, W. Yau, P. Lee and M. Naik, A new CVD process for damascene low-k applications, Semiconductor Fabtech-10th Edition, pp 285-289.

22. Q. Hong, C. M. Comris\e, S. W. Russel, J. W. Mayer, 1992Phase formation in Cu-Si and Cu-Ge, J. Appl. Phys. 7036553660

23. J. D. Mc Brayer, R. M. Swanson, T. W. Sigmon, 1986Diffusion fo metals in silicon dioxide, J. Elelctrochem. Soc. 13312421246

24. S. Y. jang, S. M. Lee, H. K. Baik, J. Mat. Sci: Mat in Elelctronics 7, p271 (1996)

25. J. O. Olowolafe, C. J. Morgah, R. B. Gregory, M. Kottke, J. Appl. Phy. 72, p 4099, (1992)

26. Anthony C. Fischer-Cripps, Nanoindentation, Second edition-2004, Springer-Verlag New York, LLC, NY.

27. Alex A. Volinsky, Mechanical reliability and characterization of modern microelectronic interconnect structures, Nano Engineering world forum 2003.

28. Y.L. Wang, C. Liu, S. T. Chang, M. S. Tsai, M. S. Fang, W. T. Tseng, Thin solid films, 308-309 (1997) 550.

29. E. Hartmannsgruber, G. Zwicker, K. Beekmann, Microelectron. Eng. 50 (2000) 53.

30. H. Yano, Y. Matsui, G. Minamihaba, N. Kawahashi, M. Hattori, Mater. Res. Soc. Proc. 671 (2001) D1.1.

31. For more details see www.mts.com

32. W. C. Oliver and G. M. Pharr, An improved technique for determining hardness and elastic modulus using load and displacement sensing indentation experiments, J. Mater. Res., Vol. 7, No. 6, June 1992, 1564-1583.

33. N. Sneddon, 1965Relation between load and penetration in the axisymmetric Boussineq problem for a punch with arbitrary profile', Int. J. Eng. Sci., 3, 47.

34. K. L. Johnson, 1985Contact mechanicsCambridge, Cambridge university press.

35. B. R. Lawn, T. R. howes, 1981Elastic recovery at hardness indnetations, J. Mater. Sci., 16, 2745.

36. W. C. Oliver, J. B. Pethica, 1989Methods for continuous determination of the elastic stiffness of contact between two bodies', US patent 4,848,141, July 18.

37. J. B. Pethica and W. COliver(1989Thin films: Stresses and mechanical properties', MRS Symp. Proc. 1301323Pittsburgh, PA, Materials Research Society.

38. W. C. Oliver and G. M. Pharr, 'Measurement of hardness and elastic modulus by instrumented indentation: Advances in understanding and refinements to methodology', J. Mater. Res., Vol. 19, No. 1, Jan2004.

39. X.Li and Bharat Bhusan, 'A review of nanoindentation continuous stiffness measurement technique and its applications' Material Characterization 48200220021136

40. E. Wu, J. Sune, W. Lai, E. Nowak, J. McKenna, A. Vayshenker and D. Harmon, Solid State Electron. 46, (11) (2002) 1787.

41. Shaohua Chen, Lei Liu, Tzuchiang Wang, "Investigation of the mechanical properties of thin films by nanoindentation, considering the effects of thickness and different coating-substrate combinations", Surface & Coatings Technology191200520052532

42. R. J. Nay, O. L. Warren, D. Yang, and T. J. Wyrobek, 'Mechanical characterization of low-k dielectric materials using nanoindentation', Microelectronic Engineering, 75 (2004) 103-110.

43. J. B. Vella, Alex A. Volinsky, Indira S. Adhihetty, N. V. Edwards, and William W. Gerberich, 'Nanoindentation of silicate low-k dielectric thin films' Mat. Res. Sym. Proc., Vol. 716, Materiual Research Society, B12.13.1-6.

44. Lugen Wang, M. Ganor, and S. I. Rokhlin, 'Nanoindentation analysis of mechanical properties of low to ultralow dielectric constant SiOCH films', J. Mater. Res., Vol.20, No.8, Aug 2005, 2080-2093.

45. D. J. Morris, S. B. Myers, and R. F. Cook, 'Sharp probes of varying acuity: Instrumented indentation and fracture behavior' J. Mater. Res., 19, 165 (2004).

46. Y. Toivola, A. Stein, and R. F. Cook, 'Depth-sensing indentation response of ordered silica foam', J. Mater. Res., 19, 260 (2004).

47. Wu Tang, Lu Shen and Kewei Xu, 'Hardness and elastic modulus of Au/NiCr/Ta multilayers on Al2O3 substrate by nanoindentation continuous stiffness measurement technique' Thin solid films2005Article in press.

48. Kaiyang Zeng, Zhi-Kuan Chen and Bin Liu, 'Study of mechanical properties of light-emitting polymer films by nanoindentation technique' Thin Solid Films47720052005111118

49. J. Wang, F. G. Shi, T. G. Nieh, B. Zhao, M. R. Brongo, S. Qu and T. Rosenmayer, 'Thickness dependence of elastic modulus and hardness of on-wafer low-k ultrathin polytetrafluoroethylene films' Scripta mater. 42 (2000) 687-694.

50. J. Vitiello, A. Fuchsmann, L. L. Chapelon, V. Arnal, D. Barbier and J. Torres, 'Imapct of dielectric stack and interfaces adhesion on mechanical properties of porous ultra low-k' Microelectronic Engineering, 2005, Article in press.

51. Lu Shen, Kaiyang Zeng, Yihua Wang, Babu Narayanan, Rakesh Kumar, Dtermination of the hardness and elastic modulus of low-k thin films and their barrier layer for microelectronic applications, Microelectronic Engineering7020032003115124

52. V. N. Sekhar, T. C. Chai, S. Balakumar, Shen. S. K. Lu, A. A. O. Sinha, Tay and Seung Wook Yoon, " Influence of Thickness on Nanomechanical Behavior of Black Diamond TM Low Dielectric Thin Films For Interconnect and Packaging Applications" J Mater Sci: Mater Electron (2009

53. Lugeb Wang et al, "Nanoindentation analysis of mechanical properties of low to ultraz dielectric constant SiOCH films" J. Mater. Res., Vol. 20, No.8, Aug 2005, 2080-2093.

54. V. N. Sekhar, Shen. Lu, Kumar. Aditya, Chong. Tai, Xiaowu. Chai, C. S. Zhang, Vaidyanathan. Premchandran, Seung. Kripersh, Yoon. Wook, H. John, Lau, " Study on the Effect of Wafer Back Grinding Process on Nanomehcanical Behavior of Multilayered Low-k Stack" IEEE Transaction on CPMT, 21Jan-2012

55. Lu Shen, Kaiyang Zeng, Comparison of mechanical properties of porous and non-porous low-k dielectric films, Microelectronic Engineering7120042004221228

56. S. Y. Chang, H.L. Chang, Y. C. Lu, S. M. Jang, S. J. Lin, M. S. Liang, Mechanical property analyses of porous low-dielectric-constant films for stability evaluation of multilevel-interconnect structures, Thin Solid Films 460 (2004) 167-174.

57. I.S.Adhihetty, J.B. Vella, A. A. Volinsky, C. Goldberg, W. W. Gerberich, Mechanical properties, adhesion and fracture toughness of low-k dielectric thin films for microelectronic applications, Proceedings of the 10th international congress on fracture, Honululu, USA, DEC-2001.

58. Y. H. Wang, M. R. Moitreyee, R. Kumar, S. Y. Wu, J. L. Xie, P. Yew, B. Subramanian, L. Shen, K. Y. Zeng, The mechanical properties of ultra-low dielectric constant films, Thin Solid Films 462-463 (2004) 227-230.

Chapter 9

Strategic Resolution of Policy, Environmental and Socio-Economic Impacts In Canadian Arctic Diamond Mining: BHP's NWT Diamondproject

William J Couch

[a] *EIA at the Międzywydziałowe Studia Ochrony Środowiska, University of Warsaw , ul. Przechodnia 2/m.1518, 00-100 , Warsaw , Poland*

ABSTRACT

BHP's Northwest Territories Diamond Project (EKATI™) illustrates how a two-step process consisting of an environmental impact assessment review followed by parallel permitting and negotiated agreements resolved an array of interdependent policy, environmental, socialimpact, legal/administrative and economic issues in the remote Canadian Arctic. The process bridged problems of intercultural communications between a multinational corporation, four small Aboriginal groups and officials working in a period of transition within government. The proposal proceeded because of participants' good faith, good science, public consultation, sound process management, mitigable impacts and shrewd ministerial handling.

Keywords: environmental impact assessment; diamond mining; Canadian Arctic; public consultation; Aboriginals

THE CASE OF Broken Hill Proprietary (BHP) Co's Northwest Territories (NWT) Diamond Project illustrates how a two-step process consisting of an environmental impact assessment (EIA) review followed by permitting and negotiated agreements together resolved an array of interdependent policy, environmental, social, legal/ administrative and economic issues in a remote, sparsely populated region. The participants held diverging world views and interests: a multinational corporation, four small Aboriginal groups seeking to defend their local traditional lifestyle from further threats, while benefiting from the development; and officials who were expected to resolve many questions in the face of large resource cuts. The story's interwoven strands included: the legacy of past Canadian Aboriginal policy; the need for a rigorous EIA to protect the fragile Northern environment and fair public consultation; averting the social havoc of previous Northern boom–bust developments; addressing Aboriginal people's present and long-outstanding concerns; responding to new policy issues; building institutions to manage the sudden arrival of a mega-project; and resolving the plethora of different economic interests. The dénouement assured that the environment would be prudently managed, Aboriginal people did benefit economically without having their lifestyle overwhelmed, precedents were created for approving subsequent diamond-mining proposals, and new policies and new institutions set in place to manage the rapidly changing Canadian North. Research was based on primary sources, that is, official and semi-official publications and interviews with central players in the process, as well as the author's own sense of events as a former official in the Canadian Federal Government.

BACKGROUND

Canada has in effect 'two environments': the South with a temperate climate; and the vast North above 600 N latitude, where the severe climate precludes a large population. The Arctic region is divided into three federal Territories: the Yukon Territory, the Northwest Territories (NWT) and Nunavut.

This very large project is in the NWT, which is on the Canadian Shield, covered by thousands of small lakes, often joined by rivers. Forests cover the land up to the tree line, north of which there is tundra. The NWT has a sparse population of 39,460 inhabiting an area of 1,183,084 km2 . Aboriginal people (48.2%) play an important role in the affairs of the NWT. Most non-Aboriginals from the South live in the capital, Yellowknife (population 17,702).

The Aboriginal people belong to one of three groups.1 The ancestors of the Indians2 and their historic rivals, the Inuit (Eskimos),3 migrated from Asia over the land bridge that once connected Siberia and Alaska, and then fanned out to settle the Western Hemisphere. Some speak only their own language. The Métis are of mixed European–Indian descent. The Indians who live in the vicinity of the Project are called the Dene.

Canada's Federal Government, its ten provinces and its three territories have a Cabinet form of government derived from the Westminster (UK) model. As Canadian provinces have constitutional control over many social policies, natural resources, Crown land and urban affairs, many environmental laws are provincial laws.

Aboriginal affairs fall under federal jurisdiction. Since there is not a sufficient tax base in the North to finance programmes that the provinces deliver in the South, the Federal Government has managed their delivery largely through the Department of Indian Affairs and Northern Development (DIAND).

There was a need to clarify overlapping Northern legal and administrative régimes in order to manage the approval and regulation of this complex megaproject, to create a framework for future activities, and to do justice to Aboriginal people.

Economically viable gem-bearing diamond pipes are very rare. There were only 15 such areas known in Africa, Siberia, Australia and Brazil. In a seemingly impossible search, Canadian geologists Charles (Chuck) E Fipke and Dr Stewart L Blusson criss-crossed the Arctic by foot and light aircraft for a decade looking for diamonds until they found indicator minerals near Lac de Gras in 1989. They began to stake out mineral claims and then reached an agreement with the Australian multinational corporation BHP Ltd to mine the diamonds in 1990.

BHP's NWT Diamond Project, later named EKATI™, became Canada's first diamond mine and was the first major resource development in the NWT for a number of years. This Project and two other diamond mines that have followed it (the Diavik and Snap Lake Projects) were to have a large impact on economies of both the NWT and Canada as a whole.

Canada's aboriginal policy

Terra nullius is a legal principle that invests discoverers of an uninhabited land with sovereignty and all rights and titles to it. As English-speaking people began to settle on faraway shores in the 17th century, they broadened the concept and claimed Aboriginal lands not being used in a 'civilised' way (that is, for agriculture, industry and commerce).

Between the early 19th century and 1945, Canadian policy aimed at 'civilising' and assimilating Aboriginals, that is, treating them as racially inferior wards in need of tight paternalistic control, dispossessing them of their land, relocating them on reserves to facilitate assimilation, and obliterating their languages and sense of identity. Since there was no public interest in Aboriginal policy, the Indian Affairs Branch remained a closed, quasi-colonial government unto itself. The legacy of these events remains the backdrop for all Aboriginal issues in Canada (Royal Commission on Aboriginal People, 1996)

Change began after World War II when Canadians began to perceive their own colonialism at home. In 1960, Aboriginal people received the right to vote in federal elections. However, the big turnaround began after Pierre Trudeau became Prime Minister in 1968. In 1971, the Government created the new Cabinet post of Indian Affairs and Northern Development to manage the changes.

A major vehicle for change has been the comprehensive land claim agreement. This is a modern treaty between an Aboriginal group and the Federal Government. It is a negotiated, legally binding and constitutionally protected accord. Land claims deal with the concerns of Aboriginals, governments and third parties about who has the right to own and use the lands and resources in the area designated by the land claim agreement. Agreements may include procedures for EIA and resource management. Since 1975 there have been ten such agreements in Canada. There was only one in the vicinity of the Project, the Nunavut Land Claims Agreement (1993).

There has been a steady devolution of financial and executive powers on to the NWT's Assembly. In 1975, the NWT Council became fully elected with a majority of Aboriginal members. The Council was soon called the Legislative Assembly (1976) and began to choose a Premier (1980). Six of the NWT's eight premiers have been Aboriginals.

Major highly publicised public EIA reviews in the subsequent years have offered yet another podium from which Aboriginals voiced their concerns and won their national acceptance.

EIA's role in the Canadian Arctic

When considering pipeline construction down the Mackenzie Valley corridor, the Trudeau Government appointed Mr Justice Thomas R Berger (March 1974) to head an inquiry into the regional social, environmental and economic impacts of pipeline construction. The Berger Inquiry set many important precedents for EIA and general governance in the Canadian North and called for a new social contract with Aboriginal people.

Berger interpreted very liberally his terms of reference so as to have an inquiry "without walls". He held two types of hearing: formal, for experts; and informal community hearings for local people to state their concerns directly to him. He gave funds to Northern groups so they could participate on a more equal footing with industry. By arranging extensive media coverage in English and Aboriginal languages in the North and the South, he brought the issues of Northern hydrocarbon development and Aboriginal policy to the attention of all Canadians (Berger, 1977; Gamble, 1978).

The openness of Berger's procedures, the breadth of issues considered and the inclusion of all Northerners created a model that came to be expected thereafter for all future impact assessments of Northern developments, including the BHP NWT Diamond Project.

Berger's approach sprang from Canada's political culture and, in turn, significantly influenced it. Canada has a different planning tradition from Europe. In Europe, an EIA may be only one study among others that the competent authority may use when granting approval to a project. Since EIA in Canada is a planning tool and not a permitting process per se, Canadian practice tends to be more flexible. An EIA's content can be expanded when necessary to fill administrative lacunae and deal with a wider range of issues — notably the socio-economic impacts of large projects.

EIA has proven effective in the undeveloped and sparsely populated regions of Canada's North. By default, the EIAs of mega-projects have had characteristics ascribed to strategic environmental assessment (SEA) and social impact assessment (SIA). They have been thrust into the breech to fill gaps for policy and/or regional planning of areas larger than many European countries, and have served as a catalyst for broader government initiatives. As public participation elicits discussion of socio-economic issues and EIA reviews have on occasion been the 'only show in town', the assessment of social impacts is usually central to EIA in the North.

Although the proposal of BHP Diamond Inc is called a "project", it has these features of SEA and SIA. The EIA identified and responded to key policy and planning issues, and led to the creation of new institutions of governance in the NWT. The NWT's area is very large — comparable to that of France, Germany and Great Britain combined, or three times the size of Japan. Lastly, even though the potential physical environmental effects were important, it was the social impacts that were central to the Project's assessment and approval (Couch, 2000a).

PROJECT AREA

Geography

The Proponent's holding has an area of 3,400 km2 . It is located 300 km northeast of Yellowknife (Figure 1). Most of the planned developments are in the Koala watershed whose water flows into Lac de Gras and thence northward via the Coppermine River to the Arctic Ocean.

The project area lies in the harsh Low Arctic Ecoclimate region, where the summers are short and cool and the winters are long and extremely cold. The average annual temperature is –11.80 C. In the summer, daily temperatures may reach 250 C. Winter temperatures are often below –300 C. Precipitation is sparse, averaging 300 mm annually and coming mostly as snow (BHP, 1995, page 23).

The EKATI™ claim is located 100 km north of the tree line in the tundra region. One third of its area is covered by 8,000 lakes. The land has continuous permafrost with 250 metre deep permanently frozen subsoil and rock, overlaid by a one metre active layer that thaws during the summer (BHP, 1995, page 23).

Figure 1. Maps of the EKATI™ and Diavik project areas, NWT and North America

Diamonds are crystallised carbon formed more than 150 km below the earth's surface. Kimberlite is a very rare igneous rock, which, during a volcanic eruption, passed through the diamond-forming layer, carried the diamonds to the surface, and then cooled to form carrot-shaped volcanic cones called 'pipes'. In the Canadian North, the surfaces of soft kimberlite pipes were scraped deeper by glaciers, so that, when the glaciers receded, diamond deposits were located under small lakes. There are 136 known pipes on the EKATI™ claim block and 300 in the NWT.

FLORA AND FAUNA

The chief form of vegetation is stunted shrubs and grass tussocks. In depressions there are shrubs such as willow and scrub bush. Wetland areas have a complex of water sledges and sledge-willow communities (BHP, 1995, page 24).

The two most important mammals are the Bathurst caribou herd and the grizzly bear. The caribou herd, consisting of about 350,000 animals, lives on a range of about 250,000 km2. They winter below the tree line and in the spring migrate northwards to their calving grounds near Bathurst Inlet on the Arctic Ocean. The caribou play a central role in the physical and cultural life of Aboriginal people. About 60% of Aboriginal households get at least half their meat and fish by hunting and fishing. The caribou's harvest value was calculated to be CAN$11,200,000 annually. Its cultural value cannot be overstated (CEAA, 1996, pages 39–42, 62).

Because of its low density and rate of reproduction, the grizzly bear has been designated as vulnerable. There is extensive movement of bears in the project area (CEAA, 1996, pages 42–43).

Arctic fish have limited species diversity and slow growth rates. The lake trout population for the five impacted lakes varied from about 90 mature fish in Misery Lake, to 260 in Koala Lake, and 2,600 in Long Lake (BHP, 1995, pages 24, 32).

Participants in the EIA review

Governments in transition

A major shift in governance is occurring in the North. In the mid-1990s the Federal Government embraced the neo-liberal paradigm, which postulates that the public good can be better served by allowing market forces and the private sector to deal increaseingly with a number of human activities that had hitherto fallen to government. It continues to cut funding for programmes to the public, to reduce government investment, generally to favour deregulation, and to delegate governing downward to territorial governments and Aboriginal groups.

In the North, DIAND has a shared mandate with the Government of the Northwest Territories (GNWT) for land and water management and for regulating mining, and oil and gas rights. The federal Department of the Environment (DOE) administers various environmental laws, and the Department of Fisheries and Oceans (DFO), the important federal Fisheries Act. The GNWT is responsible for most local matters, that is, co-management with DIAND of wildlife and forest management and fire control. The Federal Government plans to devolve to the GNWT powers to manage land and natural resources such as mining, and oil and gas.

The delegation of wide responsibilities to the small, impecunious, new entities in the North poses great challenges if they are to carry out all these new roles, even though the Federal Government will continue to provide revenue to the GNWT, and Aboriginal and other Northern bodies.

Proponent

Fipke set up Dia Met Minerals Ltd in 1983 to finance his search for diamonds. When he found indicator minerals, he began to stake out mineral claims and sought out BHP Ltd to be the operator of the project. The Proponent (BHP Diamonds, Inc) was a joint venture (signed 1990) in which BHP Ltd owned 51% of the undertaking and the Blackwater Group, 49%. The Blackwater Group consisted of Dia Met, owning 29% of the undertaking, Fipke 10%, and Blusson 10%.

BHP Diamonds Inc was a Canadian subsidiary of BHP Minerals, a business group of the Australian multinational company BHP Ltd. Prior to 2001, BHP's three principal areas of business were: mineral exploration, production and processing (principally coal, copper and iron ore); hydrocarbon exploration and production; and producing most of Australia's steel. BHP Ltd had over 35,000 employees in 30 countries.

BHP Diamonds Inc extracted the first diamonds in 1991. In 1993 it opened Koala Camp and its office in Yellowknife.

Aborginal people

Aboriginal leaders tend to suspect Southerners who set before them promises of benefits coming from large new projects, because boom–bust developments have destroyed the fabric of their fragile traditional societies, leaving despair and bewilderment, family breakdown, rampant alcoholism and adolescent suicide. Unlike the recently arrived Southerners, Aboriginal people have nowhere to return to if they do not like their lot. Still, they are torn when confronted with the money economy and chronic unemployment among their demographically young population. The younger generation has accepted the new order. Older people know that they can never return completely to their traditional lifestyle based on hunting and fishing, and that a wage economy permits them to buy rifles, boats, trucks, food and clothing to supplement their needs.

Aboriginal people have produced a new generation of educated articulate leaders. They are organised and have learned their political skills in their own national organisations, managing the NWT's machinery of government, and participating prominently in important national discussions such as the constitutional debates leading up to the Constitution Act, 1982, and further such debates in the late 1980s and early 1990s. They knew how to operate the levers of power in the Southerners' world by winning popular support in the media, negotiating with senior politicians and corporate executives, and tying up government and corporate plans in the courts. As this review was a good chance to get national attention for their concerns, they were determined that their voice would be heard so asto maximise their benefits.

The Project would affect the members of four Aboriginal bodies whose members speak English, one of several Athapaskan languages, or the Eskimo–Aleut language, Inuinnaqtun.

The four Aboriginal bodies are:

- •The Treaty 11 Dogrib Dene live southwest of the project area and north and west of Yellowknife. In 1995, the group was negotiating a comprehensive land claim.
- The Akaitcho Treaty 8 Dene (Chipewyan Dene) live south of the project area and east of Yellowknife. It is a more heterogeneous and lessstructured group than the Treaty 11 Dogrib Dene. In 1995, it was negotiating a treaty on land entitlement.
- The North Slave Métis Alliance lives in the larger communities.
- The Kitikmeot Inuit Association is better organised. It had also settled its comprehensive land claim with the Federal Government — the Nunavut Land Claims Agreement, which protects water quality flowing from the project area into Nunavut, and Inuit animal harvesting rights outside Nunavut.

All four groups have traditionally used the project area for hunting and fishing. It is a highly contentious region where two separate land claim agreements were under negotiation and another was being considered. BHP Diamonds Inc and government officials often had to meetseparately with Aboriginal groups.

An important intercultural communication problem exists between the Aboriginals, whose world view and points of reference in discussions are rooted in traditional knowledge (TK); and Southerners, who try to impose science, and a belief in progress, rational analytical thinking, objectivity, reductionism and the Judeo–Christian ethic of human domination over nature.

TK is derived from life experience and observation by Aboriginal people. Elders and active hunters and trappers pass it on orally from generation to generation. Such knowledge is tested continually in daily living and survival, and only that which has value is handed down. TK is practical common sense based on knowing the land, the environment and the relationships among things. It is holistic and cannot be compartmentalised or separated from the people who hold it because it defines their identity.

Western scientists and decision makers have had a tendency to dismiss TK as anecdotal, nonquantitative, without method, and unscientific. Whereas western science generally excludes the humanistic perspective, holistic TK embraces language, culture, practice, spirituality, mythology, feelings, customs and even the social organisation of local communities. Science is but a small part of indigenous knowledge. Similarly, to suggest that TK is only the equivalent of science is to diminish TK's strength and breadth, which, in turn, often leads to basic misunderstandings and TK's trivialisation because it makes no sense without these contextual elements.

Having always lived off the land, Aboriginal people of the NWT have an intimate unique knowledge of the climate, land, water, and animal behaviour and plant life. During the EIA process, they stressed, "this land is ours" and said they were the "guardians" of the land, and not just "landlords" (CEAA, 1996, page 11; BHP, 1995, page 35).

THE PROJECT AND ITS IMPACTS

Mine, infrastructure and transportation

BHP Diamonds Inc planned to mine five pipes under existing lakes. The sequence of their development would be: Panda, Misery, Koala, Panda underground, Fox and Koala underground and Leslie (Figure 2). Panda, Koala, Fox and Leslie pipes are near each other. Misery is 29 kms. to the southeast, adjacent to Lac de Gras. Seven lakes with an area of 890 hectares would be lost: five for mining, one for tailings disposal (Long Lake), and Airstrip Lake to be used as an aggregate source for construction. After the open pits were exhausted, an underground mine at Panda would be constructed. It would last 5 years. There would then be a second underground mine at Koala.

The ore would be processed at a processing plant near the Koala pit that would handle 9,000 tonnes daily, and, after ten years, 18,000 tonnes. Ore would be crushed and diamonds separated from the ore by physical rather than chemical means (BHP, 1995, pages 14–15).

BHP Diamonds Inc chose Long Lake to be a basin for tailing impoundment because of its size and its small watershed area. Dams at the outlet and around the perimeter would increase its capacity. The basin would be divided into five cells by rock dikes. As the tailings in each cell consolidate and convert to permafrost, rocks and soil would cover it. The soil would be revegetated and the basin converted into a wetland (BHP, 1995, pages 17–18; CEAA, 1996, pages 24–26).

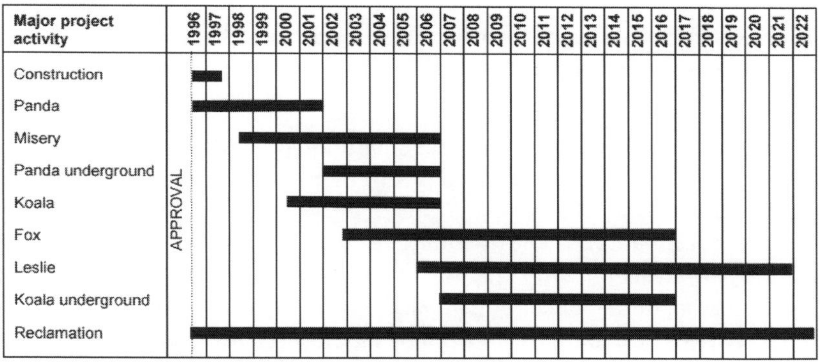

Figure 2. Schedule of major project activities

The work schedule to construct these facilities would be based on a two-weeks-in/two-weeks-out, 70 hour/week rotational work schedule. During operation there would be a permanent camp to house the 400-person rotational work force at Koala and a smaller camp for the Misery pipe.

BHP Diamonds Inc did not propose to build an all-weather road. During the winter, it would use 476 km of the winter ice road across frozen lakes and rivers, which Echo Bay Mines Ltd operates between Yellowknife to its gold mine at Lupin north of the Project. During the construction period, 2,250 truckloads of equipment and consumables would pass over the road, and, during operation, 2,000 truckloads of fuel would pass. The Proponent would construct an all-weather road to the Misery pit to haul ore for processing (BHP, 1995, page 12; CEAA, 1996, pages 27–28).

Air transport would be essential. The existing airstrip could accommodate Hercules C130, and Boeing 727 and 737 jets to carry all personnel to and from the site, and to bring food in and ship diamonds out.

Environmental issues

Water quality and the management of tailings water from Long Lake were central issues. DIAND and DOE concluded that the environmental impact statement (EIS) had addressed adequately the most important issues. BHP Diamonds Inc would revisit certain specific issues not definitively resolved in the EIA review during its application for a water licence from the NWT Water Board (CEAA, 1996, pages 34–36).

Threats to the caribou could come from disruption of movements and migration corridors, deaths caused by vehicles and changes to water quality. The risks were small because the project area represents less than 0.01% of the herd's range. It was later found that less than 2% of the herd approached the project area. As for the bears, between 1991 and 1995, 58 were killed in the Coppermine/Slave area, with only six deaths being ascribed to industry (CEAA, 1996, page 43; IEMA, 1998, page 22).

Obstacles to reclamation include the cold environment, limited topsoil, low soil moisture and the short growing season. BHP Diamonds Inc planned on-going reclamation during the life of the Project. A combination of good mining practice and reclamation would minimise vegetation loss. Decommissioning and closure would begin after all the ore has been processed. With prior progressive reclamation, decommissioning would only involve removing all structures, burying foundations and removing culverts to restore water flow patterns.

Financial and occupational issues

The total project capital expenditure of EKATI™ was an estimated CAN$1.2 billion (CAN$1.00=US$0.73 on 1 January 1997). The Project would contribute CAN$6.2 billion to the Canadian GDP (gross domestic product) and would generate CAN$400–500 million annually. The direct, indirect and induced benefits of the Project to the NWT would be CAN$2.5 billion, with 60% coming from wages and benefits. The income from the Project would be considerably greater than government expenditure for the expanded physical and social infrastructure. For every $1 accrued by Canada from the Project, the Federal Government and the GNWT would pay out $0.05 (BHP, 1995, pages 42–43).

The Project would have very wide economic and social effects on Northerners, notably on Aboriginal people. It would employ twice as many people as any other mine at that time in the NWT, that is, 830 over the mine's 25-year-life span. Yellowknife would benefit both from direct employment of residents and the purchase of goods and services. During operation 70 cents of each purchase dollar would be spent in the North, with 60% of that in Yellowknife (BHP, 1995, pages 40–41).

The Proponent's hiring policy would reduce severe northern unemployment. It would hire first NWT Aboriginal people, then non-Aboriginal NWT residents, and then other Canadians. It would base its selection process on personal aptitude, rather than standard employment criteria, such as formal schooling, experience and qualifications. When Aboriginal people lacked skills (for instance, literacy and technical) the Proponent would train them.

It would give preferential treatment to Aboriginal businesses, establish scholarship programmes and on-the-job training for Aboriginal students, crosscultural training in the work place and bring in Aboriginal elders, youth and organisations to share their traditional skills and customs to solve community problems. There would be a drug- and alcohol-free work environment with offenders being fired without exception (BHP, 1995, pages 10–11, 37, 39–40, 50; CEAA, 1996, pages 47ff, 50, 53; IEMA, 1998, page 22).

Table 1. Aboriginal population of communities and their distance from the project

	Community	**Population**	**Distance from project** (km)
Treaty 11 Dogrib Dene	Rae-Edzo	1,600	330
	Wha Ti	415	370
	Rae Lakes	255	310
	Snare Lake	135	180
Akaitcho Treaty 8 Dene (Chipewyan Dene)	Łutselk'e	300	250
	Dettah	190	310
	N'dilo	150	310
North Slave Métis Alliance	Yellowknife	1,200	310
	Rae-Edzo	50	330
Kitikmoet Inuit Association	Kugluktuk (Coppermine)	1,200	400
	Umingmaktok	50	350

Yellowknife was the designated point of hiring. BHP Diamonds Inc would fly employees from Yellowknife to the mine. It would also bring people in at no cost from tiny outlying communities, that is, Snare Lake, Rae Lake, Wha Ti,

Łutselk'e and Kugluktuk (Table 1). The rotational work shifts would permit Aboriginal employees both to participate in a wage economy and to maintain traditional lifestyles and contact with families and communities (CEAA, 1996, page 50).

BHP Diamonds Inc did not plan to do any diamond processing or manufacturing. It would sell 35% of the diamonds to De Beers and the rest at its own sales office in Antwerp (BHP, 1995, page 16).

Traditional knowledge

BHP Diamonds Inc faced difficulties with TK. Since Aboriginals use TK when negotiating claims and view it as their intellectual property, they did not want to make TK public. They insisted that its use and management should remain in their hands. As the Dene and Métis each have their own TK, the data did not always coincide. Furthermore, the information was scattered, and there was no documented baseline of TK, single protocol for standards or research methods (CEAA, 1996, page 17).

In spite of the problems, BHP officials could not simply dismiss TK as being irrelevant because the local people used it as their point of reference when voicing their concerns. Nor could the company risk being accused of tokenism. It did use TK when collecting data for archaeological and wildlife studies. It also used Dene/Métis maps (BHP, 1995, page 34).

Monitoring

BHP Diamonds Inc proposed the creation of an Environmental Advisory Group to assist in monitoring programmes for air, water, land and socio-economic effects. Its tasks were to include (CEAA, 1996, page 18):

- gathering information for regulatory compliance;
- measuring of operational performance and effectiveness of mitigation strategies;
- monitoring both natural change and that caused by the Project;
- assessing the validity of predictions; and
- initiating a response to unexpected adverse impacts.

PUBLIC REVIEW BY EA PANEL

BHP's pre-hearing consultations

BHP Diamonds Inc followed the policy of its parent firm BHP Ltd that being a 'good neighbour' is good for business. In 1992, it began public consultations in what proved to be a successful corporate public consultation programme in a remote place. The costs of meeting the affected public and acceding to reasonable requests were tiny compared to the potential costs of delays. A one-month delay would effectively impede the Project for up to one year because winter is the only time to transport equipment, fuels and other essentials, and summer the only time for major construction and lake dewatering. In turn, a stoppage would translate into very large sums of money from opportunities lost, interest on investments and credibility among investors. Furthermore, BHP Diamonds Inc would benefit from access to Northerners' unique environmental knowledge (Couch 2000b).

Between 1992 and the beginning of the panel review, BHP Diamonds Inc visited all communities in the project area at least twice. It targeted local and regional residents and organisations, especially Aboriginal people, and government bodies. It used public presentations, field trips, community meetings, open houses, cultural exchanges and joint workshops, and made audio and videotapes. To allow the Aboriginal leaders to observe first hand the quality of its relations elsewhere with Aboriginal people, BHP Diamonds Inc took a group to three mines owned by its parent company in New Mexico where 75% of the work force is Aboriginal.

PANEL REVIEW

During the quarter century following the Berger Inquiry, officials carried out a list of EIAs for large Northern proposals using procedures bearing Berger's clear influence. The review of the BHP NWT Diamond Mining Project is an example in this line of reviews.

Since the Project could have potentially significant adverse environmental impacts, the federal Minister of DIAND, as the minister responsible for the Project's proceeding, referred it to the federal Minister of the Environment for a public review by an independent Environmental Assessment Panel (26 July 1994), as required under the federal Environmental Assessment and Review Process Guideline Order, 1984 (see Table 2).

Table 2. Chronology of events in BHP's NWT Diamond Project

Year	Event
1983	Fipke set up Dia Met
1989	Fipke and Blusson found indicator minerals at Lac de Gras
1990	August – BHP Ltd. and Dia Met formed a joint venture
1991	September – Fipke and Blusson found first diamonds
1992	BHP Diamonds Ltd. began public consultations
1993	October – BHP Diamonds Ltd. opened the Koala camp
1994	26 July – Minister of DIAND referred Project for a Panel review
1995	July - BHP Diamonds Ltd. submitted its EIS to the Panel
1996	22 January – 23 February – Panel held public hearings
	21 June – Panel submitted its Report to the Ministers
	8 August – Minister of DIAND replied to Report
1997	February – Final approval obtained
	May – Construction began
	28–30 May – Inaugural meeting of IEMA's Board of Directors
1998	June – IEMA's first Annual Report
	15 October – Official opening of the ETAKI™ mine
1999	January – First diamonds sold in Antwerp
2001	June – BHP Billiton bought Dia Met

The Minister of the Environment appointed the Panel to review the Project (9 December 1994). Its members needed the appropriate professional expertise and credibility among all review participants, and to be free from any economic conflict of interest or membership of a political party. They came from outside government service and were paid on a per diem basis during the review, whereupon their connection with the Project ended. It had four members (CEAA, 1994):

- the chair was a lawyer based in Calgary (Alberta) with 24 years' experience in dealing with NWT Aboriginal people, governments, small businesses and environmental groups;
- a retired professor of geology who had spent most of his career working on Northern issues;

- a Dene who was a communications specialist with Aboriginal people; and
- an economic consultant based in Yellowknife and specializing in NWT resource and environmental issues.

The Canadian Environmental Assessment Agency (CEAA) is the Canadian federal agency responsible for overseeing EIA. CEAA recommended to the Minister the names of people who could serve on the Panel and provided the Panel's Secretariat. It worked out of CEAA's Vancouver Regional Office, but set up a communications office in Yellowknife. The Secretariat had close ongoing dealings with DIAND, the GNWT, the Proponent and Aboriginal groups (CEAA, 1996).

To ensure fairness, CEAA announced (14 December 1994) that the Federal Government would provide participate funding to help the affected public take part in the public review in a meaningful way, that is, to review the EIS for its adequacy, and to attend and prepare presentations at the public hearings. Recipients had to have a firm plan on how they would spend the money in ways related to the review, and show that they could later account for its proper use (CEAA, 1995e).

The Panel issued Operational Procedures (23 January 1995). The Panel would accept no information unless it was also available to the public. All documents would be placed in a public file in Yellowknife. People wishing to participate were to inform the Secretariat's Executive Secretary for inclusion on the mailing list for important announcements. Like Berger over 20 years earlier, it had formal procedures for the presentation of scientific information and social concerns, and informal procedures for community meetings. Information received at both types of gathering would be treated equally (CEAA, 1995a)

The Panel issued to BHP Diamonds Inc the Draft Guidelines for the Preparation of the EIS and announced a 70-day review period (31 January 1995). In February 1995, CEAA awarded CAN$105,000 to 14 groups to participate in the scoping exercise. Between 14 March and 8 April 1995, the Panel visited eight NWT communities and held scoping meetings to define the content and scope of the final guidelines. The Panel heard from 125 people (CEAA, 1995b).

The Panel issued to BHP Diamonds Inc its 19- page Final Guidelines for the Preparation of the EIS (23 May 1995). It provided that (CEAA, 1995c):

"For many issues that are raised in the Guidelines, traditional knowledge will have as important a contribution to make as scientific and engineering knowledge. The Proponent should fully consider local traditional knowledge and expertise when preparing the EIS."

The Guidelines appeared in English, the Indian languages Dogrib, Chipewyan and North Slavey, and Inuinnaqtun.

CEAA announced a second award of CAN$150,000 to 12 groups to help them take part in the public review (7 July 1995) Recipients included five Aboriginal groups, the NWT Chamber of Mines, the Yellowknife Chamber of Commerce, the NWT Federation of Labour, and the NWT Construction Association (CEAA, 1995e).

BHP Diamonds Inc submitted its EIS to the Panel (25 July 1995). It had four volumes: Project Description; Environmental Setting; Environmental Impacts and Mitigation; and Summary of the EIS. In December 1995 and January 1996, BHP submitted two information responses to the Panel (Tailings Management Plan and Preliminary Design of Retention Structures) and 12 baseline study updates.

The summary, which was for general distribution, appeared in English, Inuinnaqtun, Dogrib, Chipewyan and North Slavey. BHP Diamonds Inc prepared audiocassettes based on the summary in Dogrib, Chipewyan and North Slavey. In all, BHP Diamonds Inc presented to the Panel documents that placed side by side would be one metre in extent.

The Panel held 18 days of public hearings in nine NWT communities (22 January–23 February 1996). There were ten days of technical sessions in Yellowknife to review issues pertaining to TK, environmental management plans, water, wildlife and vegetation and socio-economic effects. The Panel received 75 written submissions and heard about 260 presentations (CEAA, 1995d).

The sequence for oral presentations was: the Proponent explained its position on the session's designated topic; Government agencies presented their comments and technical reviews; then anyone else could speak who had signed up in advance. All participants could pose issues of BHP Diamonds Inc and each other. There was no cross-examination. When an intervener gave a written text to the Secretariat in time, it made copies for distribution at the hearings.

During the community meetings the Panel made special efforts to accommodate Northerners who were unfamiliar with, or felt uncomfortable participating in, public meetings. These meetings were in fact discussions ordered along the lines of other Aboriginal meetings. Translation services were available in all remote communities. The Panel showed deference to the customs of local communities by inviting an Aboriginal elder to begin each meeting with a traditional prayer.

At the end of the hearings, the Panel retired to write its report behind closed doors.

Panel's report

The Panel presented its report to the federal Ministers of the Environment and DIAND on 21 June 1996. Although it concluded that the environmental impacts were mitigable and the Project could proceed, it made 29 recommendations. They dealt with:

- water quality and management;
- the speedy settlement of outstanding land claims;
- the need for new policies, and monitoring and management plans to deal with the loss of fish habitat, air quality, the risk of trucks' spilling fuel, caribou, birds, protection of the interests of Northern business, and the exploration of archaeological and historical sites;
- Aboriginals' hunting and fishing rights on the Proponent's claim area; and
- the Federal Government's developing a capacity for diamond valuation in Canada prior to their export and sale.

The Panel recommended that there should be a policy for TK's inclusion in EIA developed by the GNWT, Aboriginal people and industry, with the most immediate need being TK guidelines and standards for developers preparing an EIA (CEAA, 1996).

During BHP's application for a water licence, the NWT Water Board would consider outstanding issues such as: the risk from tailings and other sources (such as the contamination of waste-rock drainage by nitrates or nitrites from ammonium nitrate-based explosives) and potential impacts on Inuit rights under the Nunavut Land Claims Agreement.

The report appeared in English and French, and its executive summary in Inuinnaqtun, Dogrib and Chipewyan.

EIA's cost

Preparing an EIA in the North is more expensive than in the South. The cost of living is higher because there is no inexpensive surface transport for bulk goods. Most products from the South must be flown in. Travel outside Yellowknife must be done by airplane or on foot. The dearth of 'off-the-shelf' baseline information, the larger areas to be studied and the extreme climate impose a heavier burden on proponents.

The Panel review cost CEAA about CAN$1,000,000 plus CAN$255,000 for participant funding. This sum does not include the costs of DIAND, DOE, DFO, the GNWT and other government agencies. The environmental studies cost BHP Diamonds Inc more than CAN$10,000,000.

Yet, in comparison to the Project's capital cost, the anticipated profits to BHP Diamonds Inc and the tax revenue to governments, this outlay was very small. Furthermore, much of the data in the Proponent's environmental studies would be used to address other technical and engineering matters.

Agreements to manage Project's impacts

Minister's announcement of 8 August 1996

The Hon Ronald Irwin, the Minister of DIAND, began the second phase of the process on 8 August 1996 when he announced his acceptance of the Panel's report and the next steps in the approval process and negotiations, which led to the creation of new institutions, practices and interjurisdictional relationships to deal with environmental and social issues.

As some of the Panel's recommendations fell outside the scope of existing laws, and there were no agreements with the Aboriginal people, there was a risk that the Project could be tied up by endless wrangling. The Minister surprised everyone by giving participants 60 days to achieve "satisfactory progress" in the negotiation of agreements(DIAND, 1996).

The chief components of the new régime were: an environmental agreement, which included the creation of the Independent Environmental Monitoring Agency (IEMA); a socio-economic agreement; obtaining a licence from the Water Board; authorisation for the loss of fish habitat under the Fisheries Act; land leases; and impact and benefits agreements between BHP and the Aboriginal groups.

The need for the Minister's conditional approval, the tight time frame, the specificity of some of his goals and the ministerial discretion he allowed himself, together put pressure on everyone. Yet it was in the vital interest of each to bargain in good faith and to reach an accord. For the Proponent, a delay could mean the loss of time and money. The Aboriginal groups saw they had a defined window of opportunity to get what they wanted from BHP Diamonds Inc. The Minister did send a special envoy to Yellowknife who helped facilitate negotiations and served as a conduit for information and directions to and from his office in Ottawa.

Participants recall the intense activities up and down the corridors of Yellowknife's Explorer Hotel, as parallel negotiations were taking place on the separate, but closely interrelated agreements.

Environmental Agreement

The Environmental Agreement is a legally binding accord that consolidates the numerous environmental issues contained in the Panel's recommendations. It aimed at harmonising the activities of many government bodies, serving notice of the Federal Government's commitment to sound environmental management, and assuring on-going Aboriginal participation in the monitoring process. The Federal Government, the GNWT and BHP Diamonds Inc negotiated the Agreement with the active participation of the four Aboriginal groups. The final signing took place on 6 January 1997 (DIAND, 1996; Kennett et al, 1997, pages 19–25; IEMA, 1997; IEMA, 1998)

The Agreement required BHP Diamonds Inc to

- prepare a plan for environmental management during the Project's construction and operation for IEMA's review and the Minister's approval;
- submit annual reports on its environmental management plan to DIAND, the GNWT, the IEMA and Aboriginal groups;
- prepare an impact report every three years on the Project's environmental impacts;
- set up a monitoring programme for air and water quality, and wildlife;
- submit its reclamation plan for the Minister's approval;
- establish a security deposit (CAN$11,075,000) for potential land impacts and give a guarantee of CAN$20,000,000 for potential water impacts: the Minister could draw down on deposits to enforce compliance;
- incorporate TK into all environmental plans and programmes where it would receive full consideration with scientific knowledge. The Proponent would complete its study of TK to identify where it could be incorporated into its activities. Aboriginal people would play a major role in its design and implementation. TK would remain the property of Aboriginal people with the Proponent's staff being forbidden to disclose proprietary information to outsiders.

The Agreement created the IEMA to be the independent public watchdog for monitoring — the most neglected aspect of EIA. The IEMA's advent came as a surprise because the Panel had not recommended the creation of such a body. The IEMA's tasks would be to:

- integrate the various aspects of the Environmental Agreement;
- prepare annual reports on the Project's environmental implications; • review impact reports;

- participate as an intervener in legal and regulatory processes on environmental issues;
- provide a public document repository in its Yellowknife office; and
- inform BHP Diamonds Inc, governments, Aboriginal people and the general public about the Project, its monitoring and its regulation.

The IEMA's Board of Directors has seven members: four appointed by Aboriginal groups and the rest jointly by Federal Government, the GNWT and BHP Diamonds Inc from outside their own ranks. The Federal Government and the GNWT would help the Proponent with funding during the first two years, but thereafter, the IEMA would negotiate its annual budget with BHP Diamonds Inc.

Socio-economic Agreement

The GNWT and BHP Diamonds Inc negotiated the Socio-economic Agreement. It dealt with economic benefits and social impacts on all NWT residents and established contractual commitments for activities outside existing laws. Unlike the Environmental Agreement, this Agreement did not have noncompliance penalties. It was signed on 22 October 1996 (Kennett et al, 1997, pages 25–26).

The Agreement provided for preferential hiring, recruitment criteria, specific employment targets, employment of contractors, training programmes, and employment support (i.e. orientation, cross-cultural training, counseling, safety). It set targets for giving contracts to, and purchases from Northern businesses and employment of Northerners and Aboriginals; and reporting indicators on community health.

Licence from NWT Water Board

The NWT Water Board had nine members appointed by the Minister of DIAND from federal agencies and Aboriginal groups (Kennett et al, 1997, pages 28–33; IEMA, 1998, pages 14–18; DIAND, 1997).

BHP Diamonds Inc applied for a Type A water licence for water use and waste disposal. The Board held public hearings (9–10 September 1996). Since BHP Diamonds Inc had considered the Panel's recommendations to be an agreement-in-principle, it did not anticipate that the Board would request the amount of detailed information that it did

Because of the complexity of some issues, the unpreparedness of BHP Diamonds Inc and government officials, and the heavy scrutiny by some

Aboriginal interveners and their advisors, the Board convened a second phase of hearings (21–22 October 1996). These meetings were more conciliatory because of the progress made in parallel negotiations, the use of interrogatories and informal meetings between interested parties. The Minister approved the licence on 5 February 1997.

The licence was the most comprehensive and detailed ever issued by the Water Board. It contained special provisions that had never before been put in a Class A licence. These included: the need for the monitoring of aquatic effects, kimberlite's toxicity, waste rock and mill tailings; and reclamation needs. BHP Diamonds Inc gave a deposit as security to cover the full costs of reclamation. The term of the licence was from 1 January 1997 to 31 December 2004.

Authorisation of fish habitat destruction

The Fisheries Act is an important federal environmental law. As noted, in the South, many environmental laws are provincial laws. A key federal power comes from this Act's provision for the protection of fisheries habitat management. Since fish live almost everywhere in Canada, the Act permits the Federal Government to become involved in fisheries and environmental issues anywhere in the country. It has special significance in the North because fish are so important in the residents' diet (Kennett et al, 1997, page 33)

Although DFO has a policy of no net loss for fish habitat, the Project Area's harsh climate made it unrealistic to compensate for the loss of habitats by creating new habitats as in the South. In a compromise, BHP Diamonds Inc gave CAN$1.5 million to DFO to create the Fisheries Rehabilitation Fund for creating new fisheries habitat and habitat enhancement, with those used by Aboriginal people receiving priority. The authorisation was signed on 7 January 1997.

Land leases

Crown land for mining in the NWT cannot be bought, but can be leased for a specified period. The Lands Advisory Committee and the Regional Environment Review Committee reviewed applications and made recommendations to DIAND for its decision. The Committees consisted of federal and GNWT officials, and representatives of Aboriginal and public organisations (DIAND, 1997; Kennett et al, 1997, pages 9–10, 34).

The Minister issued six leases to BHP Diamonds Inc for 30 years that authorise it to occupy and use the lands for the open pits, the mine mill, the airstrip, an on-site road and the camps at the Koala and Misery sites. The Environmental Agreement and the water licence set out the conditions. Should the Minister decide BHP Diamonds Inc is causing damage to the environment, he may retain a portion of the security deposited under the Environmental Agreement, or in exceptional cases, suspend operations or even terminate the lease.

Impact and benefit agreements

The impact and benefit agreements (IBAs) are private, bilateral, confidential agreements between BHP Diamonds Inc and each of the four Aboriginal groups. Since there was no settled land claim agreement and BHP Diamonds Inc wanted good relations, it took the initiative to begin negotiations (1994). They reflect the general view that Aboriginal people would benefit from Northern development. (Kennett et al, 1997, pages 26–28, 89–97)

IBAs included employment practices and targets, business opportunities, training, scholarships, and transportation to and from the communities, as well as cash payments to the Aboriginal groups during the life of the mine.

CONCLUSIONS

BHP Diamonds Inc began the construction of its plant facilities in May 1997. It started production in October 1998 and began selling diamonds in January 1999. Canada is already the world's sixth largest producer of diamonds both in terms of value and weight. When all NWT diamond producers come on line, Canada will mine about 12% of the world's diamond output by value.4

The Project contained the seeds of disagreement. The huge amount of money could have led to clashes over its distribution among any combination of players. There was a potential discord arising from the cultural world views of a multinational corporation and of local Aboriginal people. Also the Aboriginal groups had conflicting interests among themselves.

Yet in the end, all the players reached an agreement and there has been no contention that the agreement was basically unfair. There were various reasons for this success.

- BHP Diamonds Inc followed the corporate policy of avoiding a confrontation that could ignite conflict and lead to significant financial losses. It began consulting Aboriginal groups and government officials early. It cooperated with the Panel and permitting agencies when directed to prepare complex and costly studies. BHP Diamonds Inc got what it wanted asthe Project proceeded as planned.
- The Aboriginal groups recognised that the moment was ripe to maximise their benefits. With adequate resources and legal representation, they too approached the bargaining tables with confidence and in good faith. As provided for in the Panel's report and the environmental, socioeconomic

and impact and benefit agreements, they too got what they wanted — economic benefits, defence of and respect for their needs and values (for instance, TK's acceptance), and an important role in future decision-making. By 2000, 78% of EKATI™ employees were Northerners, of whom 40% were Aboriginals.

- The success of the review process was critical. The public review was properly managed. Panel members had no economic interest in the Project and when the review ended so did their connection with the Project. It was in the interest of each member's professional career to act, and to be seen to act, fairly. The Panel Secretariat faced the same challenge. Everyone came to accept the Panel and its Secretariat.
- The Federal Government's funding of small local participants ensured there would be a more level playing field and acceptance of the EIA process.
- The Panel's report synthesised in a short, easyto-read document, an enormous amount of complexly interwoven printed scientific and social information, along with the views given orally by experts, Aboriginals and the general public. It focused governments' responses by making concrete recommendations for action that were sufficiently flexible to give participants room to manoeuvre during their implementation. These included policy/planning, social impact and site-specific matters and the forming of new interjurisdictional relationships. The review process and the report helped to depoliticise the final decision by placing in the hands of the Federal Cabinet — the ultimate decision-makers — a document that defanged anyone who may have wanted to use the Project's approval for partisan political reasons.
- There was also the potential for endless differences over details during the permitting and negotiation phase. By setting of a very tight time frame and the targeting of discussion on specific solutions, the Minister of DIAND forced all parties to reach a quick accord on the specifics in a list of thorny policy, regional and site-specific matters, or risk losing the very large benefits that the Project held for all.
- The two-step process set a precedent for the review and approval of diamond-mining proposals that shortly followed.
- Creating the IEMA assured independent, ongoing, and rigorous and adequately resourced monitoring. The Proponent rather than the Canadian taxpayer is now paying for it.
- The IEMA's Board of Directors held their inaugural meeting on 28–30 May 1999. Its first Annual Report (1997–1998) made 18 recommendations to BHP Diamonds Inc and regulatory bodies to give coherence to the various

regulatory requirements. These included: - determining TK's role; - assessing the wildlife-monitoring plan, DFO's management of the fish habitat compensation fund, and the management of tailing and kimberlite wastes; and - a monitoring programme for new cumulative effects in the Coppermine River's watershed. (IEMA, 1998).

- Subsequent Annual Reports have reported on and updated these initiatives, as well as other emerging issues.

The sum of money spent was a bargain for both the taxpayer and BHP Diamonds Inc when it is compared with the enormity of the capital costs and the anticipated profits, the size of the area studied, the fragility of the Arctic environment, and the need for real and enduring justice towards Aboriginal people who had been the victims of 'progress' over the generations.

There has been a major change in the ownership of EKATI™, as the control of diamond mining in the NWT in general passed from regional Canadian firms into the hands of multinational corporations.

In June 2001, the parent firm BHP Ltd merged with another multinational, Billiton, to become a world industry leader in the extraction and production of many natural resources.

In early 2000, Charles Fipke agreed to transfer his sharesin Dia Met to his estranged wife as part of their divorce settlement. In November 2000, she and another major shareholder made known their desire to sell their controlling interests. BHP Billiton bought Dia Met in June 2001. It now owns 80% of EKATI™, with Fipke still holding 10% and Blusson 10%.

BHP Diamonds Inc abandoned the Leslie pipe as unviable, but proposed three others on their claim area (the Sable, Pigeon and Beartooth pipes). Matters of concern were similar to those discussed during the first review. The newly created Mackenzie Valley Environmental Impact Review Board reviewed the proposal and agreed that these plans could proceed (BHP, 2000; MVEIRB, 2001).

Two other players followed BHP Diamonds Inc into NWT diamond mining through their own EIAs and permitting. The multinational Rio Tinton plc owns 60% of the NWT's second diamond mine, Diavik Diamond Mines Inc and a Canadian company, Aber Resources Ltd, owns the rest. Diavik located four economic pipes 10 km southwest of EKATI™'s Misery Pit, carried out extensive public consultations and environmental studies and received the perquisite permissions (1999). It should begin production in 2003.

Two Canadian companies discovered the third economically viable diamond deposit at Snap Lake, 230 km northeast of Yellowknife and 100 km south of

EKATI™. In 2000, De Beers acquired the Snap Lake development. Production should begin in 2005.

Diamonds from EKATI™ are from a high to very high quality. One kilogram of rough stones has an average value of US$500,000 and in a retail jewelry store would fetch US$4 million. The GNWT is trying to get a share of the large value-added profits by fostering local diamond processing and related manufacturing activities. BHP Diamonds Inc and Diavik will sell some of their diamonds locally. Three Canadian firms have opened cutting and polishing facilities in Yellowknife. They are either owned by, and/or work closely with, Aboriginal groups.

NOTES

1. Canada has 779,790 Aboriginal inhabitants, or 2.7% of its total population. They consist of: 529,035 North American Indians (1.9%) who are spread across the country; 40,225 Inuit (Eskimos) (0.1%) who live North of the tree line, along the sea coast in the Northwest Territories, Nunavut, northern Québec and Labrador; and 204,115 Métis (0.7%) who are of mixed European–Indian descent and who live in the Western Provinces of Manitoba, Saskatchewan and Alberta, and in the Northwest Territories (NWT Bureau of Statistics, 1996). There are 53 indigenous languages spoken in Canada. About 17,000 people speak one of the 15 languages in the Athapaskan family, that is, those Indian languages most frequently used in the vicinity of EKATITM.

2. There are 633 legally constituted autonomous, independently minded Indian bands ranging in size from 80 members to 16,000.

3. The Inuit are a separate people and not an Indian group that developed separately in the far North. The term 'Inuit' is now used instead of 'Eskimo', which is a pejorative expression derived from the Cree Indian language meaning 'raw-meat eater'. Inuit means 'the people' or 'real people' and comes from the Inuit–Inupiaq language. The singular of 'Inuit' is 'Inuk' and means 'a person'.

4. Forecast of diamond production by value for 2005: Botswana, 27%; Russia, 22%, Canada, 12%; South Africa, 12%; Angola, 8%; Democratic Republic of the Congo, 6%; Australia, 5%; and Others, 8% (NWT, DRWED, 2001, page 8)

REFERENCES

1. Berger, Thomas R (1977), Northern Frontier, Northern Homeland: the Report of the Mackenzie Valley Pipeline Inquiry (Minister of Supply and Services, Ottawa).
2. BHP Diamonds Inc (1995),. NWT Diamonds Project, Environmental Impact Assessment: Summary (BHP Diamonds Inc, Vancouver).
3. BHP Diamond Inc (2000), Environmental Assessment Executive Summary: Sable, Pigeon and Beartooth Kimberlite Pipes (BHP Diamonds Inc, Yellowknife, April).
4. CEAA, Canadian Environmental Assessment Agency (1994), BHP Diamond Mine Environmental Assessment Panel, Newsletter, No. 1, 12 December.
5. CEAA, Canadian Environmental Assessment Agency (1995a), "Environmental Assessment Panel reviewing BHP Diamond Mine: operational procedures", BHP Diamond Mine Environmental Assessment Panel, Newsletter, No. 2, 23 January.
6. CEAA, Canadian Environmental Assessment Agency (1995b), BHP Diamond Mine Environmental Assessment Panel, Newsletter, No. 4, 8 March.
7. CEAA, Canadian Environmental Assessment Agency (1995c), "Environmental Assessment Panel Reviewing BHP Diamond Mine: final guidelines for the preparation of an environmental impact statement", BHP Diamond Mine Environmental Assessment Panel, Newsletter No. 6, May.
8. CEAA, Canadian Environmental Assessment Agency (1995d), BHP Diamond Mine Environmental Assessment Panel, News release, "Funding for public participation in the review of the BHP Proposal", 7 July.
9. CEAA, Canadian Environmental Assessment Agency (1995e), BHP Diamond Mine Environmental Assessment Panel, Newsletter, No. 11, 13 December.
10. CEAA, Canadian Environmental Assessment Agency (1996), NWT Diamonds Project, Report of the Environmental Assessment Panel, June.
11. Couch, William J (2000a), "Oceny Oddziaływania na środowisko po obydwu stronach Atlantyku Ocena", in Jan Żelazo (editor), Oceny oddziaływania na środowisko: Praktyka polska i procedury w krajach

Unii Europejskiej (Wydawnictwo Szkoła Główna Gospodarstwa Wiejskiego, Warsaw) pages 4–20.

12. Couch, William J (2000b), "The role of public participation in environmental impact assessment", Aplinkos Inžinerija, VIII(1), pages 41–-53.

13. DIAND, Department of Indian Affairs and Northern Development (1996), "Backgrounders for Canada's Diamond Mine Project", News release, 8 August.

14. DIAND, Department of Indian Affairs and Northern Development (1997), "BHP Diamond Mine going through final regulatory steps", News Communiqué, 7 January.

15. Gamble, Donald J (1978), "The Berger Inquiry: an impact assessment process", Science, 199, 3 March, pages 946–952. IEMA, Independent Environmental Monitoring Agency (1997), Environmental agreement, available at , last accessed 1 March 2003.

16. IEMA, Independent Environmental Monitoring Agency (1998), Annual Report, 1997–1998 (IMEA, Yellowknife, NWT).

17. Kennett, Steven, Monique Ross and Owen Saunders (1997), Independent Review of the BHP Diamond Mine Process (Canadian Institute of Resources Law, University of Calgary, Ottawa and Department of Indian Affairs and Northern Development).

18. MVEIRB, Mackenzie Valley Environmental Impact Review Board (2001), Report on the Environmental Assessment of the Sable, Pigeon and Beartooth Kimberlite Pipes (MVEIRB, Yellowknife).

19. NWT Bureau of Statistics (1996), , last accessed 1 March 2003. NWT DRWED, Department of Resources, Wildlife and Economic Development (2001), Diamond Facts, 2000/01: NWT Diamond Report (NWT, Yellowknife).

20. Royal Commission on Aboriginal People (1996). Report of the Royal Commission on Aboriginal People. Vol 1 (Department of Supply and Services, Ottawa)

Chapter 10

The Hazardous Nature of Small Scale Underground Q7 Mining In Ghana

K.J. Bansah [a,*], A.B. Yalley [b], N. Dumakor-Dupey [b]

[a] *Missouri University of Science and Technology, Department of Mining and Nuclear Engineering, Rolla-Missouri, USA*

[b] *University of Mines and Technology, Department of Mining Engineering, Tarkwa, P.O. Box 237, Tarkwa, Ghana*

ABSTRACT

Small scale mining continues to contribute significantly to the growth of Ghana's economy. However, the sector poses serious dangers to human health and the environment. Ground failures resulting from poorly supported stopes have led to injuries and loss of human lives in recent times. Dust and fumes from drilling and blasting of ore present health threats due to poor ventilation. Four prominent small scale underground mines were studied to identify the safety issues associated with small scale underground mining in Ghana. It is recognized that small scale underground mining in Ghana is inundated with unsafe acts and conditions including stope collapse, improper choice of working tools, absence of personal protective equipment and land degradation. Inadequate monitoring of the operations and lack of regulatory enforcement by the Minerals Commission of Ghana are major contributing factors to the environmental, safety and national security issues of the operations.

Keywords: Small scale mining; Safety; Human health; Mercury amalgamation; Environment;Underground

1. INTRODUCTION

Small scale mining (SSM) may refer to the mining of ore deposits by individuals or groups of persons with little technical know-how and characterized by minimal or no mechanization. Some countries define SSM operations based on output and human resources needed (Coakley, 1999). However, the definition of SSM in Ghana has been based on the amount of capital and human resources needed. The Minerals and Mining Act 2006 (Act 703) of Ghana defines a small scale gold mining operation as the mining of gold by any effective and efficient method that does not involve substantial expenditure by an individual or group of persons not exceeding nine in number or by a co-operative society made up of 10 or more persons. Small scale mining is recognized as a major contributor to national income and a pillar for poverty reduction in developing countries (Hentschel, Hruschka, & Priester, 2003). Small scale mining of precious minerals such as gold and diamonds has provided employment for thousands of Ghanaians, especially indigenes of SSM communities, and has made significant contributions to the foreign exchange earnings of the country. In 2014, the sector produced 1.49 million ounces of gold representing 34.3% of Ghana's total gold output (Ntibrey, 2016). The sector has also helped in stemming rural-urban migration, and provided raw materials for both foreign and local mineral industries. As a result SSM is recognized by government as a cornerstone of a multimillion dollar industrial sector, the products and sales from which are controlled. The government of Ghana under the auspices of a German Non-Governmental Organization (NGO), Gesellschaft Technishe Zusannebarbeit (GTZ), and the World Bank has undertaken a number of initiatives to formalize and regularize resident SSM operations (Hilson, 2001).

In Ghana, SSM has been carried out for hundreds of years. It is currently and widely operated in the country by both licensed operators and unlicensed miners popularly known as *galamsey* operators. According to the Ministry of Lands, Forestry and Mines of Ghana, the number of small scale miners increased rapidly by 941.73% from 1984 to 2004 after its legalization following the promulgation of the Small Scale Gold Mining Law, PNDC Law 218 of 1989. The legalization was to revive the SSM sub-sector, facilitate supervision and minimize associated environmental hazards.

Although the efforts by government and the NGOs have noticeably improved the efficiency of operations, certain serious concerns continue to be largely ignored by the miners, the Minerals Commission and government, and have increasingly become unmanageable. Ground failures resulting from weak unsupported or poorly supported stopes have led to fatalities and various degrees of injury in recent times. Dust and fumes generated from chiseling, drilling, blasting, grinding and crushing of ore are potential health threats. Most of the stopes worked in by the miners are accessed by adits without any adequate ventilation systems in place leading to the accumulation of dust and fumes in the underground workings. Health hazards related to dust and fumes are well documented in literature (Dockery et al., 1989, Bascom et al., 1996, Gielen et al., 1997,Pope et al., 1999, Yu et al., 2000, Dockery, 2001, Gan et al., 2004, Gauderman et al., 2004, Jansen et al., 2005, Colucci et al., 2005 and Bansah and Amegbey, 2012). SSM activities can generate high levels of noise which can result in hearing impairment, as the use of hearing protection is largely ignored by the miners. Other problems related to noise include stress related illnesses, high blood pressure, speech impairment and lost productivity as described by (Pulles et al., 1990, Maschke et al., 1998, Maschke et al., 2002, Maschke et al., 1995 and Ising and Kruppa, 2004).

Small scale mining operations endanger the environment by inducing land degradation and contaminating surface and groundwater resources. Wide open excavated areas are left unreclaimed while heavy metals, total suspended solids, dissolved solids, and other water contaminants are introduced into water bodies by mining and mineral extraction activities of the small scale miners. Mercury amalgamation technique, which is heavily relied on for gold extraction can pose serious health threats and is deleterious to a wide range of ecological entities (Harada, 1995).

Even though, SSM in Ghana is by law limited to only Ghanaians, the last decade has seen a large increase in involvement by foreign nationals, mostly Chinese miners and migrants from neighboring Togo, Burkina Faso and Ivory Coast and others from western and non-western cultures. The small scale mining activities of these foreign nationals have involved the destruction of cocoa farms, wide areas of land and protected forests, and reported as security threat to the people of Ghana (Al-Hassan & Amoako, 2014). In 2013, an inter-ministerial taskforce (drawn from the military, immigration and police) was set up by the president of Ghana to crackdown on these illegal miners. During that year, the Ghanaian authorities arrested and deported over 4500 illegal Chinese miners (The Guardian, 2013). The Chinese foreign minister visited Ghana in 2014, and pledged Beijing's support to help tackle the illegal small scale mining issue (Agence France-Presse, 2014). Although government's interventions have reduced the scale illegal mining activities by foreign nationals, there are still some foreign nationals operating in remote mining areas (e.g. Manso Nkran) of the country.

The influx of migrant miners has led to a higher level of mechanization of the operations (use of excavators, trucks, dredging machines, crushers, etc.) and has increased the scale of mining. Consequently, this has increased the levels of land degradation, rechanneling of river/stream courses, and contamination of surface water bodies in terms of increased turbidity, total suspended solids (TSS), and total dissolved solids (TDS) to unacceptable levels. Bansah and Bekui (2015) report unacceptably high turbidity levels of water samples from the Bonsa River in the Western Region of Ghana, and attributed the levels to pronounced SSM activities in and around the river. The colour of most streams in the vicinity of the operational areas can also be described as aesthetically objectionable.

It is publicly known that small scale miners in Ghana operate in unsafe conditions, and pose serious threats to other land users and the environment. However, very little research has been conducted to ascertain and address the safety and environmental issues related to small scale underground mining in Ghana. This paper therefore identifies the safety issues in small scale underground mines in Ghana and suggests methods to improve the safety of operations. Field visits were made to four different small scale underground mines; Dakete Mining Limited, Mohammed & Co. Small Scale Mining, Johnson Mining Company Limited, and Stejoan Mining Group, around Tarkwa in the Western Region of Ghana to assess the conditions of operations. Mine owners and mine operators were interviewed on site while the underground workings were accessed for relevant first hand information. Visits were also made to the office of the Minerals Commission, a regulatory body, to interact with personnel on their views on small scale underground mining in Ghana and the need for health and safety training for the mine operators to mitigate the dangers associated with SSM operations in Ghana.

2. LEGAL AND REGULATORY FRAMEWORK OF SMALL SCALE MINING

According to the World Bank, small scale mining is widespread in developing countries in Africa, Asia, Oceania, and Central and South America (World Bank, 2013). Small scale mining (both surface and underground) is practiced in Ghana and is also reported to occur in South Africa, China, Zambia, Congo, Rwanda Philippines, and India (Bugnosen, 2001, CIFOR, 2012, Chakravorty, 2001, Gunson and Yue, 2001, Hentschel et al., 2003 and Kambani, 2003). In an effort to manage and promote an efficient SSM sector, the South African, Chinese, Indonesian and Ghanaian Mining Laws, for example, have provisions for small scale mining. The legal requirements while sharing some similarities, vary from one country to the other. This section briefly reviews the relevant legal and regulatory framework regarding SSM in South Africa, China, Indonesia and Ghana.

2.1. South Africa

Small scale mining in South Africa is regulated by the Mineral and Petroleum Resources and Development Act (MPRDA) 2002. The Department of Mineral Resources (DMR) established the Directorate of Small-Scale Mining to develop and address the challenges faced by the small-scale mining sector in South

Africa (DMR, 2012). Mutemeri, Sellick, and Mtegha (2010) report low level of compliance as one of the challenges facing the SSM sector. They further indicated that small scale miners do not comply with the requirements to submit monthly production reports, making it difficult for government to accurately quantify the mineral production and other statistics of the sector.

The SSM sector in South Africa is categorized into: (i) artisanal or subsistence mining operations (new entrants); (ii) sub-optimal formal mining operations; and (iii) entrepreneurs with upfront capital (DMR, 2012).

The Mine Health and Safety Act (MHSA), 1996 and amendments regulate health and safety in small scale mining. The Mine Health and Safety Council was set up to advise the Minister on issues of legislation, research and promotion of occupational health and safety in the mining industry. It also oversees research in the mining sector pertaining to issues of health and safety (Mutemeri et al., 2010). The MHSA (Act 74, 2008) provides for safety, health and environment (SHE) training, and requires employers to keep training records for individual employees. Environmental management is regulated by National Environmental Management Act (NEMA) 1998 by guiding the environmental provisions provided under the Mineral and Petroleum Resources and Development Act (MPRDA) 2002.

2.2. China

The mineral industry in China is governed by the 1986 Mineral Resources Law and the 1996 Mineral Resources Law. According to Shen, Dai, and Gunson (2009) substantial body of law relating directly or indirectly to the artisanal and small scale mining sector exists in China. These laws, regulations and measures are grouped into four categories (Andrews-Speed, Yang, Shen, & Cao, 2003) as follows: (i) laws and regulations of right to mineral resources; (ii) regulations of coal mining operations; (iii) laws, regulations and measures of environmental protection and land management; and (iv) regulations of township and village mines (TVMs).

Shen et al. (2009) indicate that the first category was promoted by the restructured former Ministry of Geology and Minerals Resources (now Ministry of Lands and Natural Resources). The Ministry implements the rules and issues exploration and mining licenses. The second category covers mine development, operations, production, management, marketing and safety. The third category, which relates to environmental protection and management, is administered by the Ministry of Environmental Protection. Laws and regulations regarding the protection of soil and water ecosystems are covered under this category. The final category of regulations is specific to specific to small scale mines. The regulations in this category address a wide range of issues from engineering and safety standards, to inspection procedures, role of different levels of government, responsibilities of mine managers, employee contracts, and the legal status of township and village mines.

2.3. Indonesia

The Indonesian Mineral and Coal Law (2009) allows Indonesian citizens, as individuals or cooperatives to apply for a license to conduct small scale mining in designated areas. According to Macdonald, Lund, Blanchette, and Mccullough (2014), a mechanism of licensing, permitting, management and control of small scale mining is not clearly stipulated in the law. However, the management and control of SSM is fully decentralized to regional governments. The decentralization of authority is recognized as a significant contributor to unintended growth in illegal small scale mining in Indonesia (Gita, Primanti, Zaki, & Ismawati, 2012). As a signatory to the 2013 UNEP International Treaty on Mercury, the use of mercury in mining is illegal in Indonesia.

It is reported in the literature that Indonesia lacks the institutional and technical capacity to provide adequate assistance to assess impacts or enforce compliance, especially at the local and regional levels (Sousa, Veiga, Meech, Jokinen, & Sousa, 2011). According to Macdonald et al. (2014), the sheer numbers of artisanal and small scale miners and locations, combined with poorly understood temporal and spatial variability of impacts on aquatic ecosystems complicate efforts of local and regional environmental managers to regulate activities.

2.4. Ghana

In 1989, the Small Scale Gold Mining Law (PNDCL 218), Mercury Law (PNDCL 217) and the Precious Minerals Marketing Corporation (PMMC) Law (PNDC Law 219) were promulgated after the government recognized the contribution of SSM to Ghana's development. The Minerals Commission of Ghana, which is responsible for assisting the Minister of Mines with small scale licensing procedures and monitoring of operations, established seven small scale mining district centers (in Tarkwa, Dunkwa-on-Offin, Bibiani, Asankrangwa, Assin Fosu, Akim Oda, and Bolgatanga) to provide technical extension services to the miners and improve supervision. In particular, the district centers were required to perform the following functions: (i) compile a register of all small scale gold miners and prospective small scale gold miners specifying such particulars as may be determined by the secretary; (ii) supervise and monitor the operation and activities of the small scale gold miners and prospective small scale gold miners; (ii) advise and provide such training facilities and assistance as may be necessary for effective and efficient small scale gold mining operations; and (d) submit to the Minerals Commission in such form and at such intervals as may be directed by the commission, reports or other documents and information on small scale gold mining activities within the district (Small Scale Gold Mining Law, PNDCL 218 of 1989). Currently, there are nine district centers that include the Wa and Konongo district centers to provide such services to the small scale mining community.

Under the law, no person shall engage in or undertake any small scale gold mining operation unless granted a license. The law allows provision of mining license to only Ghanaian citizens who are 18 years or over and registered by the district center in the designated area. The procedure by which individuals have to obtain small scale mining licenses are described by the miners as tedious and expensive, hence most small scale miners operate illegally. Institutionally, the Minerals Commission is responsible for policymaking and regulatory activities in the mining industry. A 2002 report by the Commission indicates that 420 SSM concessions were licensed in Ghana. Of these, nine were diamond licenses and 411 were gold. One thousand four hundred and thirty six (1436) licenses have been issued to small scale miners in Ghana as of April 2015, (Ntibrey, 2015).

The Ghanaian laws legalize the use of mercury in SSM; "a small scale gold miner may purchase from any authorized mercury dealer such quantities of mercury as may be reasonably necessary for the purposes of his mining operations" (Small Scale Gold Mining Law, PNDCL 218 of 1989). The Minerals and Mining Act, 2006 permits the use of explosives by small scale miners with the written permission of the Minister on the recommendation of the commission (Parliament of the Republic of Ghana, 2006).

Unlike China and South Africa, the Ghanaian laws have no specific regulations or measures for ensuring safety and the protection of health and the environment in the small scale mining sector. Under section 11 of PNDCL 218, "a person licensed to mine gold under this Law may win, mine and produce gold by any effective and efficient method and shall in his operations observe good mining practices, health and safety rules and pay due regard to the protection of the environment". This provision is vague, as neither "effective and efficient method" nor "good mining practices" are defined, giving room for miners to operate in the manner they like, and posing a challenge to the monitoring officers in effectively discharging their duties. The Minerals and Mining (Health, Safety and Technical) Regulation, LI 2182 offers regulations regarding health and safety in the mineral industry of Ghana. This regulation however, lacks specific and detail requirements for health and safety issues in small scale mining operations. Of significant mention is the absence of specific regulations regarding small scale underground mining in Ghana.

There are over 18 legislation and reforms (ICMM, 2015), of which 10 have been repealed. These legislations and reforms are however more geared towards the large scale mining. The few available regulations are not enforced to the letter. According toMacdonald et al. (2014), law enforcement in artisanal and small scale mining in Ghana is poor and often unevenly applied.

3. SMALL SCALE MINING METHODS

The nature of the mineral deposit generally dictates the mining method to be employed for exploitation. Historically, SSM methods for exploiting mineral deposits have involved the *anomabo*, chisel and hammer, underground "ghetto", and the dig and wash methods, described below. More recent methods that involve the use of simple inexpensive Chinese-made equipment or more expensive and sophisticated machinery are being practiced. These recent methods brought about as a result of influx of Chinese nationals into Ghana are locally referred to as "changfa", "more blade", dredge, and alluvial washing plant methods in resident communities. The following sections briefly discuss the SSM methods adopted in Ghana.

3.1. Anomabo method

This method is a dredging method used in the mining of gold bearing gravel from river beds. In this method, a miner dives and scoops the gold bearing gravel from the river bed. The gravel is put in a bucket, hoisted up onto a raft and then washed onshore. This method is employed after the rainy season when loose materials are eroded and deposited on the river beds (Yamoah, 2002, pp. 18–19). Typically, a crew of two persons is involved in this mining method.

3.2. Chisel and hammer method

According to Tepkor (2005, pp. 5–9), this method is employed for the mining hard rock (lode) formations occurring mainly as outcrops. The vegetation and the top soil are removed either manually or mechanically to expose the rock. Chisels and hammers are used to fragment the gold bearing rock. The broken rock is transported to be crushed and ground by mechanical crushers or manually with metal mortars and pestles into fines. Sluicing is usually used to recover the gold particles from the ground materials (Acheampong, 2009, pp. 3–22).

3.3. Underground "ghetto" method

This method involves mining of hard rock from underground workings. The workings are mostly in old and abandoned mines and are accessed mainly by shafts or adits. The fragmented gold bearing rock is carried to the surface in bags by the miners for processing. The underground method employed by small scale miners is labour intensive and also involves high risks due to possibility of stope collapse in poorly supported stopes.

3.4. Dig and wash method

This method is used for mining alluvial gold deposits that occur on the banks of rivers, in old valleys, on terraces, or in the tailings dumps of old mine workings. Tools such as spades, shovels and pickaxes are used to dig the materials. The dug material is then transported and usually stockpiled along the banks of nearby streams or water bodies where they are subsequently washed.

Figure 1. Artisanal Small Scale Mining Methods: (a) Changfa in Operation, (b) "More Blade" Method, and (c) Typical Dredge Method: Changfa motors (arrowed red), suction hose (arrowed blue) and working platform (arrowed dark orange).

3.5. Changfa method

The changfa method originated from the name of a Chinese-made (changfa) diesel powered rock crusher that is used for crushing excavated auriferous deposits. The unit is mostly fabricated and welded or assembled locally in Ghana by Chinese entrepreneurs. The method typically involves excavating ore deposits manually with simple tools such as pickaxes, shovels or spades. Head-loads of the materials are then carried (by women in most instances) or in wheelbarrows (by men) to the changfa location for processing. The materials are re-handled by shoveling into the crusher for comminuting. A water pump connected to the crusher continuously pumps water, mostly from a sump onto a sluice board (usually lined with a blanket) connected to the crusher chute for washing. After a while the changfa motor is shut down and the blanket is removed and washed in a drum containing water to obtain the trapped gold for further processing. Typical duration of operation is 8 h/day with the labourers earning between US$25 and US$40 a day.Fig. 1a shows a changfa used in Tarkwa for crushing rocks at a SSM site.

3.6. "More blade" method

This method is a modified form of the dig and wash method. It involves the use of excavators for mining in the pit. It is usually done close to a stream or river where the water can be directed into the pit. The site owner hires a gang of typically 50 people. A 10-person crew is assigned a sluice board in the pit. The excavated products are re-handled with shovels onto the sluice boards where the materials are washed, as some members of the crew manually fetch water and pour it onto the sluice board. Other members of the group may also use pans to wash the materials, complementing the sluice board washing. It is labour intensive and the employees typically work for 6 h a day with an average income of US$25 per person for every 3 days. Fig. 1 (b) shows the "more blade" method of mining.

3.7. Dredge method

This mining method is similar to conventional methods using suction dredges. However, the suction dredge in this method is locally made from discarded metals and powered by two changfa motors. The gold-bearing sediments are excavated directly by suction from the stream/river bed and transported hydraulically onto sluice boards mounted along the bank of the stream/river for washing (see Fig. 1c). This method is less labour intensive than the changfa and "more blade" methods. It typically involves a crew of 5 persons who together, usually earn a little over US$500 over a period of three weeks, if yields are high.

3.8. Alluvial washing plant

This method is quite mechanized and is less labour intensive. It involves the use of two excavators and a mini washing plant. The two excavators work in tandem: the first, excavating the auriferous deposit and stockpiling it adjacent to the second excavator; and the second, re-handling the excavated materials onto the washing plant for washing. This method has high output and typically involves a 9-person crew.

4. SOCIO-ECONOMIC IMPACT OF SMALL SCALE MINING

There is a wide range of published literature by researchers and international organizations such as the World Bank, International Institute for Environment and Development (IIED), and the International Labour Organization (ILO) on the socio-economic significance of small scale mining (World Bank, 2013, Kitula, 2006, Hilson, 2001, Bryceson and Jønsson, 2010, Chakravorty, 2001 and Mwaipopo et al., 2004). The World Bank describes the artisanal and small scale mining sector as an important livelihood and income source for the poverty affected local population. According to the Bank, the artisanal and small scale mining sector ensures the existence for millions of families in rural areas of developing countries. It estimates that about 100 million people (workers and their families), depend on artisanal and small scale mining compared to about 7 million people worldwide in industrial mining (World Bank, 2013). The sector has made immense contributions to the production of minerals in the world.

The small scale mining sector contributes largely to the economy of host countries. In China, for example, the artisanal and small scale mining sector employed over 5 million people and produced over half of the mineral production in 2006 (Shen et al., 2009). The small mining sector in Ghana is reported to contribute significantly to the growth of the country (Amankwah et al., 2015, Aryee et al., 2003, Hilson, 2006 and Yakovleva, 2007). It offers direct or indirect employment to mostly unskilled or less-skilled individuals from the rural communities where they are practiced, as watchmen, mechanics, electricians and equipment operators, porters, food sellers etc. Although, there is no precise data on SSM population in Ghana, the Minerals Commission estimates that more than 1, 00,000 persons directly engage in SSM in Ghana.

Ghana earns foreign exchange from the export of minerals. According to the Minerals Commission of Ghana, the small scale gold mining sector contributed 7% of the $694,970,543 gold revenue to the government in 2002. According to Ntibrey (2016), artisanal and small scale mining contributed 34.3% (1.49 million ounces) of total gold production in 2014, and 33.9% (1.44 million ounces) in 2013. All the diamonds produced in Ghana since 2008 came from artisanal and small scale mining (Ntibrey, 2016). The industry also provides some of the raw materials for mineral-based domestic industries. These include kaolin for the manufacture of local paint and cosmetic powder, salt for pharmaceutical products, mica for the ceramic industry and gold for the manufacture of jewels (Acheampong, 2009). According to Oblokuteye (2010, pp. 15–17), the commencement of SSM activities in rural areas has helped in the minimization of the problems of rural-urban migration. Indeed, according to the World Bank, small scale mining helps to stem rural-urban migration (World Bank, 1995).

However, the sudden influx of miners to new areas, in an attempt to escape poverty and get rich, introduces different cultures sometimes to the disadvantage of the indigenes. The dominance of male miners creates fertile grounds for prostitution and sexual promiscuity. There is also the danger of increased levels of truancy, school dropout and rising illiteracy levels as school-age children get engaged in SSM activities (Al-Hassan & Amoako, 2014).

5. ENVIRONMENTAL AND SAFETY ISSUES IN SSM

SSM over the years has resulted in a wide range of negative impacts on the environment (Kitula, 2006, Aryee et al., 2003 and Bonzongo et al., 2003). Tarras-Wahlberg, Flachier, Lane, and Sangfors (2001) studied the environmental impacts and metal exposure of aquatic ecosystems in rivers contaminated by small scale gold mining in the Puyango River basin of Southern Ecuador. Their study found that the discharge of cyanide, mercury and metal rich tailings into rivers of the Puyango catchment area from small scale mining were causing considerable environmental impacts. Cyanide and metal levels in the rivers were found to exceed environmental quality criteria.

In Ghana, the environmental impacts of small scale mining (mostly surface operations) have included water contamination, destruction of fauna, land degradation, and mercury contamination of soils and water (Aryee et al., 2003, Babut et al., 2003, Hilson, 2002 and Bonzongo et al., 2003). There are also reported cases of disregard for basic safety protocols resulting in death, damage to property and to the environment (Al-Hassan & Amoako, 2014). The following sections discuss the various environmental and safety concerns associated with SSM in Ghana.

5.1. Effect on land

Coomson (2004, pp. 5–9) describes land degradation as the major impact of SSM on the environment in Ghana. Flora and fauna are destroyed in the process of mining. The dimensions of surface SSM openings vary from shallow (<30 m) to deep depths (>30 m). The mining of deep deposits typically produce wider openings than shallow deposits. The openings are often left unreclaimed after the ore extraction. Small scale underground openings are commonly not backfilled after mining. These openings have the tendency of collapsing and trapping or killing farmers, hunters, and animals. Water may also fill abandoned stopes or pits, serving as potential breeding zones for mosquitoes.

Examples of deep trenches and pits left after mining by small scale miners in Ghana are shown inFig. 2a.

Figure 2. Environmental Impacts of SSM: (a) Abandoned deep pits and trenches, (b) Excavated river bank.

5.2. Effect on water

SSM of alluvial gold is believed to be the major cause of river water pollution in Ghana (Al-Hassan and Amoako, 2014 and Aryee et al., 2003). Due to the dredging activities and the washing of alluvial gold in the water, siltation is common in major rivers and streams where the miners operate. The operations have also changed some water courses of streams and rivers (Oblokuteye, 2010), depriving downstream users of their only source of water. Fish and other aquatic organism die-off can occur, altering the food web significantly. Individuals depending on the rivers for fishing are forced to abandon their source of livelihood resulting in less food availability in the communities. The Bonsa River, for example, which was relied on by farmers and some community members for their source of drinking water and fish has been affected by illegal small scale mining.Fig. 2 (b) shows a section of the Bonsa River where the mining activities occur. Piles of excavated materials have been heaped along the river bank with trees felled into the river. The impact of small scale mining on aquatic ecosystem in other parts of Ghana is reported by Babut et al. (2003) and Donkor, Bonzongo, Nartey, and Adotey (2006a).

5.3. Air quality, noise and poor ventilation

Air pollution resulting from mining related activities comes from the generation of dust and emission of mine gases especially during drilling, blasting, grinding and crushing of ore. Al-Hassan and Amoako (2014) discuss ambient air quality deterioration by fine particulates released from the sieving of crushed rock obtained by small scale mining. In surface operations, dust and emissions are diluted by the wind. Due to the confined nature of small scale underground mining (SSUM) operations, dust generated in the stopes accumulates, serving as potential health threat to the miners. Most of the stopes are accessed by adits (same opening for entry and exit) without any ventilation system in place. This issue is typically as a result of ignorance on the part of the operators. Some of the small scale mining operators interacted with were unaware about the existence of fans that can supply the underground workings with adequate air.

The small scale underground miners typically resume working in the stopes after blasting, when they no longer see suspended dust particles, or smell the blast fumes. However, this approach can be misleading since fine inhalable particles can remain suspended and not seen with the eye. Mohammed & Co. small scale underground mine attempts to solve the air quality problem in their stopes by installing a ventilation fan to decontaminate the underground workings.

Small scale mining activities can generate loud noise that can result in hearing impairment of the miners and nuisance to the residents of SSM communities. The changfa crushers produce loud noise that can affect the hearing of the operators who often operate the machines without any form of hearing protection. People who live in close proximity to the operations see the small scale mining activities as major source of nuisance. For example, residents at Nkanponase, a community located opposite to the University of Mines and Technology in Tarkwa complain about nuisance from blasting activities by small scale miners who operate in close proximity to the community. According to the 67 year old queen ("ohemaa") of the community- Nana Afua Nakra, the sudden blasting activities conducted by the miners frighten the community members who believe that such practices could endanger their health:

"They don't tell us about the blasting; we only hear of sudden loud noise and ground vibrations from their blasting work. We get frightened; sometimes it happens even when we are sleeping, and personally, it makes my heart beats faster. The last time for example, I had all my louver blades come off the building".

To support some of her claims, the queen mother granted access for visual inspections of the interior walls of her bedroom where cracks attributed to the blasting were discovered in the building. A 70-year old carpenter and other community members shared similar complaints. According to the community members, all efforts to address the issues have proven futile. Fig. 3 shows structural deformations in building structures at Nkanponase that are attributed to blasting activities by the small scale miners.

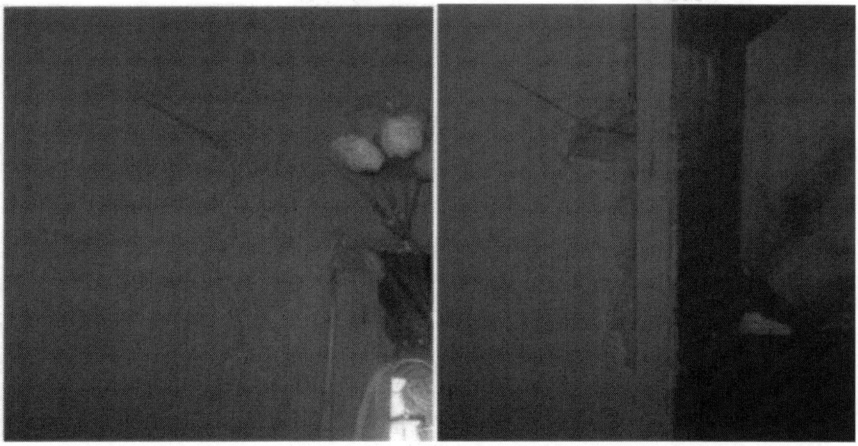

Figure 3. Cracks in the walls (bedrooms) of Building Structures at Nkanponase. *Residents attribute the cracks to blasting activities by small scale miners.*

5.4. Ground failures

Ground failures resulting from unsupported or poorly supported stopes and poor pit design have led to fatalities and various degrees of injury in recent times. In early 2016, six small scale miners were killed at Kyekyere in the Ashanti Region of Ghana when a pit wall collapsed. A collapse that occurred in 2015 killed 17 people at a small scale mine in the Central Region of Ghana. And at least 45 people were killed in 2010 when an illegal gold mine collapsed after heavy rains, while at least 18 people (including 14 women) were killed in November, 2009.

Opoku-Antwi (2010) indicates that small scale miners invariably do not report accidents that occur at the mining sites because of fear of drawing public attention. Indeed the public only gets to know of accidents that result in fatalities at the small scale mining sites. Non-fatal accidents are kept hidden from the general public. Hand and feet injuries that ever occurred in the small scale mines (the authors visited) were never reported to authorities or made public.

In the authors' opinion, the ground failures can be attributed to lack of planning, ignorance about the nature and types of rock (in terms of strength and stability), and inappropriate choice of mining methods, occasioned by lack of technical expertise of the operators. This problem can be overcome by involving professional engineers (including mining and geological engineers). However, most of the professional engineers interacted with do not find small scale mining dignifying, besides the low incomes and therefore do not want to associate themselves with such operations.

5.5. Impacts of mercury amalgamation technique

Although the use of mercury for processing of ore has serious adverse effects on human life and the ecosystem, not much attention has been given to mercury contamination in Ghana (Donkor, Nartey, Bonzongo, & Adotey, 2006b). Unfortunately, there is no reliable data to assess the extent of mercury contamination in the communities in which mercury is used. Literature available on mercury contamination by small scale mining in Ghana include Bonzongo et al., 2003, Babut et al., 2003, Serfor-Armah et al., 2005 and Kwaansa-Ansah et al., 2010 and Donkor et al., 2006a and Donkor et al., 2006b. Babut et al. (2003), for example, in a United Nations Industrial Development Organization (UNIDO) study to determine the environmental impacts of mercury prior to the introduction of mercury retorts, conducted analysis on river water, soil and fish samples obtained from Dumasi, (a small scale mining village with about 2000 people) in the Western Region of Ghana. The results showed significant contamination of soil sediments. Most of the fish fillets were also found to have accumulated mercury levels that exceed the United States Food and Drug Agency (US-FDA) action level. The fish from the rivers were reported to be unfit for consumption.

Mercury amalgamation technique is heavily relied on for processing gold-bearing ores because mercury is inexpensive and readily available to the miner. All the small scale mines visited utilize mercury for extracting the gold. The technique simply involves mixing of washed ore with mercury. The gold bearing concentrate combines with the mercury to form an amalgam. The mercury amalgam is heated to get rid of the mercury to obtain the gold. A careful observation of the amalgamation process by the authors at some of the small scale mines revealed the high possibility of mercury losses. Apart from the evaporation of the mercury vapour into the atmosphere (because the retorting is done openly), the mercury sometimes pours onto the ground. The mercury is typically kept in small dispensing bottles. At one of the mines, a miner inadvertently stepped on one of these bottles, forcing its contents out onto the ground. Even though the mercury was recovered, there were still some losses. These findings were consistent with that ofBabut et al. (2003) that mercury losses in SSM mainly occur during amalgamation.

Also, the miners do the amalgamation without any form of personal protective equipment (PPE) such as gloves, overalls and nose protection. This makes them susceptible to mercury poisoning, or agents of mercury transfer to other systems (such as food and drinking water), as small scale miners commonly share food or drinking water with their colleagues and friends. It is not uncommon to find a small scale miner holding an amalgam with the bare hands.

6. MATERIALS AND METHODS

To ascertain the safety and environmental issues in small scale underground mining in Ghana, field visits were made to four licensed small scale underground mines at Tarkwa in the Western Region of Ghana. Visits were also made to the office of the Minerals Commission to interact with the district officer and his assistant (secretary) for relevant information such as number of licensed operators and logistics, and in general, their views on the small scale underground mining operations. The underground workings of the small scale mines were accessed for first hand information about the operations, while mine owners and operators were interacted with on the safety and environmental issues of their operations. Rock samples from the stopes of each mine was taken for rock strength analysis to determine the uniaxial compressive strength of the rocks in which the miners operate. The following sections describe the study site, methodology of study, and the mining and safety practices at the small scale underground mines in Tarkwa, Ghana.

6.1. Study site and methodology

Tarkwa which is host to eight licensed small scale underground mines is a historical mining community with series of old abandoned underground mine openings. It currently hosts three large scale mining companies including Goldfields Ghana Limited, AngloGold Ashanti Iduapriem Limited, and Ghana Manganese Company Limited, (all operating surface mining). Goldfields Ghana Limited and AngloGold Iduapriem Limited produce gold, while Ghana Manganese Company produces manganese. Tarkwa is the capital city of the Tarkwa-Nsuaem Municipal Assembly and is located on latitude 5.3000° N and longitude 1.9833° W. The city is 300 km west of Accra (the capital of Ghana) by road. The 2013 census data put the population of Tarkwa at 34,941, with the main occupation of the people being mining.

The city has a small scale mining district support center to provide technical and extension services to the small scale mining sector as required by law. According to the district officer at the Minerals Commission, over 300 licenses have been issued to small scale miners to conduct small mining in the Tarkwa district. However, there are only 15 licensed small scale mines currently in active operations; five are surface operations and 10 are underground operations, of which four are in Tarkwa, and the others in the Nsuaem area of the Western Region of Ghana.

These four licensed small scale underground mining operations that produce gold in Tarkwa include Dakete Company Limited, Johnson Mining Company Limited, Stejoan Mining Group, and Mohammed & Co. Small Scale Mining. Together, these mines employ 507 permanent and casual workers: Dakete, 162; Johnson, 200; Stejoan, 100; and Mohammed & Co., 45. The industry is male dominated with the average age of the workers being 35 years. The income of the employees ranges from US$ 62 to US$ 400 per month for 8-h shift, depending on the nature of work. For example, a woman carrier who is paid a daily wage of 10 Ghana cedis (US$ 2.57) earns a monthly income of US$ 77 if she works every day for 30 days, while a miner, who chisels the rock, loads and carries the materials himself to the surface for processing may earn US$ 300 per month for 8-h shift.

Due to poor record keeping, statistical data at the mines are sketchy. However, it is estimated that more than 60% of the workers are indigenes of Tarkwa and have typical family sizes of five. The others include Ghanaians from other parts of the country and foreign nationals. The highest educational qualification of a worker is a three-year junior high school certificate. The typical working duration is 8 h; 8:00 am to 5:00 pm, with an hour break from 12:00 pm to 1:00 pm. The workers come to the surface to eat their lunch during the break time. Except Johnson Mining Company Limited which operates three shifts per day (24 h), all the other mines run one shift per day.

6.2. Mining process

The processes of mining at the small scale underground mine are captured under exploration, mine development, rock fragmentation, material handling, processing and marketing. Exploration involves the work done to gain knowledge about the grade, volume, orientation and all important information about the mineral to be mined. It helps to determine the economic viability of the proposed mining project. Development basically refers to making access to the mineral of interest in preparation for mining (extraction from the ground). In materials handling, fragmented or excavated materials are transported to a suitable location for processing. While processing involves the use of either physical or chemical methods (or both) to extract the mineral of interest from associated waste (gangue) materials, marketing basically involves the sale of the extracted mineral. The processes are described as follows:

(i)Exploration: Small scale miners generally do not conduct exploration. Where exploration is conducted, it is typically done on trial and error basis. The miners rely on their instincts, and experience gathered over the years by operating in a particular geological setting to identify mineral bearing rocks. They sample the materials (or rock) arbitrary with simple hand held tools such as chisel, hammers, picks and shovels into sacks for a confirmatory test. The test involves pounding the samples with metallic mortar and pestle into fines. The fines are then transferred into rubber pans (made from discarded car tyres) and water is added. The mixture is panned several times to ensure that the gold settles at the bottom. The suspension is decanted, leaving the gold concentrate (if any) at the bottom of the rubber pans. To the miner, obtaining some gold at the end of the washing is enough confirmation of the presence of gold in that area.

(ii)Development: Access to the orebody is through adits abandoned by old mechanised mining operations, or ones that have been constructed by the small scale miners themselves. The access to Stejoan is an abandoned adit with a ladder-way equipped with a wooden ladder for passage and iron beams for support. Johnson is accessed via an abandoned adit and a manually constructed adit. The abandoned adit is equipped with metal rails for support. Dakete and Mohammed are currently accessed by two abandoned adits. However, Dakete in previous years was accessed by manually constructed adit. Some of the mine accesses are shown inFig. 4(a) and (b). The development of adits by the miners is labour intensive since simple tools are used. The process involves loosening the in-situ formation with metallic hammer and chisel, and mucking the fragments into sacks using hand shovels, and manually transporting the load to the surface. These accesses are sometimes risky and dangerous as the miners have to descend into the stopes by holding onto ropes. At Dakete for example, a guide rope is used to descend into the stope by holding onto the rope as the miner descends into the stope. In narrow adits, the miner crawls into and out of the stope.

Figure 4. (a) Adit, (b) Descending into a stope using rope, (c) Miner carries broken rocks on the back, and (d) Wagon for carrying materials by pushing.

(iii)Rock Fragmentation and Material Handling: The principal cycle of operation in small scale underground mine operation involves: fragmentation of in situ rock; loading of ore into sacks; and hauling to the surface. These are similar methods to those applied during the development stage. That is, the orebody is fragmented with chisel and hammer, and loaded into sacks with shovel and carried to the surface for processing. When it becomes difficult to use rudimentary tools to break the rock, blasting with the use of explosives is adopted to break the rocks. This is done by drilling one to 2 m deep blast holes into the working face and charging with either dynamite or ammonium nitrate fuel oil (ANFO). The explosives are detonated to effect the fragmentation.

After the blasting, the rocks are loaded into sacks and carried on the heads, shoulders or the backs to the surface (see Fig. 4c). A sack of load is typically 30 kg–50 kg depending on the strength of the miner and how quick he wants to meet his target for the day. Sometimes, the sacks are tied to a rope and pulled to the surface by dedicated crew locally called *loco boys* (after locomotive). Some of the underground mines combine both manpower and semi-mechanised operations for material handling (see Fig. 4d). At Johnson for example, the broken rock is manually loaded into wheelbarrows and pushed to discharge its content through a chute into a wagon that is anchored to a winder on the surface. The material is hoisted to the surface when the load in the wagon reaches the expected level for processing.

6.3. Auxiliary operations/activities

The auxiliary operations include support systems, use of personal protective equipment, ventilation, lighting, and dewatering. These activities/operations are very necessary since the accidents that occur in small underground mining are as a result of not observing such activities. The following sections describe the auxiliary activities at the four small scale underground mines. Some of the safety failures that have led to fatalities and various degrees of injuries and continue to pose safety threats are also described.

(a)Support systems and ground failures: Stress redistribution can cause ground failures when an excavation is made into the ground. According to Mireku-Gyimah (2015, pp. 10–52), weak to medium strength orebody and country rock are susceptible to caving. It therefore becomes important to support underground openings to ensure that the openings are safe and accessible during the mining period.

The small scale underground miners typically operate more than one active stope that may either be supported or unsupported. In the supported stopes, waste rock, wooden pillars, bamboos and timbers are used to support the underground openings. At Dakete and Stejoan, sacks are filled with sand and or waste rocks, and then packed to reach the roof of openings to provide support.

Timber logs are also used for supporting underground openings at Dakete. However, most of these timber logs show signs of deterioration (see Fig. 5a). Mohammed & Co. combines bamboo, wooden pillars and timber for supporting the stopes. Unlike Dakete, the timber supports at Mohammed & Co. do not show significant deterioration. Freshly installed bamboo at Mohammed & Co. at the time of this study, aims to prevent loose rocks from falling from the roof of the drives and walkways (see Fig. 5b). Johnson and Stejoan utilize wooden pillars, bamboos and waste rocks for supporting the drives.

Figure 5. (a) Deteriorated timber support, (b), Bamboo support, (c) Wide Stope without any form of support, and (d) Wooden and timber support in a wide opened working area.

In the unsupported stopes, the voids are left open without any form of support. At one of the stopes in Dakete, for example, a 30 m wide stope was left open without support, while loose materials were found to have been left unsupported (see Fig. 5c). Pillars left as supports in old abandoned mine workings were gradually robbed by the small scale underground miners. This practice has the potential of causing the stopes to collapse, and injuring or killing the miners in the process. In 2010 for example, a stope collapse killed one miner at the Mohammed and Co. site. The miner according to the workers was working in an unsupported stope when the stope collapsed and killed him instantly. Fig. 5(d) shows wooden and timber support in a wide open working area.

To understand the conditions of the rocks (in terms of strength) worked on by the small scale underground miners, two rock samples from the roof, footwall and orebody were obtained from unsupported active stopes of each mine for strength analysis. Compressive strength tests were conducted on the samples at the geotechnical laboratory of the University of Mines and Technology in Tarkwa, Ghana. The testing procedure involves cutting (sawing) the samples into cubes of size 50 mm, and loading axially using a uniaxial compressive strength test machine until the samples fail. The load at which the samples fail is recorded and used to compute the uniaxial compressive strength of the rock by dividing that load by the surface area of the sample. The results of the rock strength test are given in Table 1. It is observed that the country/host rocks are weak to moderately strong. These rocks require artificial supports to make the workings stable and safe.

(b)Personal protective equipment (PPE): The Occupational Safety and Health Administration requires employers to provide their employees with PPE and also ensure its use when engineering, work practice and administrative controls are not feasible or do not provide sufficient protection (OSHA, 2003). According to thePersonal Protective Equipment at Work Regulations, 1992 (as amended), PPE should be regarded as the last resort to protect against risks to health hazards. The regulations prioritize the consideration of engineering controls and safe systems of work. In Ghana, the Minerals and Mining (Health, Safety and Technical) Regulation, LI 2182 requires the use of PPE by underground miners to minimise risk and the severity of injuries to them. As described earlier, the small scale underground miners do not have adequate engineering and administrative controls and safe systems in place, due to the fact that the operations are mostly done by unskilled labour. One important engineering control that is adopted by all the operators is the use of support systems in some of the drives and walkways to prevent ground collapse.

Table 1. Mean uniaxial compressive strength of rocks at the small scale underground mines.

Mine	Source of sample	Compressive strength, σ (MPa)	Interpretation of compressive strength, σ (MPa)
Dakete Mining Limited	Footwall	133.93	Very strong
	Orebody	213.69	Very strong
	Hanging wall	97.41	Strong
Stejoan Mining Group	Footwall	52	Strong
	Orebody	32	Medium strong
	Hanging wall	43.2	Medium strong
Johnson Mining Company Limited	Footwall	18.8	Weak
	Orebody	69	Strong
	Hanging wall	42.4	Medium strong
Mohammed & Co. Small Scale Mining	Footwall	24	Weak
	Orebody	52.8	Strong
	Hanging wall	52	Strong

Criteria: Extremely strong >250; Very strong 100–250; Strong 50–100; Medium strong 25–50; Weak 5–25; Very weak 1–5; Extremely weak 0.25–1. Values are in MPa.

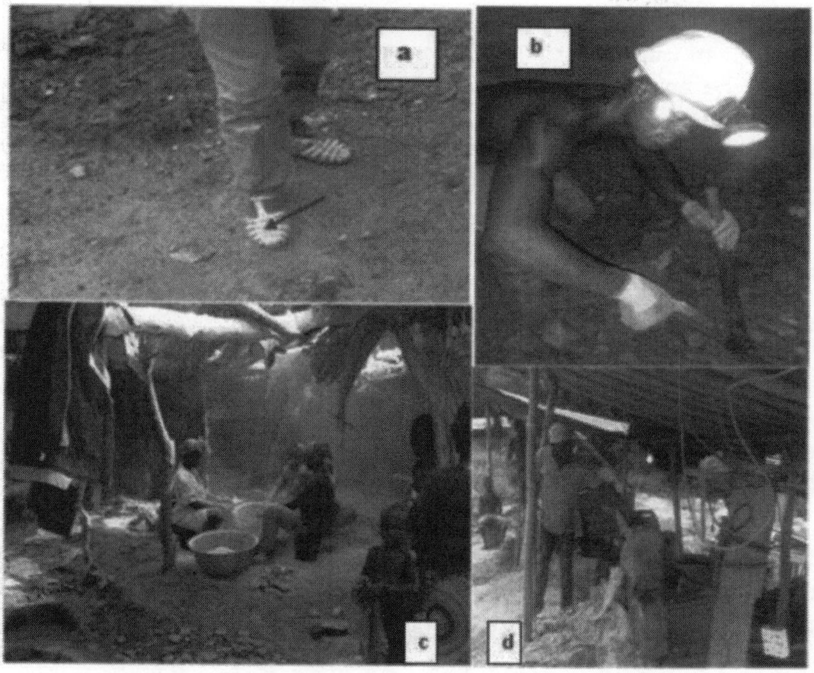

Figure 6. Some Safety Conditions in Small Scale Mining in Ghana: (a) Miner going underground in rubber sandals, (b) Much equipped miner breaks rocks using pickaxe, (c) Dust from sieving milled ore, and (d) Changfa operators without PPE.

However, due to the manual and hazardous nature of the activities at the mine, it is not uncommon for the miners to experience bodily injuries from rock falls, sharp (rock) edges, working tools, etc. Suitable PPE for use in the underground mining operations that can minimize the severity of injuries may include safety boots, protective helmets, nose mask, safety glasses, reflective overalls, ear plugs, lamps and oxygen supply set. Although, the Personal Protective Equipment at Work Regulations, 1992 (as amended) and the Occupational Health and Administration require employers to provide their employees with suitable PPE, the employers of small scale underground mines often ignore the regulatory requirement. The small scale miners generally operate without personal protective equipment. All the underground miners (over 60 people) encountered at Stejoan, Dakete and Mohammed & Co. operate without the use of PPE. According to the miners, any miner who ever used safety boots in any of these mines had obtained it as a gift from a friend who works in the large scale mining company. Some of the miners have never used safety boots. Three persons who have personally acquired their safety boots but do not use them indicate that they feel uncomfortable in them. Rubber sandals (see Fig. 6a) are worn by most of the miners who believe that such sandals provide them with firmer grip on the ground and prevent them from falling. Some of the small scale miners do not recognize the importance of PPE. They refer to their sandals as the "best and most comfortable PPE". The predominant use of dry cell battery-powered torches which sometimes fail as a result of battery die-off can pose serious safety risks to the miners as they have to walk or work in the dark places.

At Johnson Mining, however, the mine owner provides the miners with PPE such as helmet, safety boots, safety glasses, dust mask and rechargeable lamps. Even though these may not be adequate, the situation is improved compared to what pertains at the other mines. The mine owner ensures that every miner who goes underground uses the PPE provided. In general, safety concerns regarding the use of PPE and save systems of work by the underground miners are totally ignored. The story as told at Dakete Mining about a miner who was severely injured in the left foot is reported as follows:

The miner was originally contracted as a loader in 2012 to load broken rocks using hand shovel into pans for women carriers to convey the load to the crusher station. As a loader, his wage was Ten Ghana Cedis (US$ 2.60) for 3-h working duration, when the women carriers have met their target for the day. On 5th November 2012, according to Dakete, when the mine was about commencing drilling and blasting operations, they discovered that the loader had knowledge in drilling (because he was an underground miner at the erstwhile Amalgamated Banket Area Underground mine, now Goldfields Ghana Limited, Tarkwa). He was absorbed into the mainstream employment of the company as drill help with a daily wage of Fourteen Ghana Cedis (US$ 3.64) and then increased to Sixteen Ghana Cedis (US$ 4.16) in December, 2012. In March 2013, he anchored a drilling bit to the drill stem and begun drilling into the rock of the working face. Suddenly, the drill bit got disengaged from the drill stem and without much control the powered drill stem landed on his left foot and pierced through the foot, injuring him severely.

Even though safety failures such as lack of training and failure to secure are of significance in this case, the degree of injury could have been reduced if the miner had worn safety boots. Three years after the incident, the underground miners operate without the requisite training and personal protective equipment. Fig. 6b shows an underground miner digging rocks with a pickaxe.

(c)Dust, fumes and poor ventilation: Dust and fumes (e.g. carbon monoxide and oxides of nitrogen) are generated during chiseling, drilling, blasting and processing. Due to poor ventilation, dust generated in the stopes accumulate and can endanger the health of the miners. Except Mohammed & Co., all the other mines do not have ventilation systems to augment the natural ventilation and clear blast fumes and dust from the working areas. Mohammed & Co. uses a ducted ventilation fan which pushes the air towards the working face. This however, can cause trapping of mine gases and blast fumes in fractures or crevices in the working areas. The fumes and gases can be released into the workings in the event that the fan ceases to operate or is shut down, a situation that can be injurious to the health of the miners as they will unknowingly inhale harmful gases. None of the small scale underground mines conducts environmental monitoring. For example, there is no dust or air quality monitoring conducted to determine the concentrations of dust or toxic gases that are released through the activities.

Processing of materials obtained from the underground workings is typically done on the surface. The ore material is crushed, ground and then sieved to obtain the fines. These processes generate dust at each level. Persons involved in such activities (who usually do so without any form of nose protection) are exposed to inhalable dust particles. Fig. 6c shows a dust laden environment where women sieve the milled ore (Al-Hassan & Amoako, 2014), while Fig. 6d shows a changfa operator without PPE.

(d)Drainage systems: It is important to prevent the working areas from flooding to ensure safe operations. The underground workings are typically dewatered using pumps. However, poorly selected water pumps and characteristically leaking pipes that are common in the operations make the stopes highly susceptible to flooding. The SSUM operators purchase water pumps from the market without following any pump selection procedures. The result is using cheap and low capacity pumps that cannot handle the rate of water inflow into the stopes. This together with the leaking pipes does not ensure proper drainage, hence flooding of the stopes. In 2015, the two pumps at Stejoan were overwhelmed, leading to flooding of the entire mine. Consequently, the mine ceased operations for almost 12 months until the water was pumped out of the mine with the help of pumps from Mohammed & Co. The water from underground mines is also discharged directly into the environment without conducting any test to ascertain the quality of the water.

(e)Lighting: Poor electrical connections and exposed electric wires are common and sometimes located close to leaking pipes. Dry cell operated torches used by miners are mostly substandard and their batteries often run down quickly. Experienced miners are able to walk through the darkness to safety in the event of losing their lights. However, this practice can endanger the safety of the miners due to the possibility of being electrocuted or falling into an opening. These poor electrical connections and naked electrical wires are common at Dakete. To overcome the issues with dry cell torches, Johnson and Mohammed & Co. provide their underground workers with rechargeable lamps to provide them with light throughout the shift. The underground lighting system at Mohammed & Co. is better organized, as the company has an underground transformer and with all electrical works supervised by a trained electrician. Fig. 7 shows poor electrical wiring in a stope.

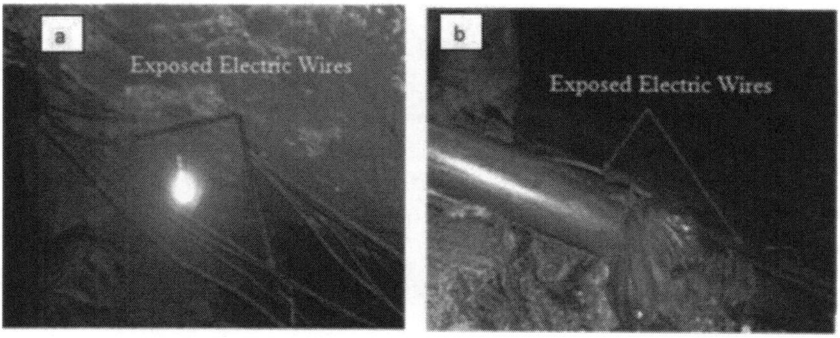

Figure 7. Poor Electrically Connected Stope: (a) Naked electrical wires (b) Naked electrical wires close to corroded leaking pipe covered with a polythene.

6.4. Processing and marketing

In processing, the broken rocks are crushed, milled and refined to obtain up to 23 carat gold. The rocks are fed into the changfa for crushing into granular sizes and further reduced into fines by milling machines. The fines are transferred into plastic bowls (containers) and mixed with water and thoroughly stirred with the bare hands until "homogenous" mixture is obtained. This is done carefully to ensure that the gold settles at the bottom of the container. The mixture is decanted to get rid of the gangue (waste), leaving the gold concentrate. Mercury is introduced into the concentrate to form mercury-gold amalgam. The amalgam is transferred onto a cotton handkerchief and squeezed until it becomes compact; this also removes excess mercury. It is then heated on charcoal fire to remove the mercury by evaporation. Nitric acid is introduced to further remove impurities. All these activities are typically done without the use of PPE. Fig. 8shows some processing activities conducted by the miners without safety precautions.

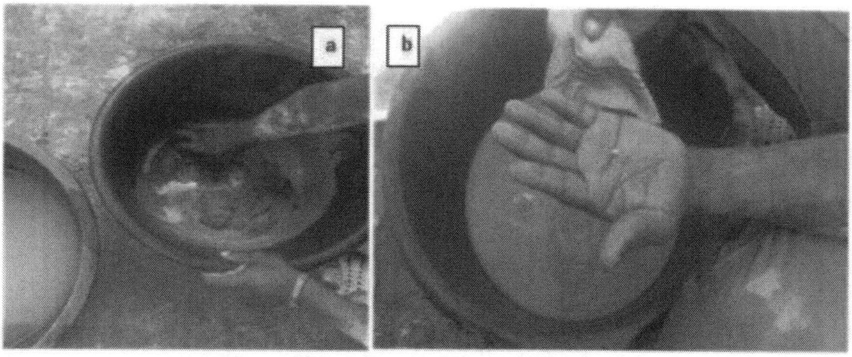

Figure 8. (a) Miner stirs slurry with bare hands, (d) Miner holds mercury-gold amalgam in the palm.

The sponged gold is subsequently fluxed with borax and smelted in a locally made refractory clay crucible. The molten gold is then poured into a mould containing palm oil for cooling and preventing the gold from getting stuck in the mould. The density and mass of the refined product are measured to ensure that it meets their set standard of 22.5–23 carat, even though the miners typically report 23 carat. The products are sold to buyers at prevailing market price. Where a miner is sponsored by a financier, the gold is sold to the sponsor at an agreed discounted price. Table 2 summarizes some of the major issues pertaining to small scale underground mining in Tarkwa, Ghana.

Table 2. Summary of Activities and safety concerns at the small scale underground mines.

Item		Dakete mining limited	Johnson mining company limited	Mohammed & Co. SSM	Stejoan mining group
Mode of Entry		Adit (2*)	Adit (2*)	Adit (2*)	Adit (1*)
Mining Method		Open stoping	Open stoping	Opening stoping	Open stoping
Ventilation Fan		No	No	Yes (1*)	No
Dewatering Pump		No	Yes (1*)	Yes (1*)	Yes (2*)
Illumination of Drives & Working Face		No	No	Yes	No
Safety Officer		Yes**	No	No	No
Rescue Team		No	No	No	No
Emergency Evacuation Plan		No	No	No	No
Hazard Signage		No	No	No	No
Air Quality Sensors & Alarm Systems		No	No	No	No
Fresh Air Chamber		No	No	No	No
Ladder Way		No	No	No	No
Ventilation Shaft		No	No	No	No
PPE	Reflective Vest	Yes	No	No	No
	Overall	No	No	No	No
	Steel Toe Boot	No	No	Yes	No
	Helmet	No	Yes	Yes	No
	Ear Plug	No	No	No	No
	Lamp	No	Yes	Yes	No
	Self Rescuer	No	No	No	No
	Hand Gloves	No	Yes	No	No
	Safety Goggles	No	No	No	No
	Nose Mask	No	No	No	No
Operating depth below Ground		200 m	200 m	100 m	200 m
Support Systems	Active Stopes	Unsupported	Unsupported	Unsupported	Unsupported
	Drives & Walkways	Unsupported, Timber	Unsupported	Bamboo and Timber	Unsupported
Materials Handling		Manual	Manual & semi-mechanised	Manual & semi-mechanised	Manual

* Quantity/Number of items,** Officer is a Junior High School graduate. Holds no qualification in safety management.

7. DISCUSSIONS

Small scale mining in Ghana has evidently provided employment to thousands of people, and continues to provide raw materials for both foreign and locally based mineral industries. The sector also plays a significant role in stemming rural-urban migration, as well as providing foreign exchange earnings to the government. These significant benefits, however, have been negated by the frequent environmental and safety issues. These issues are the result of lack of valuable technical input into the operations. The sector has remained unattractive to professional engineers with the technical know-how to effectively and efficiently manage the operations. The absence of these professionals have resulted in the use of unskilled workers, poor selection of mining methods, inappropriate choice of working tools, rampant disregard to standard safety protocols, nonexistent standard operating procedures, and improper handling of materials. For example, the injuries sustained by the driller at Dakete in 2013 could have been avoided if there were drilling engineers and safety professionals to ensure that the driller had received adequate training and observed safety protocols. Also, the flooding of Stejoan could have been mitigated if professional engineers were involved in the selection of the pumps and advised on installation and maintenance.

The issues confronting the SSM sector can also be attributed to lack of effective monitoring by the Minerals Commission of Ghana, probably due to inadequate staffing and logistics. These findings are consistent with Hilson (2002), who indicates that inadequate staff and a pool of highly obsolete research resources make it incapable for the Minerals Commission to facilitate sufficient environmental improvement in the sector. Indeed, at the Tarkwa small scale district office are only two inspectors; the district officer and his assistant who have to conduct at least one inspection per month at each of the 15 operating mines. Additionally, the district office has only one vehicle for their activities and a single hand held geographic positioning system (GPS) for field exercises. Moreover, at a stakeholder sensitization workshop presentation on the 23rd November, 2015 at Tarkwa, inadequate staffing and logistics was listed by the small scale district officer as one of the key challenges facing the small scale mining department of the Minerals Commission. The situation has partly contributed to lack of enforcement of policies and regulatory requirements, and has led to noncompliance by the SSM operators. This is also corroborated by Macdonald et al. (2014), that law enforcement in artisanal and small scale mining in Ghana is poor and unevenly applied.

An effective follow-up monitoring and evaluation, together with requisite training of the operators will enhance the small scale mining operations and minimize the hazards and negative impacts of the sector. This requires strengthening the capacity of the officers and the provision of logistics. There is also the need for an urgent effective policy direction and implementation to curtail the involvement of foreign nationals in SSM activities. The crackdown and deportation of over 4500 illegal Chinese miners from Ghana in 2013 was a good initiative by the government to arrest the situation. However, there are still some foreign nationals (fronted by Ghanaian citizens) that involve in the exploitation of the precious minerals.

Mitigating the involvement by foreign nationals and the negative impacts of the mining activities requires effective participation of the national and local governments, to make the necessary by-laws; security agencies to arrest and prosecute offenders; the Minerals Commission, to ensure implementation of regulatory requirements and conduct effective monitoring of operations; the Environmental Protection Agency, to educate the miners on environmentally friendly activities; the Ghana Immigration Service, to ensure that foreign nationals admitted to Ghana are in the country for the purpose to which their visas are granted; chiefs and elders of the rural communities, to avoid leasing their lands directly to the miners; and the people of Ghana, to report unregistered/unlicensed operations or offenders to authorities.

8. CONCLUSIONS

This study highlights the socio-economic, environmental, safety and security issues associated with the exploitation of precious minerals and metals in Ghana. A case study of four small scale underground mines is also presented to highlight the hazardous nature of the SSM activities.

Analysis of the results indicates that SSM of gold continues to make significant economic and social contributions to the socio-economic growth of Ghana. It provides jobs for the rural poor, raw materials for the local mineral industry, and contributes to the national income.

However, poor monitoring of the operations and lack of policy and regulatory enforcement by the Minerals Commission of Ghana have led to series of environmental, safety and national security issues and threats. The small scale underground mining sector in Ghana is inundated with unsafe acts and practices which lead to ground failures, improper choice of tools, land degradation, absence of personal protective equipment, drilling and blasting in confined spaces.

To mitigate the negative issues associated with small scale mining in Ghana, it is required of the Minerals Commission to improve its monitoring activities, enforce regulatory requirements and organize workshops for the miners on innovative and safe methods of operation. Other key stakeholders such as the Environmental Protection Agency, the security agencies, chiefs and landowners, and the local and national governments must play their roles to mitigate the negative impacts of the sector and enhance the socio-economic contributions of the small scale mining sector in Ghana.

ACKNOWLEDGEMENT

The authors would like to thank Safety & Environmental Research Consultancy for financial support for this work. Much appreciation also goes to members of the Ghana Small Scale Mining Association, mine owners, operators, academicians, the district officers of the Minerals Commission for their assistance, and the technicians at the Geotechnical laboratory of University of Mines and Technology, Tarkwa, Ghana.

REFERENCES

1. Acheampong, K. A. (2009). Development of small scale mining methods and their impacts on the environment of the Tarkwa area. Tarkwa, Ghana: University of Mines and Technology.

2. Agence France-Presse. (2014). 4,700 Illegal miners expelled from Ghana in 2013. http://www.globalpost.com/dispatch/news/afp/140123/4700-illegal-miners-expelled-ghana-2013 Accessed 17.03.16.

3. Al-Hassan, S., & Amoako, R. (2014). Environmental and security aspects of contemporary small scale mining in Ghana. In 3rd UMaT biennial international mining and mineral conference (pp. 146e151).

4. Amankwah, R. K., Frempong, V., & Niber, A. (2015). Women in Artisanal and small scale mining in Africa. National CompendiumGhana, United Nations Commission for Africa, 99 pp.

5. Andrews-Speed, P., Yang, M., Shen, L., & Cao, S. (2003). The regulation of China's township and village coal mines: a study of complexity and ineffectiveness. Journal of Cleaner Production, 11(2), 185-196.

6. Aryee, B. N., Ntibery, B. K., & Atorkui, E. (2003). Trends in the small-scale mining of precious minerals in Ghana: a perspective on its environmental impact. Journal of Cleaner production, 11(2), 131-140.

7. Babut, M., Sekyi, R., Rambaud, A., Potin-Gautier, M., Tellier, S., Bannerman, W., et al. (2003). Improving the environmental management of small-scale gold mining in Ghana: a case study of Dumasi. Journal of Cleaner Production, 11(2), 215-221.

8. Bansah, K. J., & Amegbey, N. (2012). Ambient particulate matter monitoring-a case study at Tarkwa. Research Journal of Environmental and Earth Sciences, 4(4), 419-423.

9. Bansah, K. J., & Bekui, P. (2015). Socio-economic and environmental assessments of illegal small scale mining in Ghana. In Proceedings of the 8th International African materials research society conference, Accra, Ghana (p. 276).

10. Bascom, R., Bromberg, P. A., Costa, D. A., Devlin, R., Dockery, D. W., Frampton, M. W., et al. (1996). Health effects of outdoor air pollution. American Journal of Respiratory and Critical Care Medicine, 153(1), 3-50.

11. Bonzongo, J. C., Donkor, A. K., & Nartey, V. K. (2003). Environmental impacts of mercury related to artisanal gold mining in Ghana. In , EDP sciences: 107. Journal de Physique IV (Proceedings) (pp. 217e220).

12. Bryceson, D. F., & Jønsson, J. B. (2010). Gold digging careers in rural East Africa: small-scale miners' livelihood choices. World Development, 38(3), 379-392.

13. Bugnosen, E. (2001). Country case study on artisanal and smallscale mining: Philippines. Mining, Minerals and Sustainable Development, 83, 8 pp.

14. Chakravorty, S. L. (2001). Artisanal and small-scale mining in India. Mining, Minerals and Sustainable Development, (78), 81. CIFOR. (2012). The formalization of artisanal and small scale mining in the Democratic Republic of the Congo and Rwanda. Center for International Forestry Research, 61 pp.

15. Coakley, G. (1999). The mineral industry of Ghana (pp. 17.1e17.11). Washington, DC: US Geological Survey, Minerals Information.

16. Colucci, M. E., Veronesi, L., Roveda, A. M., Marangio, E., & Sansebastiano, G. (2005). Particulate matter (PM10) air pollution, daily mortality, and hospital admissions: recent findings. Igiene e sanita pubblica, 62(3), 289-304.

17. Coomson, N. R. (2004). A review of the environmental impact management of small scale mining in Ghana. Tarkwa, Ghana: University of Mines and Technology.

18. DMR. (2012). Small scale mining. Republic of South Africa: Department of Minerals Resources. Retrieved from http:// www.dmr.gov.za/small-scale-mining.html Accessed 20.03.16.

19. Dockery, D. W. (2001). Epidemiologic evidence of cardiovascular effects of particulate air pollution. Environmental Health Perspectives, 109(Suppl. 4), 483-486

20. Dockery, D. W., Speizer, F. E., Stram, D. O., Ware, J. H., Spengler, J. D., & Ferris, B. G., Jr. (1989). Effects of inhalable particles on respiratory health of children. American Review of Respiratory Disease, 139(3), 587e594.

21. Donkor, A. K., Bonzongo, J. C., Nartey, V. K., & Adotey, D. K. (2006a). Mercury in different environmental compartments of the Pra River Basin, Ghana. Science of the Total Environment, 368(1), 164e176.

22. Donkor, A. K., Nartey, V. K., Bonzongo, J. C., & Adotey, D. K. (2006b). Artisanal mining of gold with mercury in Ghana. West African Journal of Applied Ecology, 9(1), 1e8. Paper 2 of 18.

23. Gan, W. Q., Man, S. F. P., Senthilselvan, A., & Sin, D. D. (2004). Association between chronic obstructive pulmonary disease and

systemic inflammation: a systematic review and a metaanalysis. Thorax, 59(7), 574e580.

24. Gauderman, W. J., Avol, E., Gilliland, F., Vora, H., Thomas, D., Berhane, K., et al. (2004). The effect of air pollution on lung development from 10 to 18 years of age. New England Journal of Medicine, 351(11), 1057e1067.

25. Gielen, M. H., Van Der Zee, S. C., Van Wijnen, J. H., Van Steen, C. J., & Brunekreef, B. (1997). Acute effects of summer air pollution on respiratory health of asthmatic children. American Journal of Respiratory and Critical Care Medicine, 155(6), 2105e2108.

26. Gita, A., Primanti, A., Zaki, K., & Ismawati, Y. (2012). Rapid assessment of the socio-economic impact and human rights aspect of mercury use in artisanal and small-scale gold mining hotspots in Indonesia international SAICM implementation project (p. 46). Denpasar: Bali Fokus.

27. Gunson, A. J., & Yue, J. (2001). Artisanal mining in the People's Republic of China (p. 19). International Institute of Environment and Development. Draft Report. No. 74.

28. Harada, M. (1995). Minamata disease: methylmercury poisoning in Japan caused by environmental pollution. Critical reviews in toxicology, 25(1), 1e24.

29. Hentschel, T., Hruschka, F., & Priester, M. (2003). Artisanal and smallscale mining: Challenges and opportunities (p. 94). UK: International Institute for Environment and Development (IIED).

30. Hilson, G. (2001). A contextual review of the Ghanaian small-scale mining industry. Mining, Minerals and Sustainable Development, 76 pp.

31. Hilson, G. (2002). The environmental impact of small-scale gold mining in Ghana: identifying problems and possible solutions. Geographical Journal, 57e72.

32. Hilson, G. M. (2006). The socio-economic impacts of artisanal and small-scale mining in developing countries. Taylor & Francis, 198 pp.

33. ICMM. (2015). Mining in Ghana e what future can we expect? Mining: partnerships for development. In International Council on Mining & Metals (p. 72).

34. Ising, H. I., & Kruppa, B. (2004). Health effects caused by noise: evidence in the literature from the past 25 years. Noise and Health, 6(22), 5.

35. Jansen, K. L., Larson, T. V., Koenig, J. Q., Mar, T. F., Fields, C., Stewart, J., et al. (2005). Associations between health effects and particulate matter and black carbon in subjects with respiratory disease. Environmental Health Perspectives, 1741e1746.

36. Kambani, S. M. (2003). Small-scale mining and cleaner production issues in Zambia. Journal of Cleaner Production, 11(2), 141e146. Kitula, A. G. N. (2006). The environmental and socio-economic impacts of mining on local livelihoods in Tanzania: a case study of Geita District. Journal of Cleaner Production, 14(3), 405e414.

37. Kwaansa-Ansah, E. E., Basu, N., & Nriagu, J. O. (2010). Environmental and occupational exposures to mercury among indigenous people in Dunkwa-On-Offin, a small scale gold mining area in the south-west of Ghana. Bulletin of Environmental Contamination and Toxicology, 85(5), 476e480.

38. Macdonald, K. F., Lund, M. A., Blanchette, M. L., & Mccullough, C. D. (2014). Regulation of artisanal small scale gold mining (ASGM) in Ghana and Indonesia as currently implemented fails to adequately protect aquatic ecosystems. In Sui, Sun, & Wang (Eds.), An interdisciplinary response to mine water challenges (pp. 401e405). China University of Mining and Technology Press.

39. Maschke, C., Harder, J., Hecht, K., & Balzer, H. U. (1998). Nocturnal aircraft noise and adaptation. Noise Effects' 98. In 7th International Congress on noise as a public health problem (Vol. 2, pp. 433e438).

40. Maschke, C., Harder, J., Ising, H., Hecht, K., & Thierfelder, W. (2002). Stress hormone changes in persons exposed to simulated night noise. Noise and Health, 5(17), 35.

41. Maschke, C., Ising, H., & Arndt, D. (1995). Nachtlicher € Verkehrslarm und Gesundheit: Ergebnisse von Labor-und € Feldstudien. Bundesgesundheitsblatt, 38, 130.

42. Mireku-Gyimah, D. (2015). Underground mine design and planning. Tarkwa, Ghana: University of Mines and Technology.

43. Mutemeri, N., Sellick, N., & Mtegha, H. (2010). What is the status of small scale mining? Discussion document for the MQA SSM colloquium (p. 24).

44. Mwaipopo, R., Mutagwaba, W., Nyange, D., & Fisher, E. (2004). Increasing the contribution of artisanal and small-scale mining to poverty reduction in Tanzania (p. 153). Dar es Salaam: Department for International Development (DFID).

45. Ntibrey, B. K. (2016). Small scale mining sector in Ghana & Minerals Commission's role in managing it. In Presentation at stakeholder sensitization workshop Tarkwa on November 23, 2015 (p. 43).

46. Oblokuteye, K. P. H. (2010). The effects of illegal small scale mining on the environment e A case study at gold hall Galamsey site. Tarkwa, Ghana: University of Mines and Technology.

47. Opoku-Antwi, G. L. (2010). Three essays on small-scale gold mining operations in Ghana: An integrated approach to benefit-cost analysis. Doctoral Dissertation (pp. 22e98). Kumasi, Ghana: Department of Economics, Kwame Nkrumah University of Science and Technology.

48. OSH. (2003). Personal protective equipment. OSHA 3151 e 12R 2003. Occupational Safety and Health Administration, US Department of Labor. Retrieved from www.osha.go Accessed 23.03.16.

49. Parliament of the Republic of Ghana. (2006). The minerals and mining act, 2006.

50. Personal Protective Equipment at Work Regulations. (1992). Guidance on Regulation L25 (Third Edition 2015). Health and Safety Executive. ISBN 978 0 7176 6597 6 www.hse.gov.uk/ pubns/indg174.pdf.

51. Pope, C. A., III, Hill, R. W., & Villegas, G. M. (1999). Particulate air pollution and daily mortality on Utah's Wasatch Front. Environmental Health Perspectives, 107(7), 567.

52. Pulles, M. P. J., Biesiot, W., & Stewart, R. (1990). Adverse effects of environmental noise on health: an interdisciplinary approach. Environment International, 16(4), 437e445.

53. Serfor-Armah, Y., Nyarko, B. J. B., Adotey, D. K., Adomako, D., & Akaho, E. H. K. (2005). The impact of small-scale mining activities on the levels of mercury in the environment: the case of Prestea and its environs. Journal of Radioanalytical and Nuclear Chemistry, 262(3), 685e690.

54. Shen, L., Dai, T., & Gunson, A. J. (2009). Small-scale mining in China: assessing recent advances in the policy and regulatory framework. Resources Policy, 34(3), 150e157.

55. Small Scale Gold Mining Law. (1989). PNDCL 218. Republic of Ghana.

56. Sousa, R. N., Veiga, M. M., Meech, J., Jokinen, J., & Sousa, A. J. (2011). A simplified matrix of environmental impacts to support an intervention program in a small-scale mining site. Journal of Cleaner Production, 19(6e7), 580e587.

57. Tarras-Wahlberg, N. H., Flachier, A., Lane, S. N., & Sangfors, O. (2001). Environmental impacts and metal exposure of aquatic ecosystems in rivers contaminated by small scale gold mining: the Puyango River basin, southern Ecuador. Science of the Total Environment, 278(1), 239e261.

58. Tepkor, A. S. (2005). Transforming small scale mining into a form of sustainable livelihood for the communities. Tarkwa, Ghana: University of Mines and Technology.

59. The Guardian. (2013). Ghana deports thousands in crackdown on illegal Chinese goldminers. Retrieved from Accessed 17.03.16 http://www.theguardian.com/world/2013/jul/15/ghanadeports-chinese-goldminers.

60. World Bank. (2013). Artisanal and small scale mining. Retrieved from http://www.worldbank.org/en/topic/extractiveindustries/brief/artisanal-and-small-scale-mining Accessed on 19 March, 2016.

61. Yakovleva, N. (2007). Perspectives on female participation in artisanal and small-scale mining: a case study of Birim North District of Ghana. Resources Policy, 32(1), 29e41.

62. Yamoah, S. A. (2002). Socio-economic impact of small scale mining in the Tarkwa area of Ghana. Tarkwa, Ghana: University of Mines and Technology.

63. Yu, O., Sheppard, L., Lumley, T., Koenig, J. Q., & Shapiro, G. G. (2000). Effects of ambient air pollution on symptoms of asthma in Seattle-area children enrolled in the CAMP study. Environmental Health Perspectives, 108(12), 1209.

Chapter 11

Joint Stochastic Optimisation of Short and Long Term Mine Production Planning: Method and Application in a Large Operating Gold Mine

R Dimitrakopoulos[*a] **& A Jewbali**[b]

[1] *COSMO – Stochastic Mine Planning Laboratory Department of Mining and Materials Engineering, McGill University, FDA Building, 3450 University Street, Montreal, Quebec H3A 2A7, Canada*

[2] *Newmont Mining, Denver, CO, USA*

ABSTRACT

A new multistage stochastic mine production scheduling approach is developed and tested in a large operating gold mine. The approach takes short scale orebody information in the form of grade control data into account. As simulated orebodies used in stochastic long term mine planning are based on sparse exploration data while grade control data are unavailable at the time of production scheduling, the short scale information is simulated stochastically. Stage 1 of the approach simulates high density future grade control data based on exploration data and grade control in previously mined out parts of a deposit. In stage 2, the technique of conditional simulation by successive residuals enables preexisting simulated orebody models to be updated using the simulated future grade control information. Stage 3 is based on a stochastic programming mine scheduling formulation that handles jointly multiple simulated orebody models from stage 2, and accommodates both maximising net present value (NPV) and minimising deviations from expected production targets. Stage 4 includes quantification of risk in the production schedules generated, comparisons and reporting. The application at a large operating gold mine demonstrates that the proposed approach is practical, and adds value to the operation. The approach is shown to deliver additional ore (3·6 Mt more) and metal (2·6 Mg more) which matches the mines reconciliations, unlike the conventional schedule, and results in a cumulative NPV which is on average 7·7 million AUD higher than that of a stochastic schedule without the simulated grade control data. The NPV is 230 million AUD higher compared to the NPV from the actual schedule of the mine. An additional key contribution of the proposed approach is the compliance of short- to long term production schedules and performance, leading to higher probability of meeting production targets and increased productivity. Given the capital intensity of mining projects, this contribution can be critically important to mining operations.

Keywords: Mine production scheduling, Stochastic optimisation, Grade control, Simulation

INTRODUCTION

Stochastic mine planning is a relatively recent development aimed at addressing uncertainty in ore supply from an orebody, commodity prices and metal demand, and also extends to other issues of uncertainty in mine planning. The existing work focuses largely on open pit mine design and production scheduling with uncertain grades/metal content and geological conditions, and is based on initial work and technical issues raised first in the 1970s (David *et al.*, 1974) and again in the 1990s (Ravenscroft, 1992; Dowd, 1994, 1997). It includes life of mine (LOM) production scheduling based on optimal mining rates and simulated annealing (Godoy and Dimitrakopoulos, 2004), simulated annealing with constant mining rates and assessment of various inputs (Leite and Dimitrakopoulos, 2007; Albor and Dimitrakopoulos, 2009), stochastic integer programming (SIP) formulations including a stochastic stock pile (Menabde *et al.*, 2007; Ramazan and Dimitrakopoulos, 2012), further tests of this last approach (Leite and Dimitrakopoulos, 2013) and expansion for multiple elements (Benndorf and Dimitrakopoulos, 2013), pushback design (Albor and Dimitrakopoulos, 2010), efficient SIP solutions based on metaheuristics (Lamghari and Dimitrakopoulos, 2012), and more. The two key aspects of all these approaches and example case studies are that the derived long term production schedules have substantially higher present value than the traditionally used approaches, with differences between 20 and 30%; and minimise the potential deviations from production targets set given the available drilling and orebody models. Recent work further considers integrating market uncertainty in choosing from different mine designs which also consider geological uncertainty (Abdel Sabour *et al.*, 2008). The effects of optimising pits in a new approach that assesses the combined effect of metal and price uncertainties using a stochastic network flow approach (Meagher *et al.*, 2009; Asad and Dimitrakopoulos, 2013) show once again the value of stochastic approaches and the integration of key uncertainties in mine planning. Notable also is the effort to integrate uncertainty in mining and processing costs (Richmond, 2011).

The work above is all based on two elements, in addition to new stochastic mine design and scheduling methods. The first is the ability to simulate sets of equally possible representations of the orebody being assessed. This set of orebodies is the input to the above mentioned stochastic optimisers and represents the geological uncertainty and local grade variability of the orebody, as understood from the available drillhole data; this drastically differs from the conventional single estimated (smoothed out or average type) input to conventional mine design and optimisation methods, and is the major reason for improved results from the stochastic scheduling methods discussed. The second is the ability to assess cash flows using evaluation methods based on multiple simulated realisations of the related commodity price, rather than single average and typically constant, over the LOM and metal price forecasts. Both this second element and the use of uncertainty in metal content lead to improvements in the last two studies noted above.

A limit of both stochastic and conventional long term planning and optimisation of production planning is that short term schedules may frequently deviate from the expectations of long term ones. While long term production scheduling is used to maximise the net present value (NPV) of a project, short term production scheduling focuses on meeting short term production demand within the long term plan, given the processing capacity of the mill, and managing the quality of ore being processed. Both short and long term schedulings are usually based on exploration scale orebody models, whereas the actual short term production performance is controlled by local, blasthole scale data used for grade control and ore/waste selection, usually not available at the time of scheduling. As a result, short term production sequences deviate from the long term plans and are adjusted to meet mill demand or the production performance may substantially deviate from forecasts. In short, discrepancies occur between what is forecasted and what is actually mined.

A solution to the above is to consider production scheduling approaches that are capable of reflecting short scale behaviour of the orebody beforehand, that is, at the planning stage and through grade information available before the actual grade control drilling. High density grade control information offers tighter controls on geology and mineralisation characteristics, and predicts qualities and quantities closer to what is actually being mined. Although not available before blasthole drilling, grade control data can be simulated in different ways. Simple random sampling errors added to simulated realisations from exploration data have been used to assess production schedules reported in a feasibility study (Guardiano *et al.*, 1995), and a more sophisticated approach is considered for assessing mineral resources and reserves (Journel and Kyriakidis, 2004); in operating mines, errors from field duplicates or from nearest paired grade control holes and drillholes have been used to simulate realisations of future grade control data and assess short term production schedules at Escondida Copper Mine, Chile (Khosrowshahi *et al.*, 2007). Multiple simulations of correlated sampling errors using data from previously mined parts of the deposit have been used at Morila Gold Deposit, Mali (Peatie and Dimitrakopoulos, 2009) with excellent results in reconciliations. All existing work in this area assesses potential risks in production schedules or potential reserves and actual reserves; however, it does not address the major question: How to improve or reduce deviations of forecasts from actual production at the time of planning while maximising the present value of an asset? This can be effectively addressed only by technically integrating short scale orebody variability and grade control and short and long term LOM production planning into one approach.

Short scale information is the core element in the above mentioned integration. As this information is unavailable and needs to be stochastically simulated at the time of LOM planning, the way to utilise a range of possible scenarios is through the stochastic optimisation formulations used in stochastic mine planning as reviewed above. This paper contributes a new multistage approach to production scheduling, incorporating short scale deposit information and related grade uncertainty into the scheduling process, thus allowing for the realistic integration of short and long term mine production schedules, as well as the generation of more reliable mine production forecasts. A case study at an operating gold deposit demonstrates the approach, provides comparisons between the traditional schedules of the mine, accounts for stochastic schedules before and after simulated dense grade control information and quantifies the expected monetary value of the method. In the next sections, the proposed approach is first outlined, the case study is presented, results are analysed and conclusions follow.

A NEW MULTISTAGE APPROACH TO PRODUCTION SCHEDULING

The multistage approach to production scheduling proposed here is a sequence of steps employing separate techniques for each stage and is based on the stochastic mine planning optimisation methods mentioned above (Albor and Dimitrakopoulos,2010; Ramazan and Dimitrakopoulos, 2012; Benndorf and Dimitrakopoulos, 2013; Leite and Dimitrakopoulos, 2013). The approach differs conceptually from the traditional approaches (current non-stochastic 'best practices') in many aspects. The major difference is that, similarly to previous stochastic mine schedule optimisation, the proposed multistage approach requires multiple, equally probable representations of the orebody, which are generated from spatial Monte Carlo simulation methods (Goovaerts, 1997), but

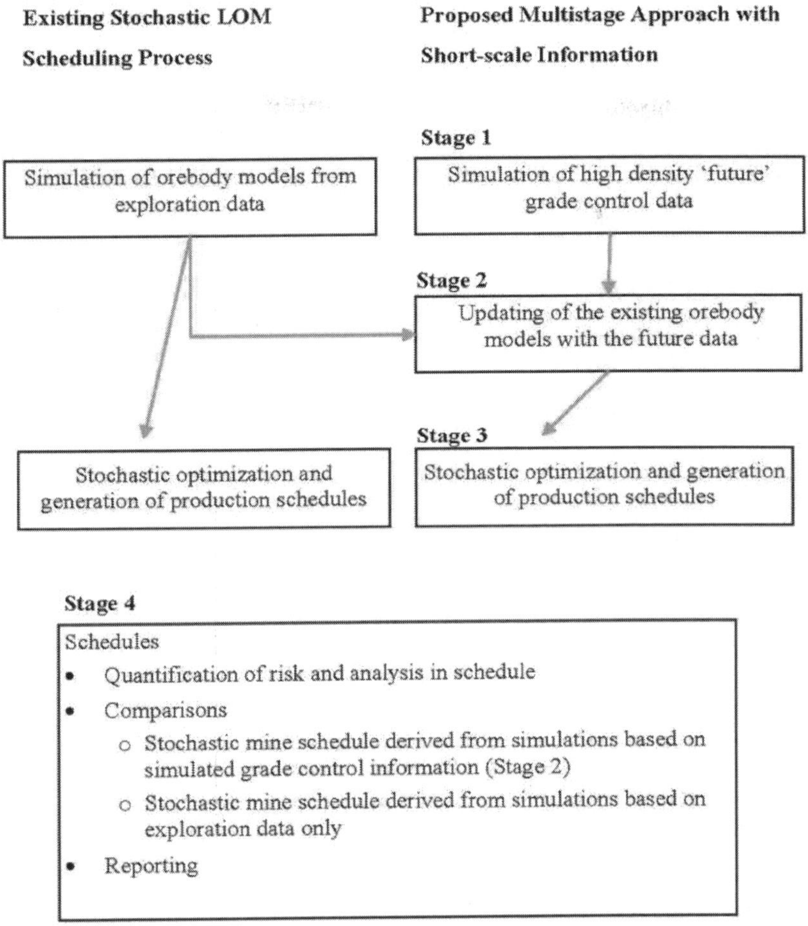

Figure 1. Schematic presentation of proposed multistage mine scheduling approach

at two different and distinct levels. The first level, and the one used in past work, is that of exploration type datasets and information; the second level is that of grade control data. This simulation based framework assists in quantifying and generating risk managing schedules as well as accommodating the interaction of information at different scales in space and time. Different scales in space refer to the local variability of the deposit and local classification of materials selected as ore and waste through a grade control process; and physical parts of the orebody that are mined over long periods of time as individual pit units (pushbacks) and parts of individual pit units that correspond to the parts of pushbacks mined over short periods of time from the same mining front. Different scales in time refer to the yearly time units on which the long term plans are reported and relatively shorter time periods in short term planning (weekly, monthly or quarterly). As a result of the ability to link information at different scales in space and time, short scale variability and short and long term schedules are integrated. The proposed approach or process has four stages shown in Fig. 1, which also shows how this compares to the steps followed in the existing stochastic LOM planning:

1. Generate high density future grade control data for incorporation into the production scheduling process. These data are not yet available at the time of planning and are simulated using stochastic methods.

2. Update preexisting simulated orebody models using the simulated future grade control information and a suitable stochastic simulation technique.

3. Generate production schedules using a stochastic optimisation method handling multiple simulated orebody models from stage 2, while accommodating the need to both maximise NPV and minimise deviations from production targets.

4. Quantify grade risk in the produced schedules, compare schedules with and without simulated short scale orebody information and report mine schedules as needed.

These four stages are discussed next in greater detail.

Stage 1: Generation of future grade control drilling data

High density future grade control data can be simulated using several approaches, such as those reviewed in the introduction. A more elaborate and effective method is outlined here and is used in the case study that follows. The method, schematically shown in Fig. 2, assumes that there are mined out parts of an orebody with grade control information of comparable quality and characteristics that can be assumed in the remainder of the orebody being studied. Exploration drilling and grade control data from a mined out section or bench of a deposit are used to quantify the special relationship between the two types of data. As it is not physically possible to have exploration and grade control information available at the same location, their spatial cross-correlation (relationship) is quantified through the so called pseudo cross-variogram (Myers, 1991). Having quantified this spatial relationship, a method such as the sequential Gaussian cosimulation of two variables (Goovaerts, 1997) can be used to generate the future grade control data in unmined sections of an open pit. The cosimulation approach ensures that the simulated future grade control data have the same distribution as past grade control information and the same auto- and cross-correlation (with exploration data).

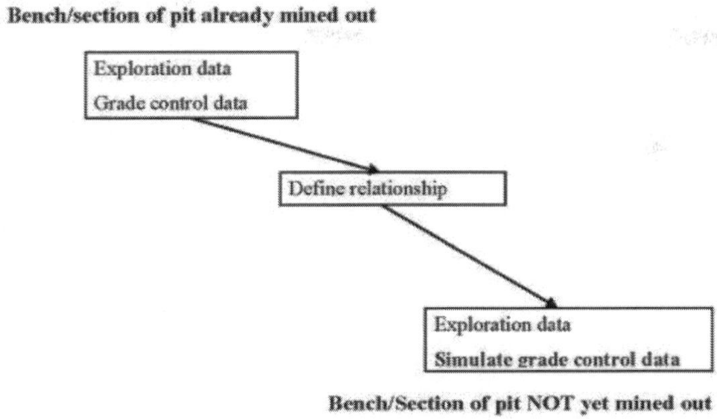

Figure 2. Simulating future grade control data

Stage 2: Updating of existing simulations with future grade control data

The second stage of the production scheduling approach presented herein involves the updating of preexisting simulated orebody models. This is because orebodies under study have frequently already been simulated for other purposes, or because there can be more than one scenario of possible sets of future data to consider, and so on. Generally the ability to update preexisting realisations of an orebody is useful in improving the efficiency of the process. The only known simulation technique that is capable of providing updating capabilities to accommodate the updating of preexisting simulations with new data such as grade control is the technique of conditional simulation by successive residuals or CSSR (Vargas–Guzman and Dimitrakopoulos, 2003; Jewbali and Dimitrakopoulos, 2011). The CSSR is a simulation method based on the column decomposition of the covariance matrix, which leads to the expression of the simulation process as a series of conditional covariance matrices.

More specifically, for the simulation of a grid G with K nodes, conditioned on a dataset D with R points (data locations), the simulation equation using the column decomposition and generating the simulated values $[\mathbf{Z}_{R+1}, \ldots, \mathbf{Z}_{R+K}]$ at the K nodes of grid G using the data $[\mathbf{Z}_1, \ldots, \mathbf{Z}_R]$ at R locations can be written as

$$\begin{bmatrix} L_{11} & & & & & \\ C_{21}C_{11}^{-1}L_{11} & L_{22} & & & & \\ \cdot & \cdot & & & & \\ C_{R1}C_{11}^{-1}L_{11} & E_{R2}^{[2]}[E_{22}^{[2]}]^{-1}L_{22} & & L_{RR} & & \\ C_{(R+1)1}C_{11}^{-1}L_{11} & E_{(R+1)2}^{[2]}[E_{22}^{[2]}]^{-1}L_{22}\cdot & \cdot & E_{(R+1)R}^{[R]}[E_{RR}^{[R]}]^{-1}L_{RR} & L_{(R+1)(R+1)} & \\ \cdot & \cdot & \cdot & & & \\ C_{(R+K)1}C_{11}^{-1}L_{11} & E_{(R+K)2}^{[2]}[E_{22}^{[2]}]^{-1}L_{22} & \cdot & \cdot & \cdot & L_{(R+K)(R+K)} \end{bmatrix} \begin{bmatrix} W_1 \\ \cdot \\ \cdot \\ W_R \\ W_{R+1} \\ \cdot \\ W_{R+K} \end{bmatrix} = \begin{bmatrix} Z_1 - m \\ \cdot \\ \cdot \\ Z_R - m \\ Z_{R+1} - m \\ \cdot \\ Z_{R+K} - m \end{bmatrix} \quad (1)$$

where $Z_{(R+1)} - m, \ldots, Z_{(R+K)} - m$ contain the simulated values, $Z_1 - m, \ldots, Z_R - m$ contain the R data values available and $W_1, \ldots, W_{(R+K)}$ contain the independent N(0,1) random numbers, and **m** is the mean. $\mathbf{C_{ij}}$ are covariance matrices. $\mathbf{L_{ii}}$ are lower triangular matrices. The $E_{ss}^{[s]}$ are residual covariances of the *s*th order and are calculated through $E_{ss}^{[s]} = E_{ss}^{[s-1]} - E_{st}^{[s-1]}[E_{tt}^{[s-1]}]^{-1}E_{ts}^{[s-1]}$ for *s*>2. For *s*=2, the residual second order covariances are calculated through $E_{ji}^{[2]} = C_{ji} - C_{j1}C_{11}^{-1}C_{1i}$.

The above equation can be seen to separate data influences which make updating of existing orebody models possible by permitting a recalculation of only those influences that change when new information becomes available. For example, if new data become available at the location of the first node, which occupies the (*R*+1)th row in the matrix

$$C_{(R+1)1}C_{11}^{-1}L_{11}W_1 + E_{(R+1)2}^{[2]}[E_{22}^{[2]}]^{-1}L_{22}W_2 + \ldots + E_{(R+1)R}^{[R]}[E_{RR}^{[R]}]^{-1}L_{RR}W_R + L_{(R+1)(R+1)}W_{(R+1)} = Z_{(R+1)} - m \quad (2)$$

Since $Z_{(R+1)} - m$ is now known, this allows for the calculation of $L_{(R+1)(R+1)}W_{(R+1)}$.

Then, the next node [(*R*+2)th row] is given by equation (3)

$$C_{(R+2)1}C_{11}^{-1}L_{11}W_1 + E_{(R+2)2}^{[2]}[E_{22}^{[2]}]^{-1}L_{22}W_2 + \ldots + E_{(R+2)(R+1)}^{[R+1]}[E_{(R+1)(R+1)}^{[R+1]}]^{-1}L_{(R+1)(R+1)}W_{(R+1)} + L_{(R+2)(R+2)}W_{(R+2)} = Z_{(R+2)} - m \quad (3)$$

By only recalculating the $\mathbf{E}^{[\mathbf{R}+1]}_{(\mathbf{R}+2)(\mathbf{R}+1)}[\mathbf{E}^{[\mathbf{R}+1]}_{(\mathbf{R}+1)(\mathbf{R}+1)}]^{-1}$ $\mathbf{L}_{(\mathbf{R}+1)(\mathbf{R}+1)}\mathbf{W}_{(\mathbf{R}+1)}$ component of the equation, i.e. the only part of the equation that changes when node 1 becomes a data location, the simulated value at node 2 is updated. The end result of the use of CSSR is a set of simulated orebody models that are conditioned on both the known exploration data and the simulated future grade control data. Note that new information for updating may also be additional data from near mine exploration drilling or in-fill drilling.

Stage 3: Stochastic integer programming formulation

To generate production schedules based on a group of simulated realisations (scenarios) of an orebody, a stochastic optimisation formulation based on SIP is used. Note that the use of stochastic optimisers allows the scheduling process to capitalise on knowing the joint local uncertainty of the economic values of the mining blocks composing the representation of the orebody at hand. The formulation, described next, is a simpler version of the one developed in Ramazan and Dimitrakopoulos (2012). The objective function of this formulation attempts to maximise expected discounted cashflows while minimising the cost of deviating from the set production targets and it is

$$\text{Max} \sum_{t=1}^{P}\sum_{i=1}^{N} \text{E(NPV)}_i^t \, X_i^t - \sum_{s=1}^{S}\sum_{t=1}^{P}\sum_{r=1}^{R} \left({}^s\text{Cu}_r^t \, \text{Yu}_r^t + {}^s\text{Cl}_r^t \, \text{Yl}_r^t\right) \quad (4)$$

where *P* is the number of periods to schedule, *N* is the number of blocks to schedule, *S* refers to the number of simulated orebody models, *R* is the number of targets (ore category 1, 2, ...), X_i^t is a binary variable indicating block *i* mined in period *t* (1=mined/0=not mined), E(NPV)_i^t is the expected NPV if block *i* is mined in period *t*, Yu_r^t is the excess amount produced compared to the target (grade/tonnage) *r* in period *t* for orebody model *s*, ${}^s\text{Cu}_r^t$ is the unit cost to penalise Yu_r^t in period *t*, Yl_r^t is the deficient amount produced in relation to target (grade/tonnage) *r* in period *t* for orebody model *s* and ${}^s\text{Cl}_r^t$ is the unit cost to penalise Yl_r^t in period *t*.

The first component in equation (4) maximises the total discounted cashflow. The expected NPV is determined by calculating economic values for each simulated model, and then averaging them. This first component reflects the conventional goal of optimising total discounted cashflow over the LOM. The second component minimises the deviations between expected productions as these are described by simulated orebody models used as input and a set of mine production targets, and minimises the risk of not meeting targets due to grade uncertainty. The deviations are calculated over the set of simulated orebody models and are penalised with costs. This second component reflects the goal of short term production scheduling and its use together with the first component provides the means to integrate long- and short term planning, in the context of the approach proposed herein. It should be noted that a novelty of this formulation is that the costs of deviating from production expectations can be discounted for each subsequent period using a geological discount rate. As a result, the cost related to not meeting mill requirements can be set higher in the first period than in subsequent periods, allowing the risk distribution between production periods to be managed. The related constraints are shown below.

Grade constraints

$$\sum_{i=1}^{N} (g_{si} - G_u) O_i \, X_i^t - Yu_r^t = 0$$
$$\sum_{i=1}^{N} (g_{si} - G_l) O_i \, X_i^t - Yl_r^t = 0 \quad (5)$$

Ore production constraints

$$\sum_{i=1}^{N} O_i \, X_i^t - Yu_r^t = PCu$$
$$\sum_{i=1}^{N} O_i \, X_i^t - Yl_r^t = PCl \quad (6)$$

Mining capacity constraints ensuring that the total tonnage of ore and waste does not exceed the given mine equipment capacity

$$\sum_{i=1}^{N} (O_i + W_i) X_i^t \leq MCap \quad (7)$$

Slope constraints ensuring that to mine a specific block, all the blocks overlying that specific block must be mined. For each overlying block of the block i and each period t:

$$X_i^t - \sum_{k=1}^{t} X_j^k \leq 0 \quad (8)$$

Reserve constraints, where a block has to be mined fully (100%)

$$\sum_{t=1}^{P} X_i^t = 1 \quad (9)$$

where g_{si} is the grade in block *i* for simulated orebody model *s*, G_u and G_l are the upper and lower grade targets, O_i is the ore tonnage in block *i*, PCu and PCl are the upper and lower limits on the tonnes mined, MCap refers to equipment capacity and W_i is the tonnage of waste in block *i*.

Stage 4: Quantification of uncertainty

Stage four aims to quantify grade risk in the produced schedules that have been generated. Any schedule can be assessed against possible scenarios of orebodies, be they constructed from exploration datasets only or in combination with future grade control data. This allows comparisons and assessments, including an evaluation of effects from potential short scale orebody variability.

CASE STUDY AT A SUNRISE DAM GOLD MINE

The deposit and mine

For this case study, data from the Sunrise Dam mine, an open pit gold mine which lies ~220 km north of Kalgoorlie and 80 km south of Laverton, WA is used. It sits on the edge of Lake Carey, in the northeastern corner of the eastern Goldfield province of Western Australia. The mine contains both the Cleo and Sunrise deposits, which are separated by an ownership boundary. The Sunrise/Cleo gold deposit lies along a shear zone and the general trend of the gold mineralisation parallel to this zone with the detailed geological description is presented in Newton *et al.* (1998). Gold mineralisation can be both secondary and primary. Secondary mineralisation is related to iron oxidation fronts and water tables and to aggregation within paleochannels, while primary mineralisation can roughly be related to high strain shear zones and low strain stockwork vein zones. Based on the deposit geology, 17 domains are inside the open pit (sections of which are shown in Fig. 3). Mining is done using four excavators which load into 25 rear dump trucks. Since opening, the mine has produced more than 2 million ounces of gold. The pit is 2·1 km long and 1·1 km wide with a final depth of 450 m to be reached at end of its life. The total amount of waste mined in 2004 was ~43 Mt, which drops to ~26 Mt in 2005 and again to 11 Mt in 2007. This will necessitate a resizing of the equipment fleet sometime in mid-2005. Figure 3 depicts the long term (yearly) production schedule in use at the mine. This is derived from an estimated model [multiple indicator Kriging or MIK (Goovaerts, 1997)] in use at the mine site. The mine's schedule proceeds one layer at a time and indicates that the pit will be exhausted by the end of the fifth year. At this mine, blasting occurs on 7·5 m benches and grade control is done using reverse circulation drilling on 5×7 m spacing. The plant capacity is 3·6 Mtpa and the mill circuit consists of crushing and grinding after which the gold is recovered through a carbon in leach circuit. A specific characteristic of Sunrise Dam mine relevant to the present study is its long history of large positive reconciliations when comparing resource models and related production forecasts to grade control, ore production and mill reconciliations. This is due to the presence of small short scale mineralised structures, too small in scale to be picked up by exploration drilling, but detected by grade control.

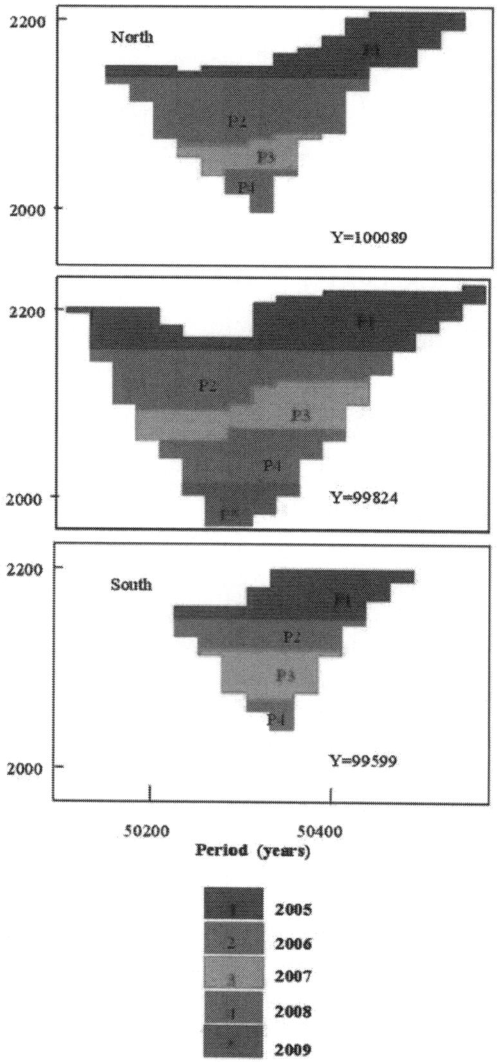

Figure 3. East–west sections of yearly schedule at Sunrise Dam Gold Mine at the time of the study presented herein

Generation of future grade control data

Within the mined out section of the pit, 75 789 grade control drillholes were available. These were separated based on the geological domain they belonged to. In addition, 3934 exploration diamond and reverse circulation holes were available within the same section of the pit. These were also divided based on geological domains. For each geological domain, these two sets of information were used to derive the relationship between exploration and grade control data. This relationship can then be extended to unmined sections of the pit where only exploration data are available. For the purposes of this study, 20 future grade control data simulations were generated, using the sequential cosimulation method discussed earlier, on 5×7×2·5 spacing in the unmined sections of the pit (the volume displayed in Fig. 3).

Figure 4 displays the histogram of grade control data for a mined out section of domain WSZ, one of the 17 domains within the pit. The figure also displays a comparison between the grade control data simulations for the unmined section of domain WSZ and the grade control data available from the mined out section of domain WSZ in the pit. As can be seen, the simulations closely approximate the grade control data histogram. Also shown is a comparison between the mean grades of the simulations and the mean grade of the WSZ grade control data. The mean grade of the simulations fluctuates around the mean of the grade control data. The grade control simulations also reproduce the grade control variogram model derived from grade control data (from mined out sections of domain WSZ). The relationship between exploration and grade control data in a mined out section of domain WSZ is given by the pseudo cross-variogram, which is reproduced by the grade control simulations for unmined sections of domain WSZ, as shown by Fig. 5. As it can be seen, statistical characteristics of the grade control data for a particular domain are derived from mined out sections of the pit and are reproduced in unmined sections of the pit for that same domain. Note that the statistical characteristics of grade control data versus exploration data are considered to be the same in the mined out part of domain WSZ and the part to be scheduled for mining; this is an assumption of the method and is validated as the operation proceeds and production reconciliation occurs.

Updating existing simulations with future grade control data

Stage 2 of the production scheduling approach involves using conditional simulation by successive residuals to update preexisting orebody models. The 20 available simulations (on 2×2×2·5 m spacing and with a total of 17 million grid nodes per simulation) conditioned on exploration data were updated with the 20 simulated future grade control data, generating a set of 400 simulations that are conditioned on both the known exploration data and the simulated future grade control data scenarios. Twenty updated simulations are chosen at random for further study.

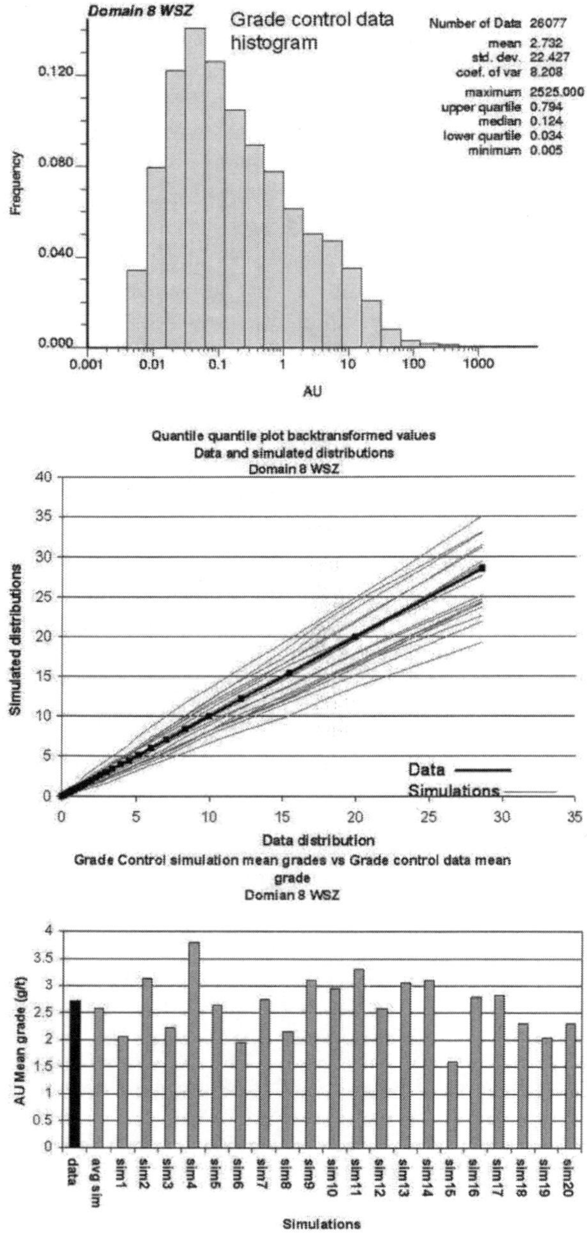

Figure 4. Reproduction of grade control data histogram

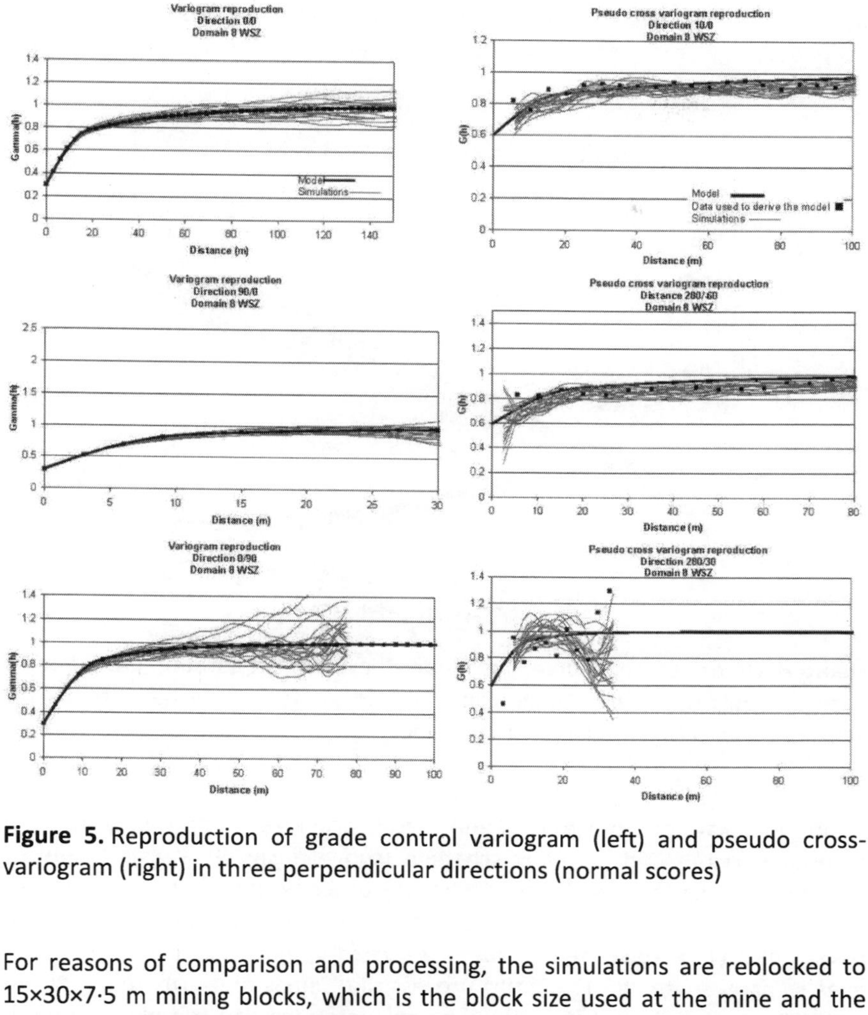

Figure 5. Reproduction of grade control variogram (left) and pseudo cross-variogram (right) in three perpendicular directions (normal scores)

For reasons of comparison and processing, the simulations are reblocked to 15×30×7·5 m mining blocks, which is the block size used at the mine and the generation of grade tonnage curves. Grade tonnage curves are shown in Fig. 6. It is apparent from the figure that for the model based on exploration data, the average grade above cutoff is higher at all cutoffs. The amount of tonnes above cutoff is lower compared to the updated models. To facilitate understanding of the grade–tonnage curves in Fig. 6, it is important to recall that over the past production time of the deposit, exploration and grade control drilling showed the grade control data as having a higher proportion of higher grades. The main reason for this is that grade control picks up small short scale mineralised structures, too small in scale to be picked up by exploration drilling. As mentioned earlier, the mine has had a long history of strong positive reconciliation between resource model and grade control/mill and the grade tonnage curves in Fig. 6 reflect that, as they should. The opposite may be observed in other mines; that is, the resource/reserve model of the orebody overstates high grades and the grade control cannot find all these high grades.

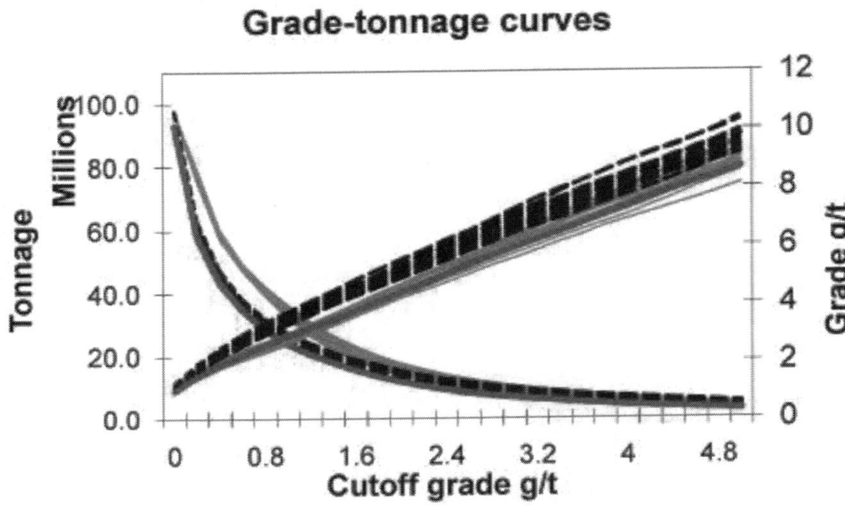

Figure 6. Grade tonnage curves for models based on exploration data (dotted lines), models updated with simulated future grade control data (thin solid lines) and mines MIK model (thick solid line)

Figure 7 displays the ore tonnes and metal recovered when mining is done according to the mine's yearly schedule (shown inFig. 3), ore is identified according to the mine's MIK model of the orebody and short scale deposit information is taken into account (simulated future grade control data). The risk profiles were generated by running the 20 updated models through the mine's yearly schedule. During this process the tonnes of ore/metal produced and NPV generated by each updated model were recorded and plotted on Fig. 7. The updated models indicated that the amount of ore tonnes produced is not enough to fill the plant to capacity (3·6 Mtpa) for all the years from 2005 to 2009, if the mine's schedule is to be implemented and ore blocks are identified by the mine's MIK model of the orebody. They also indicate a 50% chance of negative NPV in 2005 when insufficient metal is produced.

Deriving production schedules using the SIP formulation

The following section describes the process, as given in Fig. 1, used to derive the production schedules. First, the 20 simulated orebody models conditioned on exploration data are put through the SIP formulation in equation (4). Next,

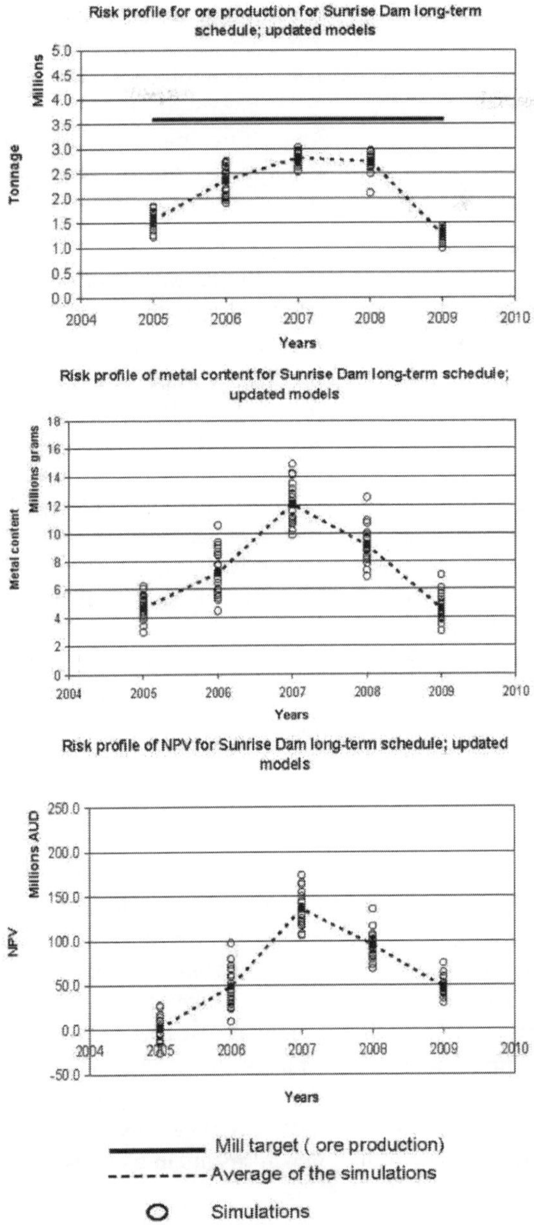

Figure 7. Performance of mine's yearly schedule when local orebody information is taken into account (simulated models updated with simulated future grade control data): note that ore blocks are defined from the mine's MIK orebody model

the 20 updated orebody models are run through the SIP formulation and a production schedule accounting for short scale orebody information is derived. This methodology allows for a comparison of the schedules derived using orebody models based on two sets of information. As a result, it provides a look at the consequences of not taking closer spaced grade control information into account when performing production scheduling. The discussion on why a group of 20 simulated models is used is deferred for a subsequent section.

The production schedules were generated for 4-month periods (short term schedules). The production capacity for each 4-month period was set at 1·2 MT and the targeted grade constraints between 2 and 7 g t^{-1}. To make the schedules practical for equipment mobility and space, the schedules were smoothed. After smoothing, the simulated orebody models are run through their respective schedules to generate risk profiles for ore, metal and NPV. The risk profiles of the production schedule derived from the 20 exploration based models for ore, metal and NPV are shown in Fig. 8. As the figure shows, the quarterly production target of 1·2 Mtpa is met for 2005–2007. Note that this schedule predicts that the pit will be exhausted by 2008. The figure also displays the performance of the mine's MIK model when mining proceeds using the schedule derived from the exploration based orebody models. The MIK model predicts a shortfall in ore production after Q3 2007. The NPV predicted by the MIK model is almost always lower or towards the lower end of the risk profile as described by the simulations. The cumulative NPV predicted by the MIK model is 364 million AUD compared to the schedule from the simulations and SIP which is 552 million AUD.

Figure 9 describes the schedule derived from the updated models. This schedule extends the LOM by another year to 2009. As the figure shows, the short term schedule has no problem delivering 1·2 Mtpa of ore in all quarters except in Q3 2009, when the last of the ore will be mined. The figure also shows that according to the MIK model, there will be a shortfall in ore production for all the remaining mine life when mining proceeds according to the schedule derived from updated simulated models. The MIK model generates a cumulative NPV of 375 million AUD, while the cumulative average of the updated simulations is 561 million AUD. A summary of the cumulative averages of ore tonnes, metal and NPV is given in Table 1, below. The schedules based on updated models produce 3·6 Mt more ore and as a consequence 2·6 Mg more metal, which results in a cumulative NPV that is 7·7 million AUD higher.

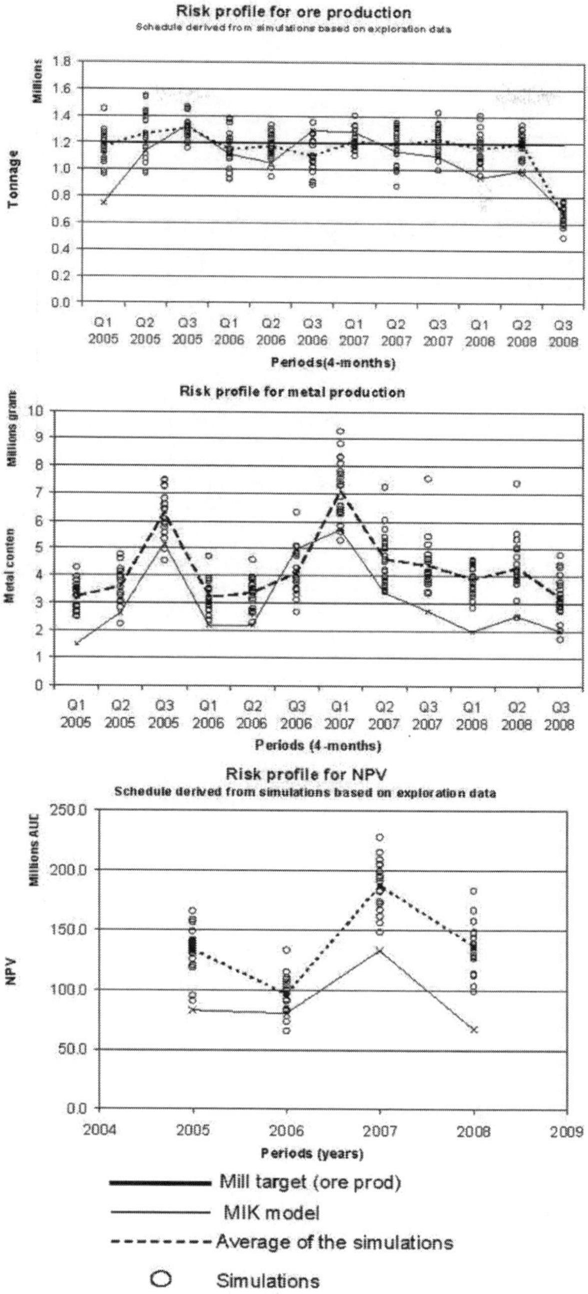

Figure 8. Risk profiles of ore, metal and NPV for production schedules derived from orebody models based on exploration data

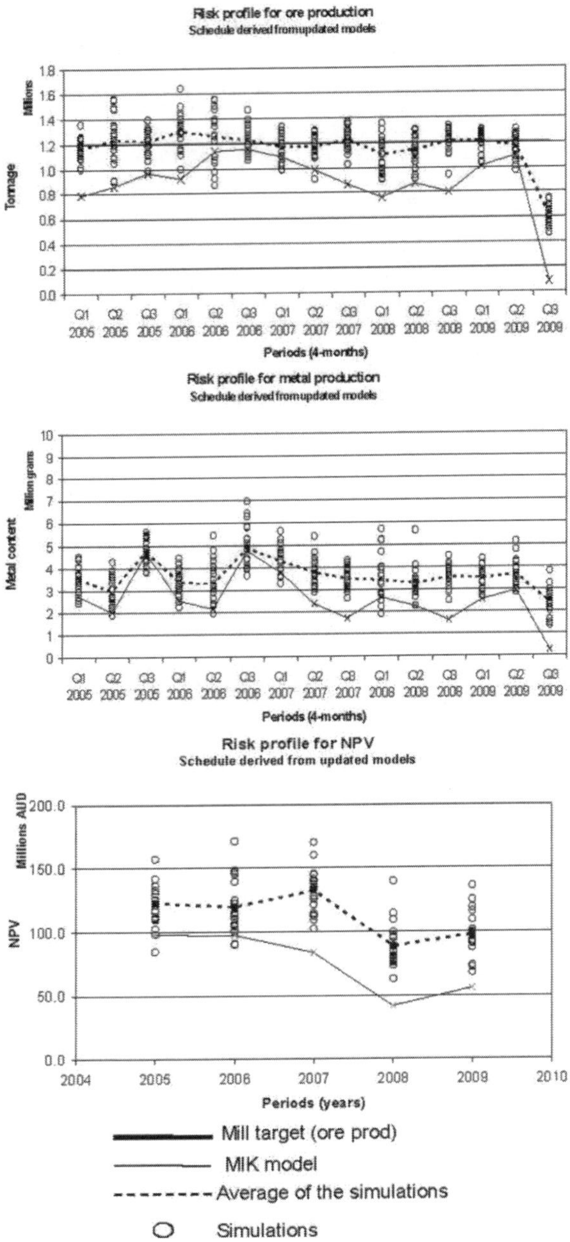

Figure 9. Risk profiles of ore, metal and NPV for mine's production schedule derived from orebody models displaying grade control characteristics

Table 1. Cumulative averages for ore tonnes, metal and NPV over all simulated orebody models (both exploration data based and updated models)

	SIP and simulations with exploration data	SIP and updated simulations with 'future' data	Mine schedule performance with updated simulations
Ore tonnes/Mt	14	18	10
Metal tonnes/Mg	52	55	38
NPV/million AUD	552	561	330

The production schedules based on orebody models derived from exploration data (that do not account for local orebody knowledge and characteristics) underestimate the actual tonnes of ore mined from the orebody. The derived exploration based schedules are a conservative option. Taking local grade control information into account indicates that there is probably sufficient ore to extend the LOM to 2009 compared to 2008 when only exploration data are used. Figure 8 displays the production schedules for the exploration based simulations and the updated simulations. As Fig. 9 shows, the two schedules are different in that different sections of the pit are mined out in different time periods. However, when compared to Fig. 3, both of the SIP derived schedules mine the pit out in a series of successive cones, while the mine's yearly schedule proceeds in layers.

Discussion on results

In assessing the results in the previous section, it is helpful to address some specific topics relevant to the methods employed herein, particularly stochastic optimisers developed for mine production scheduling such as the SIP in stage 3. The reason that 20 simulated scenarios were used in the case study at Sunrise Dam Gold Mine is that past work, for example in Albor and Dimitrakopoulos (2009), shows that starting from after ~15 simulated representations of an orebody, the stochastic schedules generated converge to a stable final schedule and using, say, 20 versus 400 will not provide different results in terms of schedules and forecasts. This is not surprizing, as it is a volume support effect: while we simulate the grades of the blocks of a given size used to represent a mineral deposit and can measure any difference one wishes between simulated scenarios, an LOM production schedule is a grouping of a few thousand of these mining blocks in just one time (mining) period under various constraints. This means that the sensitivity of a schedule is not the sensitivity one observes at the scale of individual mining blocks; the support of an LOM schedule is orders of magnitude larger and this explains the insensitivity of the stochastic schedule to additional scenarios after a point. This is valid in the case study at Sunrise Dam for both stochastic schedules; the first generated using only simulations based on the original exploration data

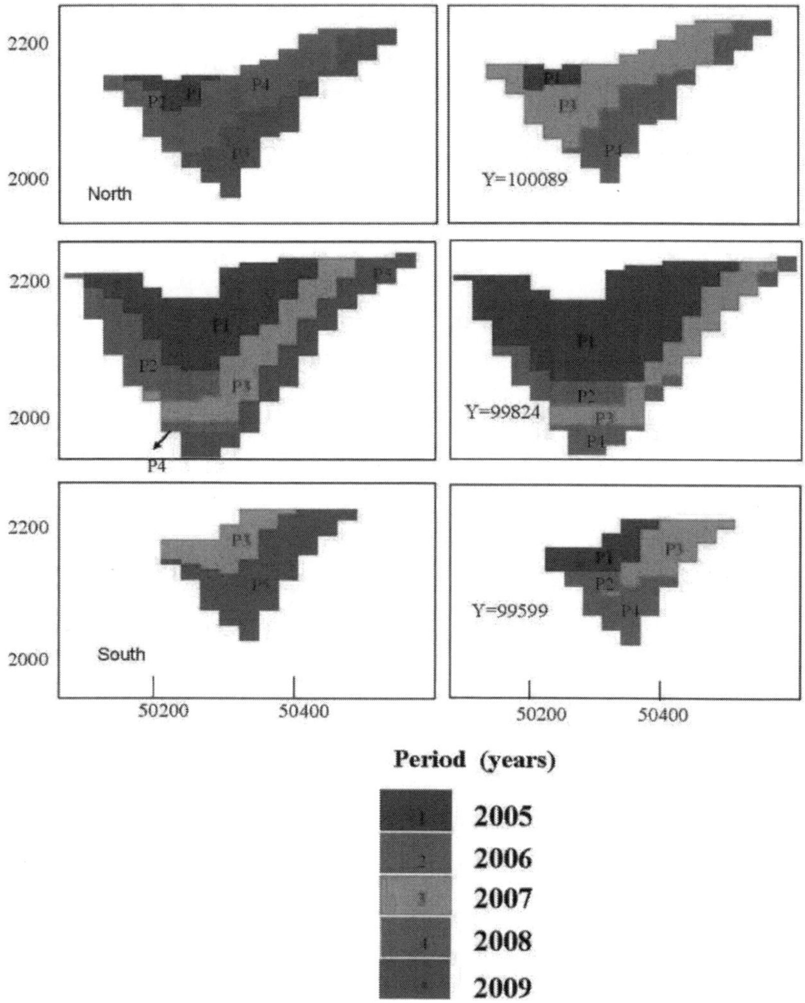

Figure 10. West–east sections of Sunrise pit displaying production schedules derived from simulation based on exploration data (right) and updated models (left)

and the second generated using the updated simulations with scenarios of future simulated grade control data. Note that the sizes of the mining blocks are the same as well as the mining capacities used for both the SIP generated and LOM production schedules.

The NPV assessments between both stochastic schedules and the mine's are not surprising despite the fact it is in the order of 50% (Table 1). One observation relevant to this difference issue is evident in past work: LOM schedules from conventional approaches and estimated orebody models, no matter what methods are being used, will have a lower NPV assessment when compared to the corresponding schedules generated from stochastic optimisation methods. A main reason for this is that the scheduling process uses the joint local uncertainty of the economic values of the mining blocks in the orebody as provided by a group of simulated deposit scenarios used as the input (see for example the second part in equation (4) and the related stochastic constraints).

There is no theoretical reason why the NPV generated from the proposed multistage approach will be higher than the corresponding stochastic schedule without the simulated grade control data. This is specific to the Sunrise Dam mine and caused due to the differences seen in Fig. 6, discussed in the related section. In a different case study, if the resource/reserve model(s) of the orebody overstates high grades (higher proportion) and the grade control cannot find these high grades (lower proportion), the exact opposite will occur. Note that the use of stochastic optimisers allows the scheduling process to capitalise on knowing the joint local uncertainty of the economic values of the mining blocks composing the representation of the orebody at hand.

The main contribution of the multistage production scheduling approach presented in this paper is its ability to integrate short and long term production scheduling through the concept of simulated future grade control data. This means that the discrepancies between the LOM schedule and the operational short term schedule will be minimised, assuming that the past grade control and production reconciliations will maintain their relation. As shown in this paper, this leads to a higher probability of meeting production targets, thus increasing productivity. It should be stressed that the value increased productivity is not represented in the standard assessment of NPV and this can be accounted for when analysing LOM production schedules.

CONCLUSIONS

A new multistage stochastic mine production scheduling approach presented herein includes four stages. Stage 1 generates high density future grade control data for incorporation into the production scheduling process. It is based on sequential Gaussian cosimulation of exploration and grade control data from previously mined out parts of a deposit. Stage 2 employs conditional simulation by successive residuals to update preexisting simulated orebody models with the simulated future grade control information from stage 1. Stage 3 uses an SIP mine scheduling formulation and balances both maximising NPV and minimising deviations from production targets. Stage 4 includes quantification of risk in the produced schedules generated and reporting.

The application at Sunrise Dam, a large operating gold mine, demonstrates that the proposed approach is practical, and adds value to the operation. The approach is shown to deliver 3·6 Mt of additional ore and 2·6 Mg more metal, which matches better with the mine's reconciliations. The approach also results in a cumulative NPV which is on average 7·7 million dollars higher than that of a comparable stochastic schedule without the simulated grade control data, and substantially higher (by 230 million AUD) when compared to the NPV from the actual schedule of the mine. It is important to note that the key contribution of the multistage production scheduling approach presented in this paper is its ability to integrate short and long term production scheduling through the concept of simulated future grade control data. This means that the discrepancies between the LOM schedule and the operational short term schedule will be minimised, assuming that the past grade control and production reconciliations maintain their relation. As shown in this paper, this leads to a higher probability of meeting production targets thus increasing productivity. It should be stressed that the value increased productivity is not represented in the standard assessment of NPV and this can be accounted for when analysing LOM production schedules. As capital intensity of mining projects is high, this last point may be critically important to mining operations. Last but not least, the additional technical work required in implementing the approach presented herein, compared to the present best mining industry practices, is insignificant when compared to the value of the operations or projects being supported. Generating mine production schedules that have a higher probability of delivering tonnes and grade is a positive contribution. Although in this case study short scale mineralised features which are picked up by grade control, but not by exploration data, lead to a history of positive reconciliation, the opposite may also happen in other mines; that is, when the resource model overstates high grades and the grade control cannot find all of the high grades.

The authors would like to acknowledge substantial financial contributions from Anglo Gold Ashanti, BHP Billiton, Rio Tinto and Xstrata Copper, as well as McGill's COSMO Stochastic Mine Planning Laboratory and NSERC Discovery Grant 239019-06.

NOTES

This paper is part of a special issue on Strategic Mine Planning

REFERENCES

1. Abdel Sabour SA, Dimitrakopoulos R, Kumral M. 2008. Mine design selection under uncertainty, *Trans. IMM A: Min. Technol.*, 117, 53–64.

2. Albor F, Dimitrakopoulos R. 2009. Stochastic mine design optimization based on simulated annealing: pit limits, production schedules,

multiple orebody scenarios and sensitivity analysis, *Trans. IMM A: Min. Technol.*, 118, (2), 80–91.

3. Albor F, Dimitrakopoulos R. 2010. Algorithmic approach to pushback design based on stochastic programming: method, application and comparissons, *Trans. IMM A: Min. Technol.*, 119, (2), 88–101.

4. Asad MWA, Dimitrakopoulos R. 2013. Implementing a parametric maximum flow algorithm for optimal open pit mine design under uncertain supply and demand, *J. Operat. Res. Soci.*, 64, 185–197.

5. Benndorf J, Dimitrakopoulos R. 2013. Stochastic long-term production scheduling of iron ore deposits: integrating joint multi-element geological uncertainty, *J. Min. Sci.*, 47, 2.

6. David M, Dowd P, Korobov S. *Forecasting departure from planning in open pit design and grade control*, Proc. 12th APCOM Symp., Boulder, CO, USA, April, 1974, Colorado School of Mines, F131–F149.

7. Dowd P. 1994. Risk assessment in reserve estimation and open pit planning, *Trans. IMM Sect. A: Min. Industry*, 103, 148–154.

8. Dowd P. 1997. Risk in minerals projects: analysis, perception and management, *Trans. IMM Sect. A: Min. Industry*, 106, 9–18.

9. Godoy M, Dimitrakopoulos R. 2004. Managing risk and waste mining in long-term production scheduling of open pit mines,*Trans. Soc. Min. Metall. Explor.*, 316, 43–50.

10. Goovaerts P. 1997. *Geostatistics for natural resources evaluation*, New York, Oxford University Press.

11. Guardiano E, Parker HM, Isaaks EH. 1995. *Prediction of recoverable reserves using condiional simulation: a case study for the Fort Knox gold project, Alaska, Unpublished report*, Mineral Resources Development Inc., Foster City, CA, USA.

12. Jewbali A, Dimitrakopoulos R. 2011. Implementation of conditional simulation by successive residuals, *Comput. Geosci.*, 37, 129–142.

13. Journel AG, Kyriakidis PC. 2004. *Evaluation of mineral reserves: a simulation approach*, New York, Oxford University Press.

14. Khosrowshahi S, Shaw WJ, Yeates G. 2007. *Quantification of risk using simulation of the chain of mining – a case study on Escondida Copper, in Orebody modelling and strategic mine planning*, (, Dimitrakopoulos R), Spectrum Series 14, 2nd edn, Carlton, Vic, AusIMM, 33–41.

15. Lamghari A, Dimitrakopoulos R. 2012. A diversified Tabu search approach for the open-pit mine production scheduling problem with metal uncertainty, *Eur. J. Operat. Res.*, 222, 642–652.

16. Leite A, Dimitrakopoulos R. 2007. A stochastic optimization model for open pit mine planning: application and risk analysis at a copper deposit, *Trans. IMM A: Min. Technol.*, 116, (3), 109–118.

17. Leite A, Dimitrakopoulos R. 2013. Stochastic optimization of mine production scheduling with uncertain ore/metal/waste supply: application at a copper deposit, *J. Min. Sci. Technol.*, under review.

18. Meagher C, Abdel Sabour SA, Dimitrakopoulos R. 2009. Pushback design of open pit mines under geological and market uncertainties, Proc. Orebody Modelling and Strategic Mine Planning Symp., Perth WA, Australia, March, AusIMM, 297–304.

19. Menabde M, Froyland G, Stone P, Yeates G. 2007. *Mining schedule optimization for conditionally simulated orebodies, in Orebody modelling and strategic mine planning*, (, Dimitrakopoulos R), Spectrum Series 14, 2nd edn, 379–384, Carlton, Vic, AusIMM.

20. Myers D. 1991. Pseudo cross-variograms, positive definitness, and cokriging, *Math. Geol.*, 14, 805–816.

21. Newton PGN, Gibbs D, Grove A, Jones C, Ryall AW. 1998. *Sunrise-Cleo gold deposit, in Geology of the Australian and Papua New Guinean mineral deposits*, (, Berkman D A, Mackenzie D H), 179–186, Carlton, Vic, AusIMM.

22. Peatie R, Dimitrakopoulos R. 2009. Forecasting recoverable ore reserves and their uncertainty at Morila Gold Deposit, Mali – an efficient simulation approach and future grade control drilling, Proc. Orebody Modelling and Strategic Mine Planning Symp., Perth WA, Australia, March 2009, AusIMM, 93–100.

23. Ramazan S, Dimitrakopoulos R. 2012. Production scheduling with uncertain supply: a new solution to the open pit mining problem, *Optim. Eng.*

24. Ravenscroft PJ. 1992. Risk analysis for mine scheduling by conditional simulation, *Trans. IMM Sect. A: Min. Industry*, 101, 104–108.

25. Richmond A. 2011. *Direct net present value open pit optimisation with probabilistic models, in Advances in orebody modelling and strategic mine planning I*, (, Dimitrakopoulos R), AusIMM Spectrum Series, Vol. 17, 285–289, Carlton, Vic., AusIMM.

26. Vargas-Guzman JA, Dimitrakopoulos R. 2003. Successive nonparametric estimation of conditional distributions, *Math. Geol.*, 35, 39–52.

Index